Drools规则引擎技术指南

来志辉 ◎ 著

北京大学出版社
PEKING UNIVERSITY PRESS

内 容 提 要

Drools 规则引擎已经有几年的发展史了,但由于学习成本较高,且国内并没有详细的中文文档,导致 Drools 规则引擎在国内市场推行缓慢。本书将对 Drools 规则引擎进行一个详细说明,共分为六篇,基石篇主要介绍规则引擎的入门知识,基础篇详细介绍规则引擎的基础语法、规则属性、关键字及错误信息,中级篇介绍规则中级语法等,高级篇介绍 Workbench、Kie-Server、动态规则、多线程中的 Drools 等高级用法,源码篇为 Drools 源码分析,扩展篇为 Drools 扩展说明。除了讲解 Drools 规则引擎的思维方式外,还在每一个知识点上辅以大量的代码案例,并且有很多实战经验及思想在里面。本书作为国内第一本 Drools 规则引擎的中文教程,Java 开发者、对 Drools 规则引擎有兴趣的软件开发人员或系统架构师都可以阅读。

图书在版编目(CIP)数据

Drools规则引擎技术指南 / 来志辉著. —— 北京:北京大学出版社,2019.7
ISBN 978-7-301-30549-2

Ⅰ. ①D… Ⅱ. ①来… Ⅲ. ①搜索引擎 – 程序设计 – 指南 Ⅳ. ①TP391.3-62

中国版本图书馆CIP数据核字(2019)第103274号

书　　　名	Drools规则引擎技术指南 DROOLS GUIZE YINQING JISHU ZHINAN
著作责任者	来志辉　著
责 任 编 辑	吴晓月　刘沈君
标 准 书 号	ISBN 978-7-301-30549-2
出 版 发 行	北京大学出版社
地　　　址	北京市海淀区成府路205号　100871
网　　　址	http://www.pup.cn　　新浪微博:@ 北京大学出版社
电 子 信 箱	pup7@ pup.cn
电　　　话	邮购部 010-62752015　发行部 010-62750672　编辑部 010-62570390
印 刷 者	山东百润本色印刷有限公司
经 销 者	新华书店
	787毫米×1092毫米　16开本　29.25印张　667千字 2019年7月第1版　2021年2月第2次印刷
印　　　数	3001-4500册
定　　　价	99.00元

未经许可,不得以任何方式复制或抄袭本书之部分或全部内容。
版权所有,侵权必究
举报电话:010-62752024　电子信箱:fd@pup.pku.edu.cn
图书如有印装质量问题,请与出版部联系。电话:010-62756370

前言

在科技急速发展的今天，各种高端技术不断涌出，常见的有大数据、人工智能、机器学习、深度学习、区块链等，这类技术已经成为各大公司的首选。如今，金融行业、医疗行业、教育行业、保险行业及传统行业都在发生着巨大的改变，都在向互联网、物联网方向进行转型，从而突出了IT（信息技术）的重要性。科技决定发展、科技改变人生已经不再遥不可及。

目前流行的行业，以金融类项目为例，风险控制系统、反欺诈系统、决策引擎等也成为常用且经常变化的业务。这类经常变更的业务让公司运营和程序员都非常头痛。按照产品开发的传统逻辑思维，基本写法就是添加If判断，或者通过SQL查询条件中的动态添加判断，如Mybatis配置文件中的<if>。这应该是大部分程序员第一时间想到的解决方案，但是依照这类传统的思维方式考虑问题会出现以下弊端。

（1）增加开发人员与测试人员的工作量。

（2）部门间需要更加频繁地进行业务沟通，时间成本增加。

（3）增加用人成本，造成公司损失。

如何解决这些弊端成了技术部门的重要任务之一。规则引擎的出现完美地解决了这个问题。我刚开始研究规则引擎时有些迷茫，所面临的问题很多，在网上搜索规则引擎时，不知道选择学习哪一个规则引擎，当时网上规则引擎有IBM ILog、URule、Rule Engine、Visual Rules、Drools。分析后，我发现只有Drools是开源项目，于是就开始了学习Drools的苦行。为什么说是苦行呢？原因很简单，Drools有一定的学习成本，而且当时（2016年）网上可以找到的教程实在是少之又少，很多人都是会用一些但并不知其道理与知识点，遇到问题连解决方案都没有，只能通过自己不断测试实践和钻研英文资料进行学习，鉴于我英文水平实在有限，就在一些QQ群和国外的论坛里找寻志同道合的人。经过一年的努力，我编写了《Drools技术指南中文教程》（电子文档），创建了有关Drools 6.4版本的博客[1]和Drools技术讨论群[2]及微信公众号[3]，为国内Drools技术填补了空白。

本教程并非翻译官方文档，虽然有些地方看起来很相似，但知识点的说明比官方更好理解，下面跟着我的思路开始学习Drools吧！

---

[1] 作者的博客地址：https://me.csdn.net/u013115157。
[2] 作者的技术讨论及教学视频QQ群：676219749。
[3] 作者的微信公众号：程序猿之塞伯坦。

本书提供以下源代码下载地址。

Drools 测试用例：https://github.com/projectLzh/DroolsTest。

Drools Spring：https://github.com/projectLzh/DroolsSpring。

Drools Spring Boot：https://github.com/projectLzh/DroolsSpringBoot。

JavaWorkbench 交互：https://github.com/projectLzh/WorkbenchJava。

读者也可以扫描下方二维码，关注微信公众号，输入书中 77 页的资源下载码，获取源代码。

目录 Contents

第一篇 基 石 篇

第 1 章 Drools 概述 002

    1.1 程序来源于生活 003

    1.2 Drools 是什么 003

    1.3 Drools 简要概述 003

    1.4 Drools 发展趋势 004

    1.5 Drools 版本 004

    1.6 Drools 新特性 005

    1.7 KIE 生命周期 006

    1.8 为什么要用规则引擎 006

第 2 章 Drools 入门实例 008

    2.1 经典 Hello World 009

    2.2 对象引用 013

    2.3 Drools 配置文件 020

第二篇 基 础 篇

第 3 章 Drools 基础语法 026

    3.1 规则文件 027

    3.2 规则体语法结构 028

    3.3 pattern（匹配模式） 028

- 3.4 运算符 030
- 3.5 约束连接 032
- 3.6 语法扩展 048
- 3.7 规则文件 drl 056

第 4 章 Drools 规则属性 057

- 4.1 属性 no-loop 058
- 4.2 属性 ruleflow-group 063
- 4.3 属性 lock-on-active 063
- 4.4 属性 salience 065
- 4.5 属性 enabled 067
- 4.6 属性 dialect 068
- 4.7 属性 date-effective 069
- 4.8 属性 date-expires 070
- 4.9 属性 duration 073
- 4.10 属性 activation-group 073
- 4.11 属性 agenda-group 076
- 4.12 属性 auto-focus 082
- 4.13 属性 timer 082

第 5 章 关键字及错误信息 085

- 5.1 关键字说明 086
- 5.2 错误信息 086

第三篇 中 级 篇

第 6 章 规则中级语法 090

- 6.1 package 说明 091
- 6.2 global 全局变量 094
- 6.3 query 查询 101

6.4 function 函数	104
6.5 declare 声明	109
6.6 规则 when	115
6.7 规则 then	146
6.8 kmodule 配置说明	150

第 7 章　指定规则名调用 153

第 8 章　Spring 整合 Drools 161
8.1 Spring+Drools 简单配置	162
8.2 Drools 整合 Spring+Web	167
8.3 Drools 整合 Spring Boot	173

第 9 章　KieSession 状态 209
| 9.1 有状态的 KieSession | 211 |
| 9.2 无状态的 StatelessKieSession | 211 |

第四篇　高　级　篇

第 10 章　Drools 高级用法 218
10.1 决策表	219
10.2 DSL 领域语言	227
10.3 规则模板	234
10.4 规则流	240
10.5 规则构建过程	272
10.6 Drools 事件监听	277

第 11 章　Workbench 283
11.1 Workbench	284
11.2 Windows 安装方式	284
11.3 KIE-WB 6.4 版本安装	287

11.4　Workbench 操作手册 291

11.5　Workbench 与 Java 交互 330

11.6　构建项目的版本控制 344

11.7　Workbench 上传文件与添加依赖关系 345

11.8　Workbench 中设置 Kbase+KieSession 349

11.9　Workbench 构建 jar 包到 Maven 私服 352

第 12 章　Kie-Server 353

12.1　整合部署 354

12.2　分离部署 362

12.3　集群部署 364

12.4　Kie-Server 与 Java 交互 380

第 13 章　动态规则 385

第 14 章　多线程中的 Drools 401

14.1　同 KieHelper 同 KieSession（有状态）... 404

14.2　同 KieHelper 不同 KieSession（有状态）... 407

14.3　不同 KieHelper 不同 KieSession（有状态），KieSession 只创建一次 ... 409

14.4　不同 KieHelper 不同 KieSession（有状态），KieSession 在线程代码中创建 411

14.5　同 KieHelper 同 StatelessKieSession（无状态）...................................... 413

14.6　同 KieHelper 不同 StatelessKieSession（无状态）.................................. 415

14.7　不同 KieHelper 不同 StatelessKieSession（无状态），StatelessKieSession 只创建一次 ... 417

14.8　不同 KieHelper 不同 StatelessKieSession（无状态），StatelessKieSession 在线程代码中创建 419

第五篇　源　码　篇

第 15 章　Drools 源码分析 424

15.1　KieServices 分析 425

15.2　KieContainer 分析 433

15.3　KieSession 分析 438

15.4	KieBase 分析	440
15.5	KieFileSystem 分析	441
15.6	KieHelper 分析	442

第六篇　扩　展　篇

第 16 章　Drools 扩展说明 446

16.1	规则引擎优化方案	447
16.2	规则实战架构	450
16.3	规则引擎项目的定位	453
16.4	规则引擎实战应用思想	454
16.5	规则引擎日志输出	455

参考文献 **458**

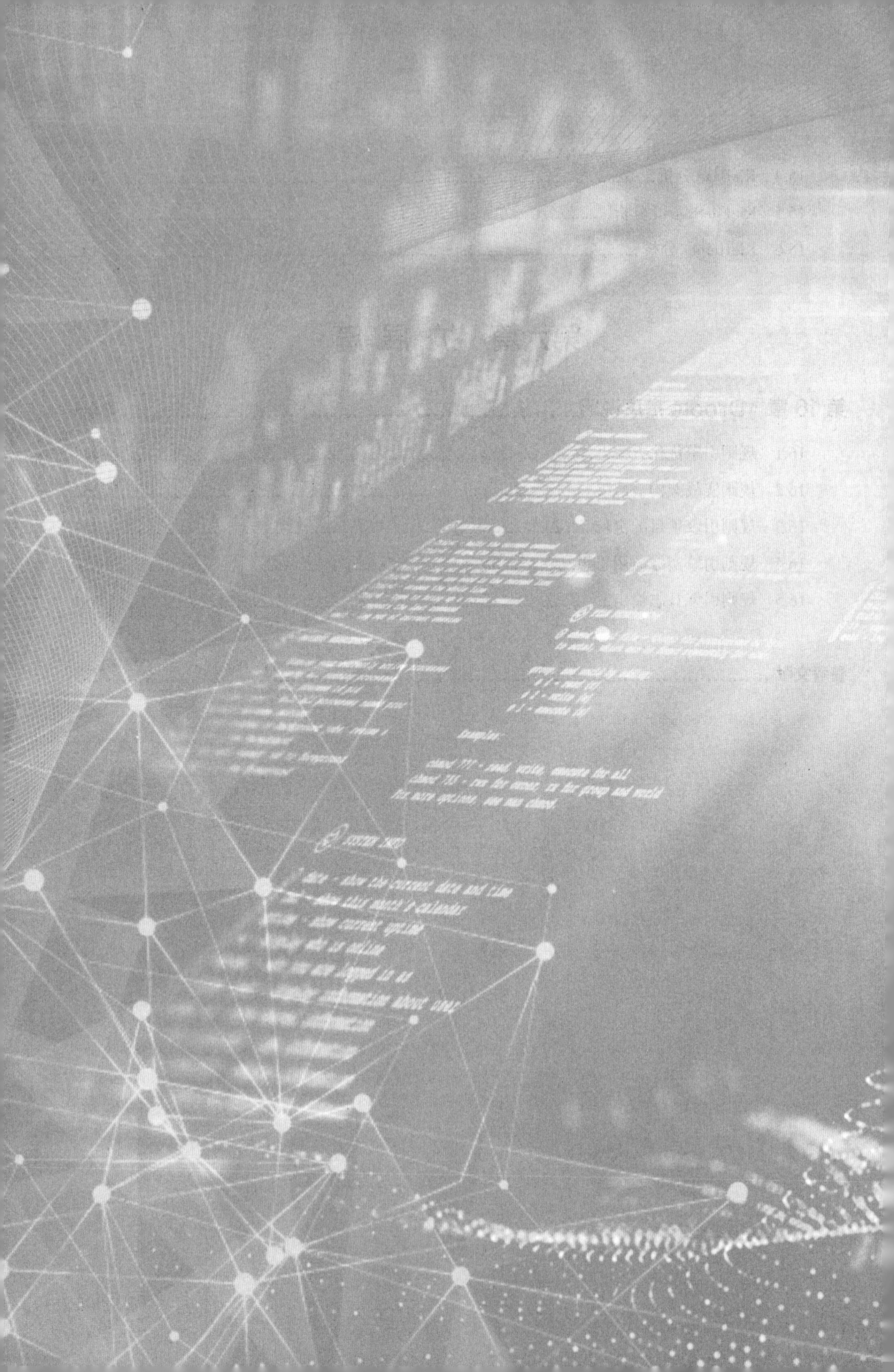

Drools 规则引擎技术指南

第一篇

基石篇

第1章 Drools概述

1.1 程序来源于生活

程序来源于生活，这一点从 Java 语言中就能深刻体会到。万事万物皆对象，各种设计模式、框架技术、核心思想都源于生活。在日常生活中处处都有规则，用特定的规则来约束人们的行为，按照这些规则人们就知道哪些应该做、哪些不应该做，甚至哪些是必须要去做的。例如，当一个路口红灯亮起时，所有人和车都是不能通过的，否则就会有安全隐患，这里的红灯就是一个规则，它使人或车停下。Drools 是一个开源的规则引擎技术，下面就针对 Drools 规则引擎进行详细说明。

1.2 Drools 是什么

Drools 是开源的项目，是具有一定学习成本的规则引擎技术，通过 Drools 特定的语法，将固定的业务、经常变的业务统一管理。它以特定的文件或数据库方式将规则内容进行存储（存放在哪里都可以，可以理解为它只需一个存储介质），以 Drools 包提供的接口对规则内容进行处理，并返回公司运营所规定的业务结果，当然这个结果也是正确的。

通过更简单的规则结构，预判业务是否正确，及时对业务进行测试，如果需要了解规则内部的执行过程，可以添加 Drools 规则引擎对外提供的事件监听功能。

1.3 Drools 简要概述

Drools 是一款基于 Java 语言的开源规则引擎，可以将复杂且多变的业务规则从硬编码中解放出来，以规则脚本的形式存放在文件或特定的存储介质中（这里可以是数据库表），使得业务规则的变更不需要修正项目代码、重启服务器就可以在线上环境立即生效。这里可以理解为动态代码（动态业务）。

规则引擎的核心目的之一是将业务决策从程序代码中分离出来，使其代码与业务解耦合。通过特定的语法内容编写业务模块，由 API 进行解析并对外提供执行接口，再接收输入数据、进行业务逻辑处理并返回执行结果。引用规则引擎后的效果如图 1-1 所示。

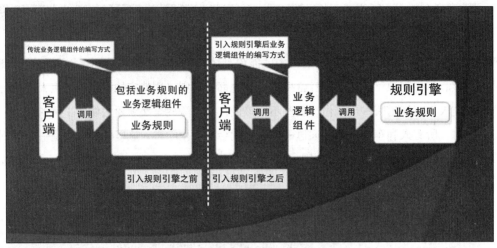

图1-1 引用规则引擎后的效果

1.4 Drools发展趋势

Drools 规则引擎是为开发人员提供的 KRR 功能程序，KRR 是人工智能的核心，所以 Drools 也是人工智能的一个分支，是专家系统的另一种展现方式。Drools 的核心之一是操作业务，与传统项目不同的是，业务由业务操作员管理，不再需要程序员去改代码，极大地减少了开发周期，使得业务员不再依赖 IT 部门就可以完成业务的变更。正是因为规则引擎有这样的优势，所以受到了风险控制类、反欺诈类、促销类项目的青睐。目前，不仅阿里巴巴、京东、智联招聘、智业软件（国内知名医疗研发公司）等公司在使用，而且国内一些商业产品的规则引擎底层也是通过 Drools 实现的。

1.5 Drools版本

在开始学习 Drools 规则引擎之前，先制订学习计划，我刚学习 Drools 时的版本是 Drools 6.4，本书主讲的版本是 Drools 7.10。从我研究 Drools 规则引擎开始到本书中主讲的 7.10 版本，我发现规则引擎的脚本并没有太多的变化，无论是 5.x 版本还是 7.x 版本，或者是 6.4 版本，在规则引擎的语法上都相差无几。发生变化的应该是性能的优化和 Drools 提供的 API 了，自 6.0 版本后 Drools 的 API 发生了巨大的变化，其相比 5.x 版本，更加简洁、容易理解。所以读者在学习和使用过程中，无须担心 Drools 规则引擎语法的使用问题。

1.6 Drools新特性

自 6.0 版本开始，Drools 推出了一个全新的概念，基于 KIE 的全新 API，其目的是为了更简单地操作规则引擎，用过 Drools 5.x 版本甚至更早版本 Drools 的人都知道，要执行一个规则文件是相当烦琐的事情。

KIE（Knowledge Is Everything）是"知识是一切"的缩写，是 Jboss 一系列项目的总称。如图 1-2 所示，KIE 的主要模块有 OptaPlanner、Drools、UberFire、jBPM。下面分别讲述这些模块的用途。

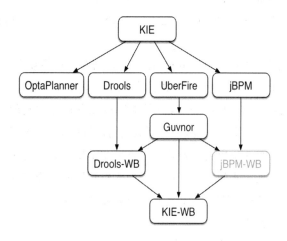

图1-2　KIE结构

OptaPlanner：一个本地搜索和优化工具，按官方的说法，OptaPlanner（以下简称为 OP）原本是 Drools 平台的组件，但由于发展的趋势，现在独立于 Drools（虽说已经独立了，但还是拥有良好的整合功能），成为与 Drools、jBPM 等同级的 KIE 组件。

Drools：规则引擎，自 6.0 版本后，Drools 提供了 Web IDE，从图 1-2 中可以看出，Drools 组件包含了 Drools-WB，Drools-WB 又指向了 KIE-WB。按照官方的说法，KIE-WB 是一个强大的 Web IDE，它结合了 Drools、Guvnor、jBPM 插件成为一个超级平台，KIE-WB 的作用将在后面进行详细说明。

UberFire：一个全新的组件，其功能类似于 Eclipse，包括插件中的样式和页面。UberFire（以下简称为 UF）独立于 Drools 和 jBPM，可以进行独立部署，对外提供功能服务。

jBPM：一个业务流管理组件，用于实现自动化业务流程和决策的工具包。jBPM 源于 BPM（业务流程管理），但它已经发展到使用户能够在业务自动化中选择自己的路径。jBPM 提供了各种功能，可以将业务逻辑简化和外部化为可重用的资产，如案例、流程、决策表等。为什么 jBPM-WB 是灰色的？按官方的说法，jBPM-WB 对于 KIE-WB 是多余的，因为 KIE-WB 完美地结合了工作流，所以将其制成灰色也不为过。

1.7　KIE生命周期

通过上述的说明，相信读者对 KIE 已经有一个大致的了解了，下面来介绍 KIE 的生命周期。

（1）创建：通过 KIE-WB 创建知识库，如 DRL、BPMN2、决策表、决策树、实体等。

（2）构建：构建一个可提供 KIE 部署的组件，简单地说，就是生成一个包含知识库的 jar 包，通过 Java 代码或 KIE-WB 提供的服务器（Kie-Server）来操作业务规则。

（3）测试：在构建部署前，对整体知识库进行测试。

（4）部署：KIE 使用 Maven 将其组件部署到应用程序上。

（5）使用：通过 KieContainer 创建 Kie 会话（KieSession），为执行提供前提条件。实际上 KieSession 是由 KieBase 创建的。

（6）执行：通过执行 KieSession 与 Drools 系统进行交互，执行规则、流程、决策表等。

（7）交互：用户与 KieSession 的交互，通过代码或页面进行操作。

（8）管理：管理 KieSession、KieContainer 等 Drools 提供的相关对象。

Drools 是很多年前就有的规则引擎技术，但随着新技术的涌现，为了适用于各式各样的场景，Drools 进行了模块化分类，使其功能更加独立、内容更加丰富、分工也更加明确。下面通过 3 种业务建模技术来实现不同应用级的业务场景，其中较为核心的是业务规则管理。

① 业务规则管理：主要以规则管理为核心进行详细的业务介绍。

② 流程管理：指规则流部分。

③ 复杂事件处理：负责事件处理功能。

1.8　为什么要用规则引擎

一般的项目中没有引用规则引擎之前，通常的做法都是使用一个接口进行业务工作。首先要传进去参数，通过 if…else 或其他方式进行业务逻辑判断，其次要获取到接口执行完毕后的结果。引用规则引擎后就截然不同了，原有的 if…else 不复存在，代替它们的是规则引擎脚本，通过规则引擎实现可动态变化的"if…else"。

规则引擎可以给项目带来什么？规则引擎的应用场景是什么？使用规则引擎的好处是什么？下面将进行详细介绍。

1. 规则引擎可以给项目带来什么

（1）给公司运营人员带来了什么？

① 将业务规则交于业务员来处理。

② 提高业务灵活性，业务员可以随时对公司业务进行修改（设计时要加权限）。

③ 增加业务处理的透明度，业务规则可以被管理。

④ 修改业务将不再通过开发人员，极大地减少了对 IT 人员的依赖。

⑤ 减少各部门之间的矛盾，各司其职。

（2）给公司 IT 部门带来了什么？

① 简化了系统的复杂度，使系统间变得简单、透明。

② 提高了系统的可维护性。

③ 减少了维护成本。

④ 规则引擎是相对独立的，只关心业务规则，并不关心与谁交接。

⑤ 减少了"硬代码[①]"业务规则的成本和风险。

⑥ 减少了与业务员的冲突。

2. 规则引擎的应用场景

（1）适用的行业分类。

① 金融行业——黑名单、白名单、风险投保。

② 医疗行业——合理输血、合理用药。

③ 电商行业——促销平台。

（2）适用的系统分类。

① 风险控制系统——风险贷款、风险评估系统。

② 反欺诈项目——银行贷款、征信验证。

③ 决策平台系统——财务计算。

④ 促销平台系统——满减、打折、加价购。

3. 使用规则引擎的好处

（1）应用概述说明。

① 应对复杂多变的业务场景。

② 快速且低成本地进行业务规则变更。

③ 业务员直接管理，不需要程序员进行干预，减少风险。

④ 平台独立化，系统迁移、系统升级都极为方便。

（2）作用与优点。

① 业务规则与系统代码分离，实现代码与业务的解耦合。

② 提供领域语言（自然语言），使业务人员更容易理解。

③ 提供了可视化页面[②] 操作，使用更简单。

④ 大大提高了对复杂逻辑代码的可维护性。

⑤ 可随时对业务进行扩展和维护。

⑥ 符合公司对敏捷性或迭代性开发的策略。

---

① 硬代码指将业务逻辑写到项目代码中，风险是维护成本高、开发难度加大、服务器重启等。
② 可视化页面指 KIE-WB，有一定的学习成本，比较适合 IT 人员使用。

第2章

Drools入门实例

2.1 经典Hello World

在讲 Drools 规则引擎的使用前，先要了解它的文件扩展名，因为这会关系到后面的知识点。Drools 规则引擎文件扩展名有很多种，最基本的是 *.drl 文件，也可以是 *.xml[①] 和 *.drls 文件，甚至还可以是 *.xls 或 *.xlsx[②] 文件。为什么会有这样多的规则文件呢？因为每一种规则文件都代表一类规则的应用。本章将用最基本也是最常用的 *.drl 文件方式对 Drools 规则引擎进行详细说明。

创建一个空项目，这里除了 JDK 1.8 外，什么都不需要引用，创建资源目录后创建 Java 的目录。在资源文件夹中创建 rules/ rulesHello 文件夹，如图 2-1 所示。

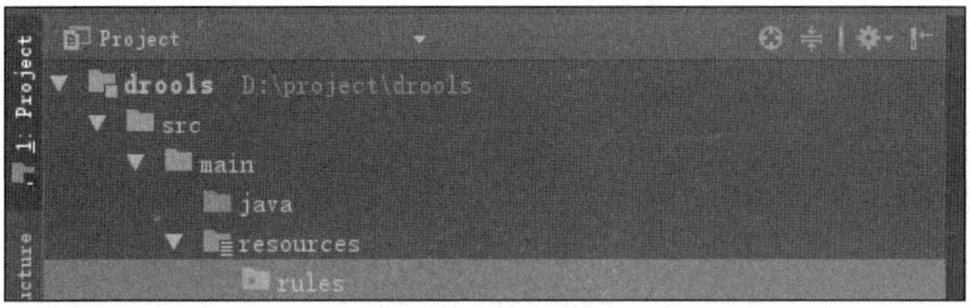

图2-1 项目结构

在当前目录下创建 hello.drl 文件，其代码为：

```
package rules.rulesHello
    rule "test001"
        when
            eval(true);
        then
            System.out.println("hello world");
    end
```

如图 2-2 所示，创建一个以 .drl 为扩展名的文件，在 IntelliJ IDEA 工具下图标都变了。当然，如果在 Eclipse 下是需要安装插件才可以使其变化的，读者不必担心为什么显现方式与本书中不同。

图2-2 规则文件内容

[①] XML 规则文件基本已经不再使用，但不代表它不可以用。
[②] *.xls 和 *.xlsx 是决策表。

hello.drl 代码就是一个标准的规则文件。其中，package 表示规则逻辑路径，rule 表示规则开始，end 表示规则结束，test001 表示规则名，when 表示规则条件，then 表示规则返回结果。在关键字 then 中的代码是 Java 的打印语句。在规则引擎脚本中可以写 Java 代码，这是因为 Drools 是基于 Java 语言开发的。

简单说明了规则文件中各行代码表示的含义后，就要执行规则了，看其能否在控制台输出 hello world。

执行规则文件，就不得不引用 Drools 相关的 jar 包。这时就将项目变成一个 Maven 项目，并引用 Drools 的相关 jar 包。pom.xml 文件内容为：

```xml
<properties>
        <drools.version>7.10.0.Final</drools.version>
    </properties>
    <dependencies>
        <dependency>
            <groupId>org.drools</groupId>
            <artifactId>drools-compiler</artifactId>
            <version>${drools.version}</version>
        </dependency>
        <dependency>
            <groupId>junit</groupId>
            <artifactId>junit</artifactId>
            <version>4.11</version>
        </dependency>
    </dependencies>
```

这里只引用 Drools 相关的 jar 包是不够的，接下来需要在资源目录中创建一个 META-INF 目录，并创建一个 kmodule.xml 配置文件，其内容为：

```xml
<?xml version="1.0" encoding="UTF-8"?>
<kmodule xmlns="http://www.drools.org/xsd/kmodule">
    <kbase name="rules" packages="rules.rulesHello">
        <ksession name="testhelloworld"/>
    </kbase>
</kmodule>
```

完成如上配置后，创建调用规则的 Java 文件。创建包名 com.rulesHello，并添加 Java 文件 RulesHello.java，其内容为：

```java
package com.rulesHello;

import org.kie.api.KieServices;
import org.kie.api.runtime.KieContainer;
import org.kie.api.runtime.KieSession;
public class RulesHello {
    public static void main(String[] args) {
        KieServices kss = KieServices.Factory.get();
```

```
        KieContainer kc = kss.getKieClasspathContainer();
        KieSession ks =kc.newKieSession("testhelloworld");
        int count = ks.fireAllRules();
        System.out.println("总执行了"+count+"条规则");
        ks.dispose();
    }
}
```

执行 RulesHello 类的主函数 main，可以看到控制台输出以下代码，如图 2-3 所示。

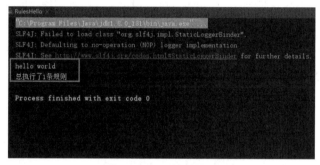

图2-3　规则执行结果

看到控制台成功输出了规则文件 then 中的内容，就证明规则测试成功。上述规则代码中只是一个非常简单的例子，只要触发了对应的 ks. fireAllRules () 方法就会在控制台输出"hello world"。虽然规则代码看似非常简单，但整个执行过程却是非常复杂的。下面对规则文件的内容进行扩展说明。

规则文件内容一般包含三大块，即包路径、引用、规则体。一个最简单的规则至少要包含包路径与规则体。规则文件还有一个重点，即规则文件名，接下来将对规则文件中的三大块及规则文件名进行阐述。

1. 规则文件名

规则文件名即 hello.drl，其命名规则不像 Java 那样要求首字母大写。虽然要求并不严格，但对规则文件命名时也最好要规范，见名知意，既方便自己，又方便他人。

2. 规则文件的内容

（1）package。关键字 package 为三大块中的包路径，这里的包路径是逻辑路径，理论上是可以随便定义的，并且是必填内容。为了更方便地开发，建议读者在编辑包路径时最好与文件目录同名，类似 Java 一样以小数点 (.) 的方式隔开，且必须用小数点隔开。这里要提醒一下读者，在规则文件中关键字 package 永远在代码的第一行（规则模板除外）。

（2）rule。关键字 rule 为三大块中的规则体，是核心内容之一，以关键字 rule 开头，以 end 结尾，每个规则文件中可以包含多个 rule 规则体，但 rule 规则体之间不能交叉使用，即一个 rule 只能对应一个 end。rule 的参数是可以随意定义的，rule 的参数指规则名，建议读者在编写规则名时以驼峰式命名,虽然有时可以不添加双引号("")[①],但还是建议每一个规则名都加上引号,以避免出现编译报错的问题。同一个规则库[②]中相同规则逻辑路径下的规则名不可以相同,可以理解为规则名即是一个 ID,不可重复。

(3) import。关键字 import 为三大块中的引用,它与 Java 引用其他类是一样的,其目的是为了对象类引用。与 Java 不同的是,在引用静态方法时,需要添加 function[③] 关键字。

3. 规则体说明

规则体是规则文件内容中的核心,分为 LHS、RHS 两大功能模块。

(1) LHS。条件部分又被称为 Left Hand Side (LHS),即规则体 when 与 then 中间的部分。在 LHS 中,可以包含 0~n 个条件(非常类似 Java 语法中的 if 判断语句),如果 LHS 部分为空,那么引擎会自动添加一个 eval(true) 条件,由于该条件总是返回 true,因此 LHS 为空的规则体也总是返回 true。其内容为:

```
package rules.rulesHello
        rule "test001"
            when
            //这里为空 则表示 eval(true)
            then
              System.out.println("hello world");
        end
```

规则体 LHS 不会如此简单,业务规则是否正确关键在业务条件,业务条件是否可以满足决定返回结果是否正确。换句话说,只要整体业务条件不满足,规则体就不会执行 RHS 模块,LHS 部分为 OR 的关系除外。所以规则体 LHS 部分与 Java 中的逻辑运算符的功能是一样的。

(2) RHS。结果部分又被称为 Right Hand Side,即一个规则体中 then 与 end 之前的部分,只有 LHS 部分的条件都满足时 RHS 部分才会被执行。这里要注意的是,在 Rete[④] 的算法中,规则在匹配时只会执行 LHS 为 true 的规则;加载规则时,会将所有规则体中的 LHS 部分先执行,即当前规则库中的 LHS 部分会被先一步加载。例如,当前规则体中的 LHS 条件为 false 时也是会被加载的,这可以理解为规则执行前的预加载功能,区别在于规则体的 RHS 部分不进行运算。只有 Fact 对象发生了改变,规则体才有可能重新被激活,之前为 false 的 LHS 就有可能变成 true。

RHS 才是规则体真正做事情的部分,即要处理和返回业务结果的部分。可以将条件满足而触发的动作写在该部分,在 RHS 中可以使用 LHS 定义绑定变量名、设置全局变量,或者直接编写 Java 代码(对于要用到的 Java 类或静态方法需要在规则文件中用 import 将该类引入后才能使用,这与 Java 文件的编写原则相同)。

规则体中的 LHS 部分是用来放置条件的,RHS 部分是编写满足条件后处理结果的,虽然 RHS

[①] 添加双引号的目的是在规则编译时将规则名转化为字符串,如果不添加,就必须定义为标准的类似定义变量名。
[②] 规则库是指规则执行必要的内存数据。
[③] function 是指规则引擎脚本中的函数。
[④] Rete 是一种网络算法,是 Drools 的核心算法之一。

部分可以直接编写 Java 脚本，但不建议在 RHS 中有条件判断。如果需要条件判断，那么要重新考虑将其放在 LHS 中，否则就违背使用规则的初衷了。

（3）Fact。上述说明中提到了一个关键字 Fact，它在规则引擎中是非常重要的，在介绍对象引用章节之前，必须先将其概念性的知识点进行一个说明。它也是一个必须要理解的概念，希望读者可以认真多读几遍加深印象。

Drools 规则引擎中传递的数据，术语称 Fact 对象[1]。Fact 对象是一个普通的 JavaBean（不只是 JavaBean 对象，也可以是任何 Object 对象），规则体中可以对当前对象进行任何的读/写操作，调用该对象提供的方法。当一个 Fact（JavaBean）插入 Working Memory（内存储存）中，规则体使用的是原有对象的引用（并不是克隆，与 Java 变量性质相似），规则体通过操作 Fact 对象来实现对应用数据的管理，对于其中的属性，需要提供 getter setter 或 "Object" 的可操作方法。执行规则时，可以动态地向当前 Working Memory 插入、删除或更新 Fact 对象。

规则进行计算时需要用到应用系统中的数据，先将这些数据设置到 Fact 对象中，然后将其插入规则的 Working Memory 中，一个 Fact 对象通常是一个具有 getter 方法和 setter 方法的 POJO 对象，由 Java 代码进行 insert 操作。通过 getter 方法和 setter 方法可以方便地对 Fact 对象进行操作，所以可以更通俗地把 Fact 对象理解为规则与应用系统数据交互的桥梁或通道。

当 Fact 对象插入 Working Memory 后，会与当前 Working Memory 中所有的规则进行匹配，同时返回一个 FactHandler 对象。FactHandler 对象是插入 Working Memory 中 Fact 对象的引用句柄，通过 FactHandler 对象可以实现对相应的 Fact 对象通过 API 进行删除及修改等操作。

在 RHS 中提供了一些对当前 Working Memory 实现快速操作的宏函数或对象，如 insert/insertLogical、update/modify 与 retract/delete。通过这些函数可以实现对当前 Working Memory 的 Fact 对象进行新增、修改或删除的操作。如果读者还要使用 Drools 中提供的其他方法，那么也可以使用其他的宏函数进行更多的操作。

2.2 对象引用

对象引用就是之前提到的 import 引用模块，是规则内容中非常重要的组成部分，即引入所需要的 Java 类或方法。关键字 import 是用来导入规则文件需要使用的外部对象，这里的使用方法与 Java 相似，与 Java 不同的是，import 不仅可以引入类，也可以导入类中的某一个可访问的静态方法[2]，其内容为：

---

[1] Fact 对象：Fact 是指在 Drools 规则应用中（是规则的因子），JavaBean 插入 Working Memory 中变成 Fact 之后，Fact 对象不是对原来的 JavaBean 对象进行克隆，而是原来 JavaBean 对象的引用。
[2] 静态方法多用于函数，是 Drools 自定义函数的两种方式之一。

```
import com.drools.demo.point.PointDomain;  导入类
import function com.drools.demo.point.PointDomain.getById;  导入函数
```

创建 Person 类文件，其内容为：

```java
package com.pojo;
public class Person {
    private String name;//姓名
    private String age;//年龄

    public String getName() {
        return name;
    }

    public void setName(String name) {
        this.name = name;
    }

    public String getAge() {
        return age;
    }

    public void setAge(String age) {
        this.age = age;
    }
}
```

重新编辑规则文件，其内容为：

```
package rules.rulesHello
import com.pojo.Person;
    rule "test001"
        when
        then
          System.out.println("hello world");
    end

rule "test002"
    when
        $p:Person();
    then
        System.out.println("输出引用对象"+$p);
end
```

关注的点是 rule "test002" 规则，包路径下引用了刚刚创建的 Person 类，现在再次执行测试类（不做任何变动），查看控制台输出，其代码如图 2-4 所示。

规则 test002 并没有输出，但规则本身没有任何问题，上述的内容中强调了一个概念性知识点——Fact 对象。rule test002 之所以没有输出，不是因为规则语法出了问题，而且规则执行时没有将实体插入工作内存中。既然规则体没有语法问题,就一定是在执行规则时出现了误差。定位到 KieSession ks =kc.newKieSession("testhelloworld") 这段代码,看看 KieSession 提供了哪些方法。

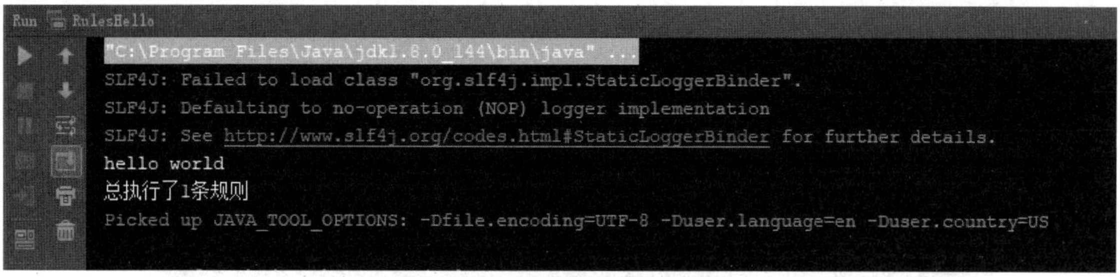

```
Run  RulesHello
"C:\Program Files\Java\jdk1.8.0_144\bin\java ...
SLF4J: Failed to load class "org.slf4j.impl.StaticLoggerBinder".
SLF4J: Defaulting to no-operation (NOP) logger implementation
SLF4J: See http://www.slf4j.org/codes.html#StaticLoggerBinder for further details.
hello world
总执行了1条规则
Picked up JAVA_TOOL_OPTIONS: -Dfile.encoding=UTF-8 -Duser.language=en -Duser.country=US
```

图2-4　规则执行结果

KieSession 是一个接口,当然这里并不是要研究它的接口,而是找它操作 Fact 对象的方法。通过 IDE 代码提示功能发现 insert(Object object) 方法,如图 2-5 所示。

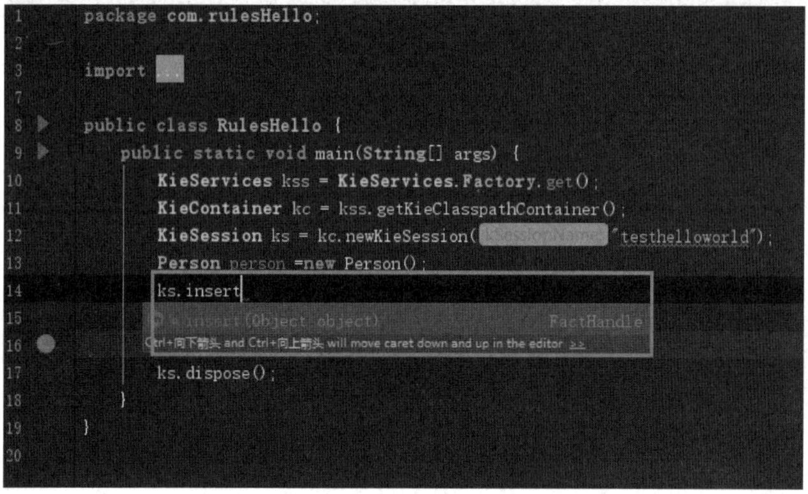

```
 1  package com.rulesHello;
 2
 3  import ...
 7
 8  public class RulesHello {
 9      public static void main(String[] args) {
10          KieServices kss = KieServices.Factory.get();
11          KieContainer kc = kss.getKieClasspathContainer();
12          KieSession ks = kc.newKieSession(SessionName  "testhelloworld");
13          Person person =new Person();
14          ks.insert
15          ⓘ ▾ insert(Object object)                    FactHandle
16          Ctrl+下箭头 and Ctrl+上箭头 will move caret down and up in the editor  ≫
17          ks.dispose();
18      }
19  }
20
```

图2-5　KieSession操作Fact对象的方法

重新编辑代码,实例化 Person 类,并使用 insert 方法再次调用,其代码为:

```
package com.rulesHello;

import com.pojo.Person;
import org.kie.api.KieServices;
import org.kie.api.runtime.KieContainer;
import org.kie.api.runtime.KieSession;

public class RulesHello {
    public static void main(String[] args) {
        KieServices kss = KieServices.Factory.get();
        KieContainer kc = kss.getKieClasspathContainer();
        KieSession ks =kc.newKieSession("testhelloworld");
```

```
        Person person=new Person();
        ks.insert(person);
        int count = ks.fireAllRules();
        System.out.println("总执行了"+count+"条规则");
        ks.dispose();
    }
}
```

执行 RulesHello 类 main 函数，执行结果如图 2-6 所示。

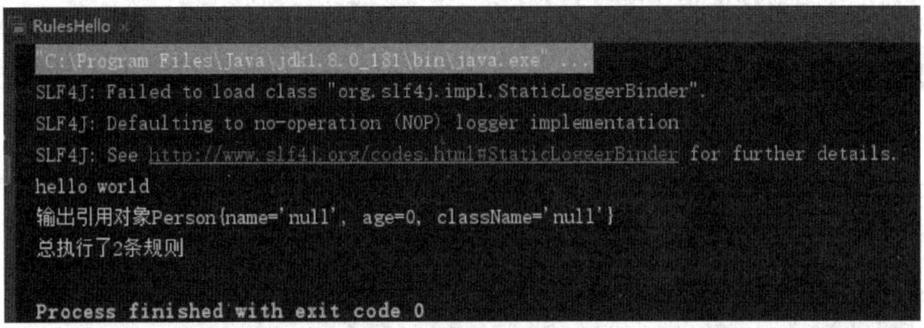

图2-6 调用规则后的结果

本节主要讲解引用对象，通过例子说明引用对象后如何使用其属性。

【例1】查看对象实例中是否有一个名字是张三且年龄为 30 岁的人。

一般 Java 代码写法，其内容为：

```
public static void main(String[] args) {
        //实际代码肯定是通过其他方法传入的，这里为了测试直接将值固定
        Person person=new Person();
        person.setName("张三");
        person.setAge("30");
        if("张三".equals(person.getName()) && "30".equals(person.getAge())){
            System.out.println("输出 传入的参数中确实有一位叫张三且年龄在30岁的人");
            //... 再处理其他业务
        }
    }
```

规则引擎写法，其内容为：

```
    rule "test003"
        when
            $p:Person(name=="张三",age==30);
        then
            System.out.println("输出 传入的参数中确实有一位叫张三且年龄在30岁的人");
    end
```

为保证代码的完整性，在 hello.drl 下方直接追加代码 rule"test003"，相应的 Java 代码也需要在规则执行的类中为 Person 实例添加属性内容，其代码为：

```
public static void main(String[] args) {
    KieServices kss = KieServices.Factory.get();
    KieContainer kc = kss.getKieClasspathContainer();
    KieSession ks =kc.newKieSession("testhelloworld");
    Person person=new Person();
    person.setName("张三");
    person.setAge("30");
    ks.insert(person);
    int count = ks.fireAllRules();
    System.out.println("总执行了"+count+"条规则");
    ks.dispose();
}
```

结果如图 2-7 所示。

图2-7 规则执行结果

图 2-7 中执行了 3 条规则，是因为 rule test002 也成功执行了，只不过 rule test002 的约束只有 Fact 对象，并没有针对 Fact 属性有额外的约束。通过上例可以反映出一个问题，一般 Java 代码的写法，如果再加一个功能，如判断传入的对象 Person 不为空，就不得不在 Java 代码中添加一个 if 比较语句进行判断。虽然该实例中只有两个约束条件，但是一个业务健全的程序又何止这些，如风控系统，其中业务场景少则几千，多则上万，如果都要写在 Java 代码中，程序员估计就要"疯"了。而规则引擎的出现恰好能解决这类问题，对规则进行分组管理，极大地提高了程序的可维护性，将业务场景做成可配置的动态规则①，真正实现业务与代码的解耦合。

认真观察规则编写的 Java 代码，做年龄判断时代码是需要加上双引号的，否则无法通过编译。而规则对于数字型字符串是会自动转化的。但要注意的是，只限于使用 == 号！

介绍规则体 RHS 部分时，提到了 RHS 是真正处理业务的部分，既然已经成功地将引用对象使用在规则中，就必须真正把 RHS 部分利用起来，下面举例说明。

【例2】将名为张三且年龄为 30 岁的人改为 40 岁。

修改规则代码为：

---

① 动态规则：泛指可动态变更的规则，俗称动态业务，第 4 章中会有详细说明。

```
rule "test004"
    when
        $p:Person(name=="张三",age==30);
    then
        $p.setAge("40");
        System.out.println("将名为张三且年龄为30岁的人改为40岁");
end
```

调用执行规则代码为:

```
public static void main(String[] args) {
    KieServices kss = KieServices.Factory.get();
    KieContainer kc = kss.getKieClasspathContainer();
    KieSession ks = kc.newKieSession("testhelloworld");
    Person person = new Person();
    person.setName("张三");
    person.setAge("30");
    ks.insert(person);
    int count = ks.fireAllRules();
    System.out.println("总执行了" + count + "条规则");
    System.out.println("输出修改后的Person age" + person.getAge());
    ks.dispose();
}
```

执行结果如图 2-8 所示。

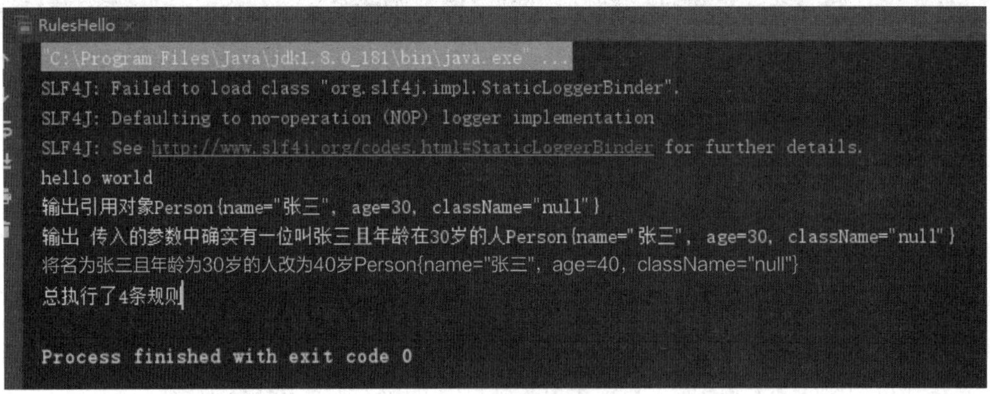

图2-8 规则执行结果

控制台中输出修改后 age=40 的信息,这一切看似没有问题,但结果真的被改变了吗? 可以做个实验,添加一个规则,其内容为:

```
rule "test004"
    when
        $p:Person(name=="张三",age==30);
    then
        $p.setAge("40");
        System.out.println("将名为张三且年龄为30岁的人改为40岁");
end
```

```
rule "test005"
    when
        $p:Person(name=="张三",age==40);
    then
        System.out.println("规则test005执行成功");
end
```

观察代码中加粗的部分,其他代码不变,再次执行调用规则的 Java 代码。再次观察控制台,输出结果与上一次是一样的,rule test005 没有被执行。规则代码在编译过程中是正确的,这在 RHS 部分就有说明,该部分是处理实际业务的,经常处理的函数如 update、insert、delete 等。那么先从 update 函数开始,将规则稍作修改,在 RHS 部分加上 update 函数操作,其内容为:

```
rule "test004"
    when
        $p:Person(name=="张三",age==30);
    then
        $p.setAge(40);
        update($p);
        System.out.println("将名为张三且年龄为30岁的人改为40岁"+$p);
end
```

执行结果如图 2-9 所示。

```
RulesHello  ×
 C:\Program Files\Java\jdk1.8.0_181\bin\java.exe ...
 SLF4J: Failed to load class "org.slf4j.impl.StaticLoggerBinder".
 SLF4J: Defaulting to no-operation (NOP) logger implementation
 SLF4J: See http://www.slf4j.org/codes.html#StaticLoggerBinder for further details.
 hello world
 输出引用对象Person{name="张三", age=30, className="null"}
 输出 传入的参数中确实有一位叫张三且年龄在30岁的人Person{name="张三", age=30, className="null"}
 将名为张三且年龄为30岁的人改为40岁Person{name="张三", age=40, className="null"}
 规则test005执行成功Person{name="张三", age=40, className="null"}
 总执行了5条规则

 Process finished with exit code 0
```

图2-9 规则执行结果

可以看出代码 rule test005 已被执行。在规则代码中,每一个 RHS 部分都输出 $p,它的作用是为了证明 Fact 对象是引用,而非克隆的,并且可以方便地对引入对象进行操作。

从控制台可以看出来,规则代码添加 update 后 rule test005 就被执行了。其实这是因为 Rete 算法的问题,简单说明一下,Rete 算法会将规则中的内容先全部加载出来,在规则中看似把 Person 的 name 属性改变了,但本质上只是引用发生了改变,Fact 对象并没有真正改变。当 Fact 对象真正发生改变时,规则将再次被执行,但这样是有风险的,容易产生死循环,其解决方案会在后面的章节中详细说明。

2.3 Drools配置文件

前面介绍了规则文件、规则文件内容，并且通过测试用例与执行结果的分析对规则引擎有了一个初步的认知。本节主要讲解 Drools 的配置文件，主要配置文件有两个，即 pom.xml 与 kmodule.xml，核心内容是 kmodule.xml 的说明。下面先来介绍 pom.xml，这里用 Maven 做管理项目，但这并不是唯一的。可以通过 Eclipse 安装插件创建项目，但要注意 Drools 的版本问题。如果是 IDEA，就会自带 Drools 插件，否则 .drl 文件的图标也不会变。

通过 pom.xml 配置文件要用 KIE 操作统一管理规则，其中 drools-compiler 是必须引用的包。这个包的主要目的是对规则进行编译、构建等。

kmodule.xml 是一个规则引擎的核心配置文件，它与 Spring 类似，都需要有相对应的配置文件做统一管理。它的主要功能是用来设置规则库名称、包路径、规则会话名称、规则会话类型等。

kmodule.xml 文件要放到 src/main/resources/META-INF/ 文件夹下，其原因通过分析如图 2-10 所示的源码可以得到。

```
public static final String               = "kmodule.xml";
public static final String               = "META-INF/" +
public static final String                             = "META-INF/kmodule.info"
public static final String               = "src/main/resources/" +
public static final String                             = "META-INF/kmodule-spring.xml";
```

图2-10 kmodule.xml 配置文件路径说明

kmodule.xml 配置文件，其内容为：

```xml
<?xml version="1.0" encoding="UTF-8"?>
<kmodule xmlns="http://www.drools.org/xsd/kmodule">
    <kbase name="rules" packages="rules.rulesHello">
        <ksession name="testhelloworld"/>
    </kbase>
</kmodule>
```

源码分析通过以下几点进行说明。

（1）一个 kmodule.xml 配置中可包含多个 KieBase，每一个 KieBase 都有 name 属性，可以取任意字符串，但不能重名。

（2）KieBase 有一个 packages 属性，它的值是一段字符串形式的路径信息，其内容是指 src/main/resources 目录下文件夹的名称，或者称之为包名，规则引擎会根据 packages 定义的内容查找规则文件。可以同时定义多个包，以逗号进行分隔。每一个 KieBase 可以包含多个 KieSession。

（3）每一个 KieSession 都有一个名称，名称可以是任意字符串，但是不能重复[1]。Java 代码中 KieSession 设置的 name 就是在执行规则代码时的 name，是用来指定操作具体规则的。

---

[1] Drools 6.4 版本中，KieBase 名称不可以与 KieSession 名称相同。

（4）packages 指定当前路径下的所有规则文件都会被读取。例如，在 rules/Hello 目录下定义了 10 个规则文件，运行时，这 10 个文件都会被加载，当然也只有满足条件的规则文件才会被执行。注意，如果在 rules/Hello 下有子目录，那么子目录中的规则文件不会被加载。

（5）避免分析（4）中的问题，解决方案有以下 3 种。

① 设计多个规则文件目录，每一种只放一个规则文件。虽然这种方式比较烦琐，但确实可以解决这类问题。

② 指定规则名，规则名称指规则文件中的 rule 参数，第 7 章中有指定规则名称的详细说明。

③ 指定规则文件名的方式，这种方式是通过 API 实现的，与第二种相似。

（6）就分析（4）而言，要注意的问题很多，同一目录下规则文件中的规则名称不能重复[①]，当目录中含有多个规则文件时，通过 insert 进行插入对象操作，则规则文件中的所有规则都将有这样的值，规则之间既相互独立，又相互依赖。

通过上述分析，编写过程中要注意的事项还是很多的。下面分析一下执行调用规则的 Java 代码，删除一些无关的代码，重点关注加粗部分。

```java
public static void main(String[] args) {
    KieServices kss = KieServices.Factory.get();
    KieContainer kc = kss.getKieClasspathContainer();
    KieSession ks = kc.newKieSession("testhelloworld");
    Person person = new Person();
    ks.insert(person);
    int count = ks.fireAllRules();
    ks.dispose();
}
```

"KieServices kss = KieServices.Factory.get();" 是指通过单例模式创建 KieServices，既可以访问所有 KIE 构建和运行时的接口，也可以理解为 KIE 的服务。

"KieContainer kc = kss.getKieClasspathContainer();" 是指创建一个 KieContainer 来读取路径中需要构建的文件，是 KIE 的容器，它提供了很多操作规则库的方法。KieServices 与 KieContainer 均属于初始化过程，可以放在静态块或其他加载时就会执行的方法中。

在运行时，KieContainer 会根据 ***Mode 对象来创建 KieModule、KieBase、KieSession 对象。一般情况下，KieModule 和 KieBase 只会创建一次，而 KieSession 则有可能被创建多次，因为 KieSession 的创建成本很低，同时 KieSession 包含了运行时的数据，所以可以进行销毁或创建若干次数。

"KieSession ks = kc.newKieSession("testhelloworld");" 是指新建一个 KIE 会话，通过参数进行获取，newKieSession 的参数正是在 kmodule.xml 中配置的名称。

"ks.insert(person);int count = ks.fireAllRules();ks.dispose();" 是指实际操作规则的代码，执行规则是 fireAllRules 方法，返回类型是数字，表示成功匹配了多少条规则。

---

① 规则名称不能重复：指在当前规则库中相同逻辑路径下的规则名不能重复。

代码执行过程中，控制台会出现如图 2-11 所示的 SLF4J 日志异常信息。

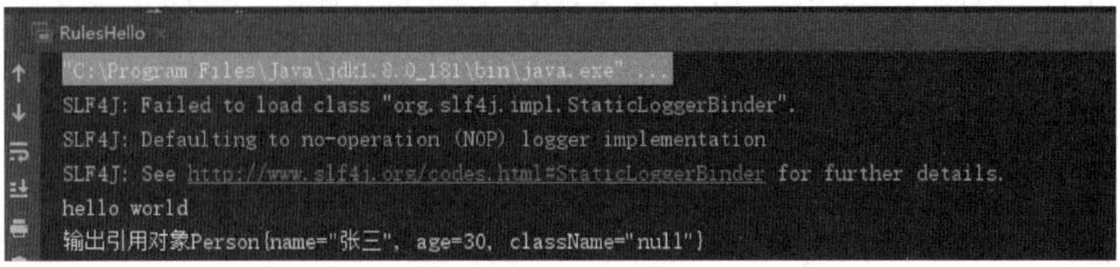

图2-11　日志问题

修改 pom.xml 配置文件，添加日志操作，其内容为：

```xml
<properties>
        <drools.version>7.8.0.Final</drools.version>
        <log4j2.version>2.5</log4j2.version>
</properties>
        <!-- start Log4j2  -->
        <dependency>
            <groupId>org.apache.logging.log4j</groupId>
            <artifactId>log4j-api</artifactId>
            <version>${log4j2.version}</version>
        </dependency>
        <dependency>
            <groupId>org.apache.logging.log4j</groupId>
            <artifactId>log4j-core</artifactId>
            <version>${log4j2.version}</version>
        </dependency>
        <dependency>
            <groupId>org.apache.logging.log4j</groupId>
            <artifactId>log4j-slf4j-impl</artifactId>
            <version>${log4j2.version}</version>
        </dependency>
        <!-- end Log4j2    -->
```

添加日志配置文件 log4j2.xml 到资源文件夹，其内容为：

```xml
<?xml version="1.0" encoding="utf-8"?>
<configuration>
    <properties>
        <property name="pattern">%date{yyyy-mm-dd hh:mm:ss} [%-5level] %logger{*.*...} - %msg%n%ex</property>
    </properties>
    <appenders>
        <console name="console" target="system_out">
            <patternlayout pattern="${pattern}"/>
        </console>
    </appenders>
    <loggers>
        <root level="debug">
            <appenderref ref="console"/>
```

```
        </root>
        <logger name="org.springframework" level="warn"/>
    </loggers>
</configuration>
```

日志文件可以根据需要进行设置，上述文件只是样板，不需要完全一样。配置完成后，再次运行 Java 代码，查看控制台显示信息，如图 2-12 所示。

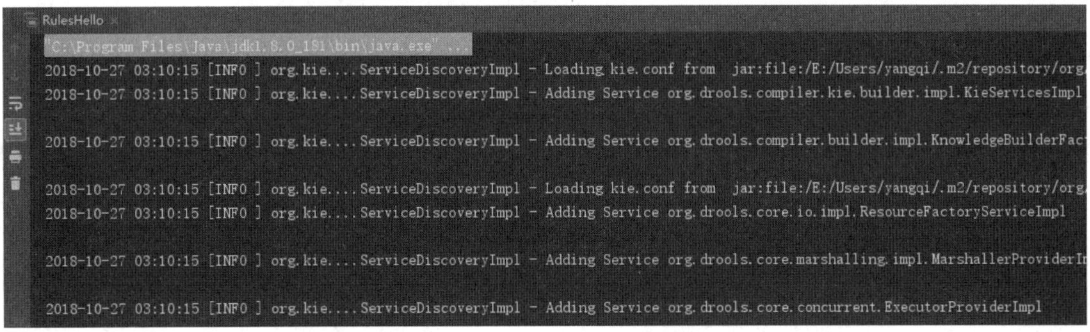

图2-12　引用日志后的结果

汇报日志异常时，因为项目代码并没有使用任何与日志相关的信息，所以异常是在调用 Drools 过程中抛出的。根据这一猜想，通过在 IDE 中打断点的方式定位问题所在，当代码执行"KieServices kss = KieServices.Factory.get();"后就会出现此问题，通过对这段代码的分析，终于在 KieContainerImpl 类中找到"元凶"，如图 2-13 所示。

图2-13　源码分析

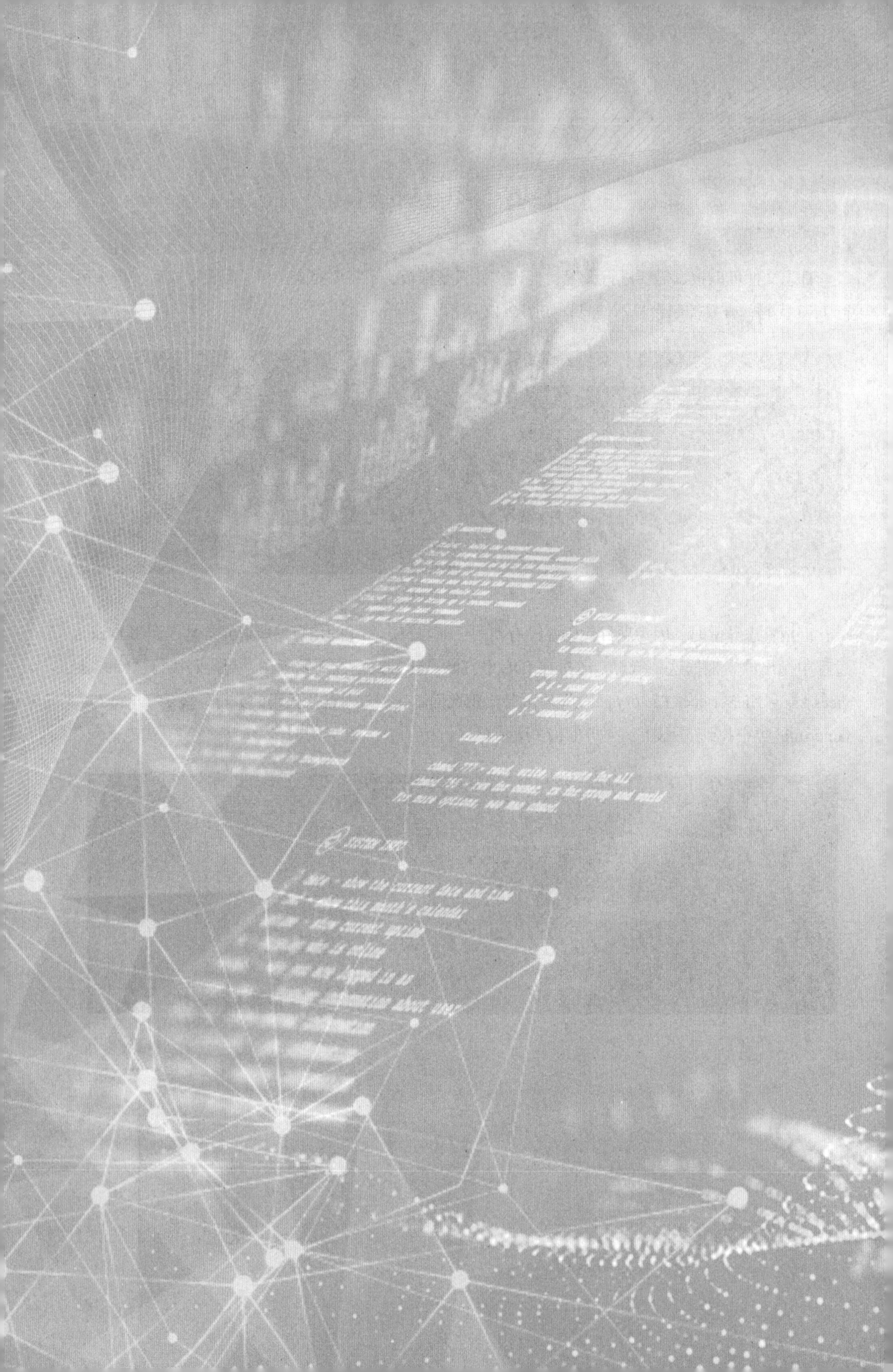

Drools 规则引擎技术指南

第二篇

基础篇

第3章
Drools基础语法

3.1 规则文件

Drools 规则引擎中，标准的规则文件就是以 ".drl" 结尾的文本文件，由于它是标准的文本文件，因此可以通过一些记事本工具对其进行查阅和编辑。规则内容[①] 是放在规则文件中的，一个规则文件可以存放多个规则体。除此之外，规则文件还可以存放用户自定义的函数、数据对象及自定义查询等。

一套完整的规则文件内容如表 3-1 所示。

表3-1 一套完整的规则文件内容

关 键 字	描 述
package	包名，只限于逻辑上的管理，若自定义的查询或函数位于同一包名，不管物理位置如何，都可以直接调用
import	规则引用问题，导入类或方法
global	全局变量
function	自定义函数
queries	查询
rule end	规则体

package：除 package 之外，其他关键字在规则文件中的顺序是任意的，规则文件中必须要有一个 package 声明，并且 package 声明要放在规则文件的第一行（规则模板[②] 除外）。规则文件中的 package 和 Java 语言中的 package 有相似之处，不同的是在 Java 文件中 package 是用来把功能相似或相关的文件放在同一个 package 下进行管理。这种 package 管理既有物理上 Java 文件位置的管理，又有逻辑上文件位置的管理；在 Java 文件中通过 package 管理文件要求文件位置在逻辑上与物理上都要保持一致。在 Drools 规则引擎的规则文件中，package 对于规则文件中规则的管理只限于逻辑上的管理，并不管规则文件所在的物理目录，这是规则文件与 Java 类文件中 package 的区别。

同一个 package 下，用户可以自定义函数、自定义查询等，不管这些函数与查询是否在同一个规则文件中，都是可以直接使用的，这与 Java 中同一 package 的 Java 类调用相似。

import：用来导入类名或静态方法。

global：又称全局变量，使用时需要单独定义变量类型。

function：自定义函数，读者可以理解为 Java 静态方法的一种变形，与 JavaScript 函数定义相似。

queries：表示查询，具体说明请查阅相关章节进行学习。

rule end：规则内容中的规则体，是进行业务规则判断、处理业务结果的部分，也是本节主讲的核心内容之一。

---

①规则内容：指符合规则语法规范的文本内容。
②规则模板：指通过嵌套模板语法，制订生成规则内容。

3.2 规则体语法结构

规则体语法结构如表 3-2 所示。

表3-2 规则体语法结构

关键字	描述
rule	规则开始，参数是规则的唯一名称
attributes	规则属性，是rule与when之间的参数，为可选项
when	规则条件部分，默认为true
then	规则结果部分
end	当前规则结束

一个规则体包含的 3 个部分，唯有 attributes 部分是可选的，其他关键字都是必填信息。属性可选并不表示没有，属性是有默认值的，如规则默认是被激活的。

3.3 pattern（匹配模式）

LHS 部分由一个或多个条件组成，条件又称为 pattern。多个 pattern 之间既可以使用 and 或 or 进行连接，又可以使用小括号确定 pattern 的优先级，默认条件是 true。

例如，第 2 章中 "Hello World" 例子规则中的 LHS 部分 "$p:Person(name==" 张三 ",age==30)"，其中 $p 为绑定变量名，其语法结构为：[绑定变量名 :Object(Field 约束)]。

pattern 的 "绑定变量名" 是可选的，当前规则的 RHS 部分需要操作 pattern 匹配的参数，若要用到某些对象，则可以通过为该对象设定一个绑定变量名来实现对它的操作。对于绑定变量的命名，通常是为其添加一个 "$" 符号作为前缀，与对象的命名方法相同；绑定变量不仅可以用在对象上，也可以用在对象的属性上，作用是方便 RHS 部分的操作，同时也避免与 Fact 对象属性的使用相混淆。

"Field 约束" 是指当前对象中属性或方法的使用，如添加条件限制 "name==" 张三 ",age==30"。

规则体中 LHS 部分绑定变量基本上有两种形式：一种是整个 Fact 变量的绑定，另一种是约束条件属性变量的绑定。

第一种方式：

```
rule "rule1"
    when
```

```
    $customer:Customer(age>20,gender=='male')
    then
        ....
end
```

第二种方式：

```
rule "rule1"
    when
    $customer:Customer(age>20,$g:gender=='male')
    then
        ....
End
```

上述两种方式没有本质的区别，主要考虑是在约束条件中使用，还是在结果设置中使用，即是在 LHS 多个匹配模式下，还是 RHS 需要单独处理某一个 Fact 对象属性。

多个匹配模式是指多个 Fact 对象的匹配，执行匹配时只有结果都为 true 才会被执行。上述的代码中，约束条件有 3 个：① 必须有 insert(Customer 对象)；② age 必须是大于 20 的；③ gender 必须等于 'male'。

多匹配模式说明的内容为：

```
rule "rule1"
    when
    $customer:Customer(age>20,gender=='male')
    Order(customer==$customer,price>1000)
    then
        ....
end
```

分析上述规则代码。

第一个：pattern(模式) 有以下 3 个约束。

① 对象类型必须是 Cutomer。

② Cutomer 的 age 要大于 20。

③ Cutomer 的 gender 是 'male'。

第二个：pattern(模式) 有以下 3 个约束。

① 对象类型必须是 Order。

② Order 对应的 Cutomer 必须是上一个匹配模式中的 Customer。

③ 当前 Order 的 price 要大于 1000。

pattern 没有符号连接，Drools 语法默认的约束为逻辑与 (and)，当该规则的 LHS 部分中两个 pattern 都满足时才会被执行。then 默认情况下，每行可以用 ";" 作为结束符（与 Java 的结束符一样），当然行尾是可以不加 ";" 的。

注意：接下来的讲述中，将 Person 的 age 属性变更为 int 类型。

3.4　运算符

运算符是在程序中最常用的计算方法，一般的运算符包括"+""-""*""/""%"等，优先级与 Java 相同。下面通过"/"运算举例说明。

编辑 hello.drl，添加 test006 规则，其内容为：

```
rule "test006"
    when
        $p:Person(age/2==20);
    then
        System.out.println("规则test006规则执行成功"+$p);
end
```

执行调用规则代码，运行结果如图 3-1 所示。

图3-1　操作除号取除结果

% 取模的操作是常用的运算方法，编辑 hello.drl，添加 test007 规则，其内容为：

```
rule "test007"
    when
        $p:Person(age%2==0);
    then
        System.out.println("规则test007规则执行成功"+$p);
end
```

执行调用规则代码，运行结果如图 3-2 所示。

图3-2　取模操作异常结果

为了验证测用例的语法是否正确，查阅官方文档后，找到取模运算符的操作，证明是可以使用的，如图 3-3 所示。

multiplicative	* / %
additive	\+ -

图3-3　官方取模说明

如图 3-4 所示，通过参考官方提供的取模测试用例进行运算时，需要先进行取模运算的比较。然而规则 test007 只有一个匹配模式，属性约束条件也只有一个，我对需要添加括号产生了疑问。

> It is possible to change the evaluation priority by using parentheses, as in any logic or mathematical expression:
>
> Person(age > 100 && (age % 10 == 0))

图3-4　官方取模用例

为了验证官方文档上提到的使用括号方式是否正确，编辑规则添加的属性约束条件，添加",age>10"。再次执行调用规则代码，结果与图 3-4 所示的效果一样。虽然测试失败了，但根据官方文档提供括号"()"的作用来看，应该可以解决以下问题。

编辑 test007 规则，其内容为：

```
rule "test007"
    when
        $p:Person((age%2)==0);
    then
        System.out.println("规则test007规则执行成功"+$p);
end
```

再次执行调用规则代码，结果如图 3-5 所示，也就是说，在 % 取模运算上，规则需要控制优先级，即先运算后匹配。

```
hello world
输出引用对象Person{name="张三", age=30, className="null"}
输出 传入的参数中确实有一位叫张三年龄在30岁的人Person{name="张三", age=30, className="null"}
将名为张三且年龄为30岁的人改为40岁Person{name="张三", age=40, className="null"}
规则test005规则执行成功Person{name="张三", age=40, className="null"}
规则test006规则执行成功Person{name="张三", age=40, className="null"}
规则test007规则执行成功Person{name="张三", age=40, className="null"}
2018-47-27 04:47:57 [DEBUG] org.drools...DefaultAgenda - State was FIRING_ALL_RULES is now HALTING
2018-47-27 04:47:57 [DEBUG] org.drools...DefaultAgenda - State was HALTING is now INACTIVE
总执行了7条规则
2018-47-27 04:47:57 [DEBUG] org.drools...DefaultAgenda - State was INACTIVE is now DISPOSED

Process finished with exit code 0
```

图3-5　通过括号取模结果

3.5 约束连接

匹配模式中可以有多种约束符的连接，常用的有"&&"（and）、"||"(or) 和","(and)。这3个连接符号如果没有用括号来显示定义的优先级，那么"&&"优先级大于"||"优先级。从表面上看","与"&&"具有相同的含义，但是在 Drools 6.4 版本中，","与"&&"和"||"不能混合使用，即在有"&&"或"||"出现的 LHS 部分，是不可以有","连接符出现的，反之亦然。可能是 Drools 规则引擎的研发人员觉得这样做并不好，所以在 Drools 7.10 版本进行测试后竟然可以同时存在了。

Drools 提供了 12 种类型比较操作符，如果进行常量比较，必须通过函数或引用比较对象属性进行比较，不能单独使用，其中 > 或 <，>= 或 <=，== 或 != 语法与 Java 是一样的。

本节的核心内容是 Drools 自带的约束条件，共有 6 种比较操作符（共三组），每一组操作相同、功能相反，其内容为：

```
contains | not contains
memberOf | not memberOf
matches  | not matches
```

1. contains 比较操作符

contains 是用来检查一个 Fact 对象的某个属性值是否包含一个指定的对象值。其语法格式为：

```
Object( field[Collection/Array] contains|not contains value )
```

创建 School.java 类，放在 pojo 包下，其代码为：

```
package com.pojo;

public class School {
    private String className;
    private String classCount;
    public String getClassName() {
        return className;
    }
    public void setClassName(String className) {
        this.className = className;
    }
    public String getClassCount() {
        return classCount;
    }
    public void setClassCount(String classCount) {
        this.classCount = classCount;
    }
}
```

修改 Person.java 类，添加属性 className，其代码为：

```
package com.pojo;
```

```java
public class Person {
    private String name;//姓名
    private int age;//年龄
    private String className;//所在班级
    public Person(String name, int age) {
        this.name = name;
        this.age = age;
    }
    public Person() {
    }
    public String getName() {
        return name;
    }
    public void setName(String name) {
        this.name = name;
    }
    public int getAge() {
        return age;
    }
    public void setAge(int age) {
        this.age = age;
    }
    public String getClassName() {
        return className;
    }
    public void setClassName(String className) {
        this.className = className;
    }
}
```

创建规则文件 contains.drl 的第一种写法，目录为 rules/constraint/isContains，并添加以下代码：

```
package rules.constraint.isContains;

import com.pojo.Person;
import com.pojo.School;

rule containsTest
    when
        $s:School();
        $p:Person(className contains $s.className);
    then
       System.out.println("恭喜你，成功地使用了 contains");
end
```

修改 kmodule.xml 配置文件，添加如下配置：

```xml
<kbase name="contains" packages="rules.constraint.isContains">
    <ksession name="contains"/>
</kbase>
```

创建执行调用规则代码 RulesConstraint 类，目录为 com/rulesConstraint，并添加以下代码：

```
package com.rulesConstraint;

import com.pojo.Person;
import com.pojo.School;
import org.junit.Test;
import org.kie.api.KieServices;
import org.kie.api.runtime.KieContainer;
import org.kie.api.runtime.KieSession;

public class RulesConstraint{
    @Test
    public void contains() {
        KieServices kss = KieServices.Factory.get();
        KieContainer kc = kss.getKieClasspathContainer();
        KieSession ks = kc.newKieSession("contains");
        Person person = new Person();
        person.setName("张三");
        person.setAge(30);
        person.setClassName("一班");
        School school = new School();
        school.setClassName("一班");
        ks.insert(person);
        ks.insert(school);
        int count = ks.fireAllRules();
        System.out.println("总执行了" + count + "条规则");
        ks.dispose();
    }
}
```

执行 contains() 方法查看输出结果，如图 3-6 所示。

图3-6　contains第一种写法测试结果

第二种写法与第一种相似，只不过使用的 Fact 对象是 get 方法，其内容为：

```
package rules.constraint.isContains;

import com.pojo.Person;
import com.pojo.School;

rule containsTest
    when
```

```
        $s:School();
        $p:Person(className  contains $s.getClassName());
    then
        System.out.println("恭喜你，成功地使用了 contains");
end
```

结果与图 3-6 所示是一样的，使用第二种写法时不能忽略 get，只写 className() 是错误的。

添加 rule containsTest002，将 contains 的第二个参数变成一个常量，其内容为：

```
rule containsTest002
    when
        $s:School();
        $p:Person(className  contains "一班");
    then
        System.out.println("规则 containsTest002 恭喜你，成功地使用了 contains");
end
```

执行调用规则代码，结果如图 3-7 所示。证明比较符是可以直接进行常量比较操作的。

图3-7 contains常量比较

2. not contains 比较运算符

not contains 的作用与 contains 相反，它是用来判断一个 Fact 对象的某个字段不包含一个指定的对象。

修改规则文件，其内容为：

```
rule containsTest003
    when
        $s:School();
        $p:Person(className  contains $s.className);
    then
        $s.setClassName("二班");
        update($s)
end
rule containsTest004
    when
        $s:School();
        $p:Person(className not  contains $s.className);
```

```
    then
        System.out.println("规则 containsTest004 恭喜你,成功地使用了 not con-
tains");
end
```

在 containsTest003 规则中修改 School Fact 对象的值,并在 rule containsTest004 中进行比较。执行调用规则代码结果如图 3-8 所示。

若删除规则体 containsTest003,则规则体 containsTest004 不会执行。

图 3-8　not contains测试结果

3. memberOf 比较运算符

memberOf 用来判断某个 Fact 对象的某个字段是否在一个或多个集合中。

memberOf 的语法为:

```
Object(fieldName memberOf|not memberOf value[Collection/Array])
```

修改 School.java 文件,添加 String[] classNameArray 属性并实现 get set 方法。

创建规则文件 memberOf.drl[①] 第一种写法,目录为 rules/constraint/isMemberOf,并添加以下代码:

```
package rules.constraint.isMemberOf;

import com.pojo.Person;
import com.pojo.School;

rule memberOfTest001
    when
        $s:School();
        $p:Person(className    memberOf $s.getClassName());
    then
        System.out.println("恭喜你,成功地使用了 memberOf");
end
```

修改 kmodule.xml 配置文件,添加如下配置:

```
<kbase name="memberOf" packages="rules.constraint.isMemberOf">
```

[①] 比较运算符的使用,两种语法都是没有问题的,第一种调用 get 方法;第二种直接使用成员变量名。注意,引用成员变量时不能加()。

```xml
    <ksession name="memberOf"/>
</kbase>
```

添加执行调用规则，其代码为：

```java
@Test
public void memberOfArray() {
    KieServices kss = KieServices.Factory.get();
    KieContainer kc = kss.getKieClasspathContainer();
    KieSession ks = kc.newKieSession("memberOf");
    Person person = new Person();
    person.setName("张三");
    person.setAge(30);
    person.setClassName("一班");
    School school = new School();
    school.setClassNameArray(new String[]{"一班","二班","三班"});
    ks.insert(person);
    ks.insert(school);
    int count = ks.fireAllRules();
    System.out.println("总执行了" + count + "条规则");
    ks.dispose();
}
```

执行 memberOfArray() 方法查看输出结果，如图 3-9 所示。

图3-9 memberOf测试结果

经过上述测试结果，添加两个规则，如 memberOfTest002、memberOfTest003，其内容为：

```
rule memberOfTest002
    when
        $s:School();
        $p:Person(className memberOf "一班");
    then
        System.out.println("恭喜你 memberOfTest002, 成功地使用了 memberOf");
end

rule memberOfTest003
    when
        $s:School();
        $p:Person(className memberOf $s.classNameArray);
    then
```

```
        System.out.println("恭喜你 memberOfTest003，成功地使用了 member-
Of");
end
```

修改执行调用规则代码并运行，其代码为：

```
@Test
public void memberOfArray() {
    KieServices kss = KieServices.Factory.get();
    KieContainer kc = kss.getKieClasspathContainer();
    KieSession ks = kc.newKieSession("memberOf");
    Person person = new Person();
    person.setName("张三");
    person.setAge(30);
    person.setClassName("一班");
    School school = new School();
    school.setClassName("一班");
    school.setClassNameArray(new String[]{"一班","二班","三班"});
    ks.insert(person);
    ks.insert(school);
    int count = ks.fireAllRules();
    System.out.println("总执行了" + count + "条规则");
    ks.dispose();
}
```

执行结果如图 3-10 所示，常量匹配是没有问题的，但使用属性对象值则不行。如果猜测正确，在使用常量情况下，memberOf 第二个参数会转换为数组。

测试完数组后，下面分别对 List、Set、Map 进行测试。

① 测试 List，修改 School.java 文件，添加"private List classNameList;"成员变量并实现 get set 方法。

图 3-10　memberOf扩展测试

添加 rule memberOfTest004，其内容为：

```
rule memberOfTest004
    when
        $s:School();
```

```
            $p:Person(className memberOf $s.classNameList);
    then
        System.out.println("恭喜你 memberOfTest004，成功地使用了 memberOf");
end
```

添加测试方法，其代码为：

```
@Test
public void memberOfList() {
    KieServices kss = KieServices.Factory.get();
    KieContainer kc = kss.getKieClasspathContainer();
    KieSession ks = kc.newKieSession("memberOf");
    Person person = new Person();
    person.setName("张三");
    person.setAge(30);
    person.setClassName("一班");
    School school = new School();
    List classNameList = new ArrayList();
    classNameList.add("一班");
    classNameList.add("二班");
    classNameList.add("三班");
    school.setClassNameList(classNameList);
    ks.insert(person);
    ks.insert(school);
    int count = ks.fireAllRules();
    System.out.println("总执行了" + count + "条规则");
    ks.dispose();
}
```

执行 Java 方法 memberOfList()，输出结果如图 3-11 所示。

图3-11 memberOfList 测试结果

② 测试 Set，修改 School.java 文件，添加 "private Set classNameSet;" 成员变量并实现 get set 方法。

添加规则 memberOfTest005，其内容为：

```
rule memberOfTest005
    when
        $s:School();
```

```
        $p:Person(className memberOf $s.classNameSet);
    then
        System.out.println("恭喜你 memberOfTest005，成功地使用了 memberOf");
end
```

添加测试方法，其内容为：

```
@Test
public void memberOfSet() {
    KieServices kss = KieServices.Factory.get();
    KieContainer kc = kss.getKieClasspathContainer();
    KieSession ks = kc.newKieSession("memberOf");
    Person person = new Person();
    person.setName("张三");
    person.setAge(30);
    person.setClassName("一班");
    School school = new School();
    Set classNameSet = new HashSet();
    classNameSet.add("一班");
    classNameSet.add("二班");
    classNameSet.add("三班");
    school.setClassNameSet(classNameSet);
    ks.insert(person);
    ks.insert(school);
    int count = ks.fireAllRules();
    System.out.println("总执行了" + count + "条规则");
    ks.dispose();
}
```

执行 Java 方法 memberOfSet()，输出结果如图 3-12 所示。

图3-12　memberOf Set 测试结果

③ 测试 Map，修改 School.java 文件，添加"private Map classNameMap;"成员变量并实现 get set 方法。

添加规则 memberOfTest006，其内容为：

```
rule memberOfTest006
```

```
    when
        $s:School();
        $p:Person(className memberOf $s.classNameMap);
    then
        System.out.println("恭喜你 memberOfTest006,成功地使用了 memberOf");
end
```

添加测试方法，其代码为：

```
@Test
public void memberOfMap() {
    KieServices kss = KieServices.Factory.get();
    KieContainer kc = kss.getKieClasspathContainer();
    KieSession ks = kc.newKieSession("memberOf");
    Person person = new Person();
    person.setName("张三");
    person.setAge(30);
    person.setClassName("一班");
    School school = new School();
    Map classNameMap = new HashMap();
    classNameMap.put("一班","1");
    classNameMap.put("二班","2");
    classNameMap.put("三班","3");
    school.setClassNameMap(classNameMap);
    ks.insert(person);
    ks.insert(school);
    int count = ks.fireAllRules();
    System.out.println("总执行了" + count + "条规则");
    ks.dispose();
}
```

执行 Java 方法 memberOfMap ()，输出结果如图 3-13 所示。

图3-13 memberOf Map 测试结果

使用 Map 是一个有争议的测试，本例就结果而言是没有问题的，但是一旦将 Map 赋值的代码变更，如将 Map 中的 Key 设置为班级编号，value 设置为班级名称，那么结果 memberOfTest006 将不会输出。即使用 memerOf 比较符操作 Map 集合时，比较是否存在 Key。如果一个 Map 的 Key 都不存在，就不用提 Value 了。

4. not memberOf 比较运算符

not memberOf 与 memberOf 的作用恰恰相反，是用来判断 Fact 对象中某个字段值不在某个集合中。

添加规则 memberOfTest007，并执行 memberOfMap 测试方法，测试结果不再阐述。

```
rule memberOfTest007
    when
        $s:School();
        $p:Person(className not memberOf $s.classNameMap);
    then
       System.out.println("恭喜你 memberOfTest007，成功地使用了 member-Of");
end
```

5. matches比较运算符

matches 用来对某个 Fact 对象的字段与标准的 Java 正则表达式进行相似匹配，被比较的字符串可以是一个标准的 Java 正则表达式。但需要读者注意的是，正则表达式字符串中不用考虑"\"的转义问题，其语法为：

```
Object(fieldName matches | not matches "正则表达式")
```

创建规则文件 matches.drl，并添加如下代码：

```
package rules.constraint.isMatches;

import com.pojo.Person;

rule matchesTest001
    when
        $p:Person(name matches "张.*");
    then
       System.out.println("恭喜你，成功地使用了 matches");
end
```

修改 kmodule.xml 配置文件，并添加如下配置：

```
<kbase name="matches" packages="rules.constraint.isMatches">
    <ksession name="matches"/>
</kbase>
```

添加测试方法，其代码为：

```
@Test
public void matches() {
    KieServices kss = KieServices.Factory.get();
    KieContainer kc = kss.getKieClasspathContainer();
    KieSession ks = kc.newKieSession("matches");
    Person person = new Person();
```

```
        person.setName("张三");
        person.setAge(30);
        person.setClassName("一班");
        ks.insert(person);
        int count = ks.fireAllRules();
        System.out.println("总执行了" + count + "条规则");
        ks.dispose();
}
```

执行结果如图 3-14 所示。

图3-14 matches匹配操作结果

简单来说就是模糊查询语法不能是 "*.三",否则会抛出语法错误。该规则用来查找所有 Person 对象的 name 属性是不是以 "张" 字开头,若条件成立,则输出正确结果。使用 $ 符号也是成立的,$ 符号是指 "." 后一位,而 "*" 则可以多位。但如果写成 "**" 就会抛出语法错误。$ 符号只会匹配一次,虽然 $$ 符号不会报错,但规则不成立。例如,将 "person.setName(" 张小三 ");" 分别测试规则为 "张 .$$" "张 .**" 两种,如图 3-15 所示。

图3-15 使用两个通配符的结果

通过上述测试得出一个结论:matches 的第二个参数是字符串,并且可以进行中文匹配。以 "." 作为分隔符,比较符第二个参数进行模糊匹配,即 "*" 部分,并且必须以 "." 的方式隔开,英文字符 "," 同理。以 "." 隔开的前面部分也必须是 insert 参数属性中的一部分。

简单地说,就是根据 insert 参数属性内容进行以 "." 为分隔符的正则匹配比较运算功能。在英文模式下使用 $ 符号并不可行。

图 3-15 所示的结果是通过测试得出的。官方解释说 matches 可以匹配任何有效 Java 正则表达

式的字段。通常 regexp 是字符串文字，但也允许解析为有效正则表达式的变量。下面通过测试进行总结。

官方例子：
Cheese(type matches "(Buffalo)?\\S*Mozzarella")

第一次测试：
修改person.setName()为person.setName("zs")，
修改规则$p:Person(name matches "(zs)");
执行matches()方法，输出结果与图3-14一样
第二次测试：
修改person.setName()为person.setName("zs")，
修改规则$p:Person(name matches "(z)");
执行matches()方法，输出结果是规则没有成立
第三次测试：
修改person.setName()为person.setName("zs")，
修改规则$p:Person(name matches "(z)[a-z]");
执行matches()方法，输出结果与图3-14一样
第四次测试：
修改person.setName()为person.setName("zs")，
修改规则$p:Person(name matches "(zs)[a-z]");
执行matches()方法，输出结果是规则没有成立
第五次测试：
修改person.setName()为person.setName("zs")，
修改规则$p:Person(name matches "(z|s)[a-z]");
执行matches()方法，输出结果与图3-14一样
第六次测试：
修改person.setName()为person.setName("zs")，
修改规则$p:Person(name matches "(z|s|l)[a-z]");与
修改规则$p:Person(name matches "(z|s|)[a-z]");
执行matches()方法，输出结果与图3-14一样
第七次测试：
修改person.setName()为person.setName("zsx")，
修改规则$p:Person(name matches "(z|s|x)[a-z]");
执行matches()方法，输出结果是规则没有成立
第八次测试：
修改person.setName()为person.setName("zsx")，
修改规则$p:Person(name matches "(z|*)[a-z]");
执行matches()方法，输出结果是规则语法错误
第九次测试：
修改person.setName()为person.setName("zsx")，
修改规则$p:Person(name matches "(z.*)[a-z]");
执行matches()方法，输出结果与图3-14一样

上述测试结果中与第一个例子明显不同，即参数加了"()"。"()"的作用类似于增加了匹配优先级，先进行正则匹配，如果正则配置成功，就会进行上述说明的 insert 参数属性内容的匹配。至于"|"符号，在第七次测试及后来的测试中，通过执行结果发现"|"与"||"和"&&"功能相似，

都是进行逻辑判断的。

6. not matches比较运算符

not matches 的作用与 matches 相反,是用来将某个 Fact 对象的字段与一个 Java 标准正则表达式进行匹配,若与正则表达式不匹配,则规则成立。

添加规则 matchesTest002,其内容为:

```
rule matchesTest002
    when
        $p:Person(name  not matches  "(zs && s.*)[a-z]");
    then
        System.out.println("恭喜你,成功地使用了 not matches");
end
```

运行 matches() 方法,结果如图 3-16 所示。

7. soundslike比较运算符

soundslike 用来检查单词是否具有与给定值几乎相同的声音(使用英语发音)。基于 Soundex 算法的语法为:

```
Object(fieldName   soundslike 'value')
```

图3-16　not matches测试结果

创建规则文件 soundslike.drl,目录为 rules/constraint/isSoundslike,并添加如下代码:

```
package rules.constraint.isSoundslike;

import com.pojo.Person;

rule isSoundslikeTest001
    when
        $p:Person(name  soundslike "foobar");
    then
      System.out.println("恭喜你,成功地使用了 isSoundslike");
end
```

修改 kmodule.xml 配置文件,并添加如下配置:

```
<kbase name="soundslike" packages="rules.constraint.isSoundslike">
  <ksession name="soundslike"/>
```

```
</kbase>
```

修改 RulesConstraint.java 文件，并添加测试方法，其代码为：

```
@Test
public void soundslike() {
    KieServices kss = KieServices.Factory.get();
    KieContainer kc = kss.getKieClasspathContainer();
    KieSession ks = kc.newKieSession("soundslike");
    Person person = new Person();
    person.setName("fubar");
    person.setAge(30);
    person.setClassName("一班");
    ks.insert(person);
    int count = ks.fireAllRules();
    System.out.println("总执行了" + count + "条规则");
    ks.dispose();
}
```

执行 soundslike() 方法，结果如图 3-17 所示。

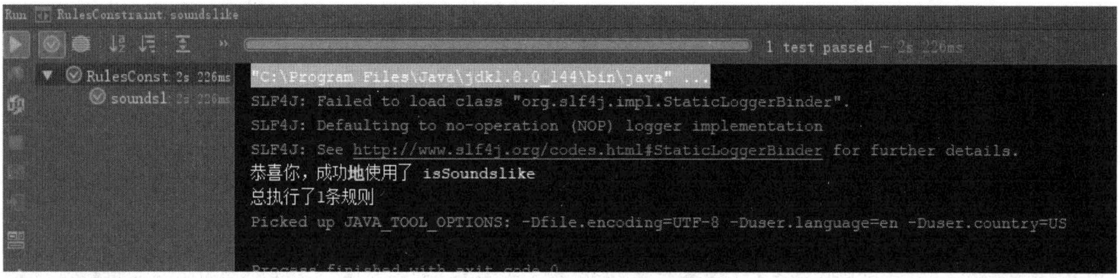

图3-17　soundslike测试结果

8. str比较运算符

str 不仅检查 String 字段是否以某一值开头/结尾，还可以判断字符串长度，其语法为：

```
Object(fieldName    str[startsWith|endsWith|length] "String"|1)
```

创建规则文件 str.drl，目录为 rules/constraint/isStr，并添加如下代码：

```
package rules.constraint.isStr;

import com.pojo.Person;

rule strTest001
    when
        $p:Person(name str[startsWith] "张");
    then
      System.out.println("恭喜你，成功地使用了 str startsWith");
end

rule strTest002
```

```
    when
         $p:Person(name str[endsWith] "三");
    then
         System.out.println("恭喜你,成功地使用了 str endsWith");
end
rule strTest003
    when
         $p:Person(name str[length] 3);
    then
         System.out.println("恭喜你,成功地使用了 isSoundslike");
end
```

修改 kmodule.xml 配置文件,并添加如下配置:

```
<kbase name="isStr" packages="rules.constraint.isStr">
    <ksession name="isStr"/>
</kbase>
```

修改 RulesConstraint.java 文件,并添加测试方法,其代码为:

```
@Test
public void str() {
    KieServices kss = KieServices.Factory.get();
    KieContainer kc = kss.getKieClasspathContainer();
    KieSession ks = kc.newKieSession("isStr");
    Person person = new Person();
    person.setName("张小三");
    person.setAge(30);
    person.setClassName("一班");
    ks.insert(person);
    int count = ks.fireAllRules();
    System.out.println("总执行了" + count + "条规则");
    ks.dispose();
}
```

执行 str() 方法,结果如图 3-18 所示。

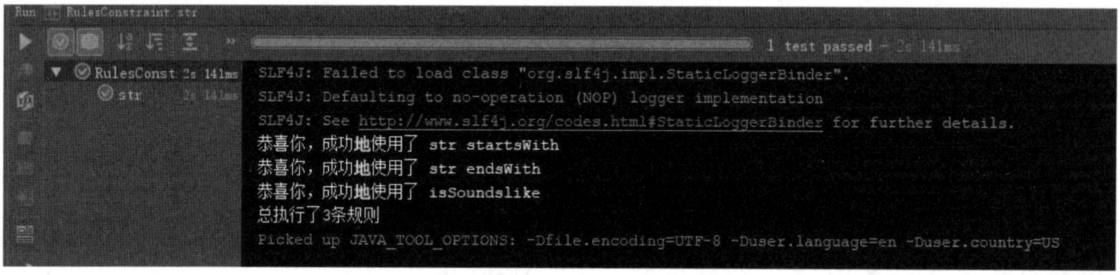

图3-18 str测试结果

3.6 语法扩展

规则中的集合处理遇到集合应该怎样获取它们的元素呢?下面针对 List、Map、Set 展开说明。

1. List元素操作

创建规则文件 collection.drl,目录为 rules/constraint/isCollection,并添加如下代码:

```
package rules.constraint.isCollection;

import com.pojo.Person;
import com.pojo.School;

rule collectionTestList
    when
        $s:School();
    then
        System.out.println("School属性classNameList的第二个元素为"+$s.get-ClassNameList().get(1));
end
```

修改 kmodule.xml 配置文件,并添加如下配置:

```
<kbase name="collection" packages="rules.constraint.isCollection">
    <ksession name="collection"/>
</kbase>
```

创建执行调用规则文件 rulesCollection.java,其代码为:

```
package com.rulesConstraint;

import com.pojo.Person;
import com.pojo.School;
import org.junit.Test;
import org.kie.api.KieServices;
import org.kie.api.runtime.KieContainer;
import org.kie.api.runtime.KieSession;

import java.util.*;

public class RulesCollection {
    @Test
    public void testList() {
        KieServices kss = KieServices.Factory.get();
        KieContainer kc = kss.getKieClasspathContainer();
        KieSession ks = kc.newKieSession("collection");
        Person person = new Person();
        person.setName("张三");
        person.setAge(30);
        person.setClassName("一班");
        School school = new School();
```

```
        List list=new ArrayList();
        list.add("一班");
        list.add("二班");
        list.add("三班");
        school.setClassNameList(list);
        ks.insert(person);
        ks.insert(school);
        int count = ks.fireAllRules();
        System.out.println("总执行了" + count + "条规则");
        ks.dispose();
    }
}
```

运行 testList() 方法,结果如图 3-19 所示。

图3-19　List在then中的使用

第二种写法是通过属性名直接引用,但使用元素需要通过"[]"来调用。添加规则文件 collectionTestList2,其代码为:

```
rule collectionTestList2
    when
        $s:School(classNameList[1]=="二班");
    then
       System.out.println("规则 collectionTestList2    School属性classNameList的第二个元素为"+$s.getClassNameList().get(1));
end
```

运行 testList() 方法,结果如图 3-20 所示。

图3-20　规则中List的第二种用法

List 语法使用总结:classNameList[1] 与 $s.getClassNameList().get(1) 是相同的,但 classNameList[1]

方式只能在 LHS 部分使用，$s.getClassNameList() 方式只能在 RHS 部分使用，且操作不当可能会出现异常情况。

2. Set元素操作

编辑规则文件 collection.drl，并添加规则 collectionTestSet，其代码为：

```
rule collectionTestSet
    when
        $s:School();
    then
        System.out.println("School属性classNameSet的第一个元素为"+$s.get-ClassNameSet().iterator().next());
end
```

编辑执行调用规则 testList() 方法，其代码为：

```
@Test
public void testList() {
    KieServices kss = KieServices.Factory.get();
    KieContainer kc = kss.getKieClasspathContainer();
    KieSession ks = kc.newKieSession("collection");
    Person person = new Person();
    person.setName("张三");
    person.setAge(30);
    person.setClassName("一班");
    School school = new School();
    List list=new ArrayList();
    list.add("一班");
    list.add("二班");
    list.add("三班");
    school.setClassNameList(list);
    Set classNameSet = new HashSet();
    classNameSet.add("一班");
    classNameSet.add("二班");
    classNameSet.add("三班");
    school.setClassNameSet(classNameSet);
    ks.insert(person);
    ks.insert(school);
    int count = ks.fireAllRules();
    System.out.println("总执行了" + count + "条规则");
    ks.dispose();
}
```

关注加粗部分，并执行 testList() 方法，结果如图 3-21 所示。

Set 集合元素操作是否也有与 List 相似的功能呢？在规则体 LHS 部分获取 Set 元素，编辑 collection.drl 规则文件，并添加 collectionTestSet2 规则，其代码为：

```
rule collectionTestSet2
    when
        $s:School(classNameSet.iterator().next=="一班");
```

```
    then
        System.out.println("规则 collectionTestSet2 School属性classNameSet
的第一个元素为"+$s.getClassNameSet().iterator().next());
end
```

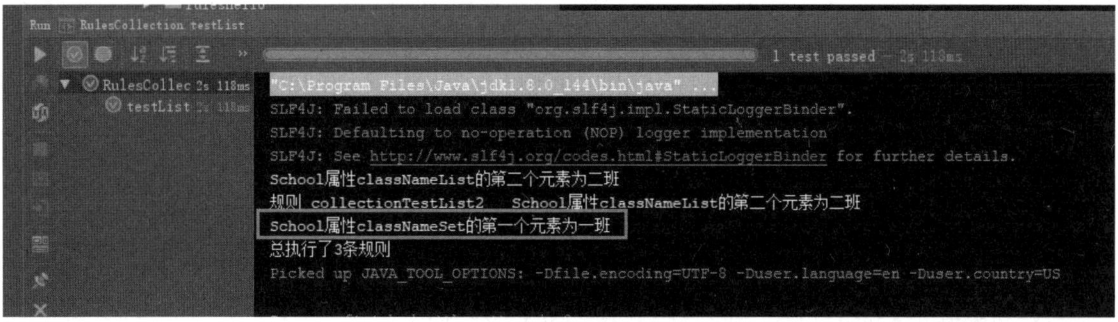

图3-21 Set集合元素操作

再次执行 testList() 方法,结果提示语法错误,使用 from 遍历 Set 集合结果也一样。由此得出结论:使用 Set 时只能通过 RHS 部分,就像是 Java 写法一样。

3. Map元素操作

编辑规则文件 collection.drl,并添加规则 collectionTestMap,其代码为:

```
rule collectionTestMap
    when
        $s:School();
    then
        System.out.println("School属性classNameMap的元素Key为一班的值"+$s.getClassNameMap().get("一班"));
end
```

编辑执行调用规则 testList() 方法,其代码为:

```
@Test
public void testList() {
    KieServices kss = KieServices.Factory.get();
    KieContainer kc = kss.getKieClasspathContainer();
    KieSession ks = kc.newKieSession("collection");
    Person person = new Person();
    person.setName("张三");
    person.setAge(30);
    person.setClassName("一班");
    School school = new School();
    List list=new ArrayList();
    list.add("一班");
    list.add("二班");
    list.add("三班");
    school.setClassNameList(list);
    Set classNameSet = new HashSet();
    classNameSet.add("一班");
```

```
classNameSet.add("二班");
classNameSet.add("三班");
school.setClassNameSet(classNameSet);
Map classNameMap = new HashMap();
classNameMap.put("一班","1");
classNameMap.put("二班","2");
classNameMap.put("三班","3");
school.setClassNameMap(classNameMap);
ks.insert(person);
ks.insert(school);
int count = ks.fireAllRules();
System.out.println("总执行了" + count + "条规则");
ks.dispose();
```

关注加粗部分，并执行 testList() 方法，结果如图 3-22 所示。

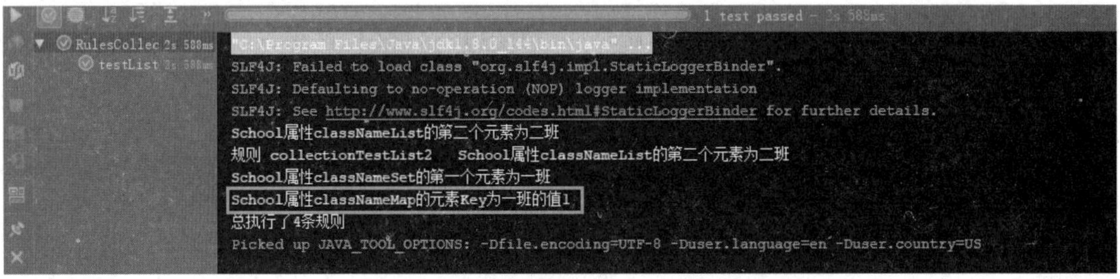

图3-22 Map元素操作

Map 元素操作是否也有与 List 相似的功能呢？在规则体 LHS 部分获取 Map 元素，编辑 collection.drl 规则文件，并添加 collectionTestMap2 规则，其代码为：

```
rule collectionTestMap2
  when
    $s:School(classNameMap.get("一班")=="1");
  then
    System.out.println("规则 collectionTestMap2 School属性classNameMap的元素Key为一班的值"+$s.getClassNameMap().get("一班"));
end
```

执行 testList() 方法，结果如图 3-23 所示。

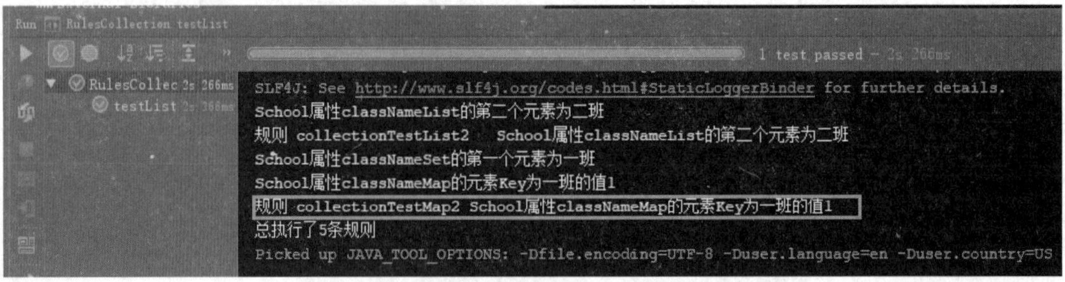

图3-23 Map元素在规则LHS的部分操作

Map 语法使用总结：classNameMap.get(key) 与 $s.getClassNameMap().get("key") 是相同的，但 classNameMap.get(key) 方式只能在 LHS 部分使用，$s.getClassNameMap().get("key") 只能在 RHS 部分使用。

在上述操作集合的例子中，是通过 JavaBean 的属性进行传值的，如果直接 insert List、Set、Map 是否可行呢？

4. List集合测试

创建规则文件 collectionInsert.drl，目录为 rules/constraint/isCollection/insert，并添加如下代码：

```
package rules.constraint.isCollection.isinsert;

import com.pojo.School;
import com.pojo.Person;
import java.util.List;

rule collectionTestList
    when
        $l:List();
    then
    System.out.println("输出List第二个元素内容为"+$l.get(1));
end
```

修改 kmodule.xml 配置文件，并添加如下配置：

```
<kbase name="collectionInsert" packages="rules.constraint.isCollection.insert">
    <ksession name="collectionInsert"/>
</kbase>
```

编辑 RulesCollection.java 文件，并添加 testInsertList() 方法，其代码为：

```
@Test
public void testInsertList() {
    KieServices kss = KieServices.Factory.get();
    KieContainer kc = kss.getKieClasspathContainer();
    KieSession ks = kc.newKieSession("collectionInsert");
    Person person = new Person();
    person.setName("张三");
    person.setAge(30);
    person.setClassName("一班");
    School school = new School();
    List list=new ArrayList();
    list.add("一班");
    list.add("二班");
    list.add("三班");
    ks.insert(list);
    ks.insert(school);
    int count = ks.fireAllRules();
```

```
            System.out.println("总执行了" + count + "条规则");
            ks.dispose();
}
```

执行 testInsertList() 方法，结果如图 3-24 所示。

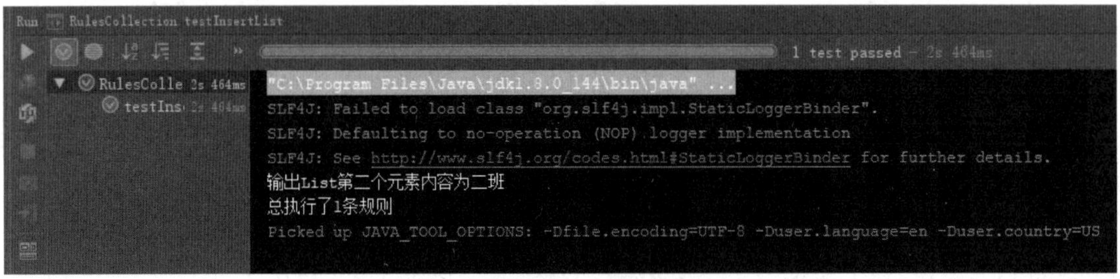

图3-24　List集合测试结果

虽然结果输出是正确的，但实际操作中肯定不会这样简单，因此提出了 3 个问题：第一，如何遍历 insert(list) 这个特殊的 Fact 对象；第二，如果同时 insert 两个 List 对象，规则又应该如何识别呢；第三，如何在 LHS 部分使用 List。针对这 3 个问题做如下说明：第一个问题，遍历 List 有两种方式，一种是在 RHS 部分通过 Java 代码进行遍历，另一种则是通过 from[①] 进行遍历；第二个问题，如果同时 insert 两个 List，规则会被激活两次，insert 的优先级以最后一次使用 insert 的参数做第一次匹配，包括同时 insert 的多个对象；第三个问题，使用 Set、Map 实验完成后再给出总结。

5. Set、Map进行测试

修改规则文件 collectionInsert.drl，并添加 collectionTestSet 及 collectionTestMap 规则，其内容为：

```
rule collectionTestSet
    when
        $s:Set();
    then
    System.out.println("输出Set第一个元素内容为"+$s.iterator().next());
end

rule collectionTestMap
    when
        $m:Map();
    then
    System.out.println("输出Map的元素Key为一班的值"+$m.get("一班"));
end
```

修改 testInsertList 方法，其内容为：

```
@Test
public void testInsertList() {
    KieServices kss = KieServices.Factory.get();
    KieContainer kc = kss.getKieClasspathContainer();
```

[①] from 是规则体 LHS 中的关键字，功能之一是遍历集合。

```
    KieSession ks = kc.newKieSession("collectionInsert");
    Person person = new Person();
    person.setName("张三");
    person.setAge(30);
    person.setClassName("一班");
    School school = new School();
    List list=new ArrayList();
    list.add("一班");
    list.add("二班");
    list.add("三班");
    school.setClassNameList(list);
    Set classNameSet = new HashSet();
    classNameSet.add("一班");
    classNameSet.add("二班");
    classNameSet.add("三班");
    Map classNameMap = new HashMap();
    classNameMap.put("一班","1");
    classNameMap.put("二班","2");
    classNameMap.put("三班","3");
    ks.insert(list);
    ks.insert(classNameSet);
    ks.insert(classNameMap);
    int count = ks.fireAllRules();
    System.out.println("总执行了" + count + "条规则");
    ks.dispose();
}
```

执行 testInsertList() 方法，结果如图 3-25 所示。

```
2018-19-28 02:19:51 [INFO ] org.drools.....ClasspathKieProject - Recursed up folders, found and used pom.xml E:\drool
2018-19-28 02:19:51 [DEBUG] org.drools.....ClasspathKieProject - Discovered classpath module com.test:drools:1.0-SNAP
2018-19-28 02:19:51 [INFO ] org.drools.....KieRepositoryImpl - KieModule was added: FileKieModule[releaseId=com.test:
2018-19-28 02:19:52 [DEBUG] org.drools...DefaultAgenda - State was INACTIVE is now FIRING_ALL_RULES
输出List第二个元素内容为二班
输出Set第一个元素内容为一班
输出Map的元素Key为一班的值1
2018-19-28 02:19:52 [DEBUG] org.drools...DefaultAgenda - State was FIRING_ALL_RULES is now HALTING
2018-19-28 02:19:52 [DEBUG] org.drools...DefaultAgenda - State was HALTING is now INACTIVE
总执行了3条规则
2018-19-28 02:19:52 [DEBUG] org.drools...DefaultAgenda - State was INACTIVE is now DISPOSED

Process finished with exit code 0
```

图3-25 Set、Map测试结果

通过图 3-25 所示结果，得出第三个问题的答案，List、Set、Map 如果单独被 insert 操作，在规则体 LHS 部分的使用与 JavaBean 中的成员变量操作集合基本相似。

3.7 规则文件drl

1. 单行注释

单行注释可以用"//"进行标记，如图 3-26 所示。

```
//规则rule1的注释
rule "rule1"
    when
        eval(true) #没有条件判断
    then
        System.out.println("rule1 execute");
end
```

图3-26　规则文件单行注释

2. 多行注释

如果要注释的内容较多，可以采用 Drools 的多行注释标记来实现。Drools 的多行注释标记与 Java 语法完全一样，以"/*"开始，以"*/"结束，如图 3-27 所示。

```
/*
规则rule1的注释
这是一个测试用规则
*/
rule "rule1"
    when
        eval(true) #没有条件判断
    then
        System.out.println("rule1 execute");
end
```

图3-27　规则文件多行注释

第4章 Drools规则属性

规则体中的属性是学习规则语法的重要组成部分，是有默认值的。它的使用直接关系到规则是否可以更好地在业务场景中起到作用，是编写良好规则的方式之一。如图4-1所示，规则属性共有12个，它们分别是 activation-group、agenda-group、auto-focus、date-effective、date-expires、dialect、duration、enabled、lock-on-active、no-loop、ruleflow-group、salience。这些属性分别适用于不同的场景，属性之间会相互制衡。

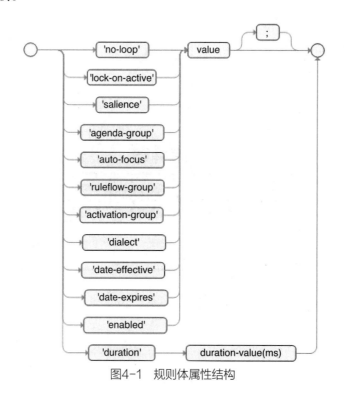

图4-1　规则体属性结构

4.1　属性no-loop

默认值： false。

类型： Boolean。

属性说明： 防止死循环，当规则通过 update 之类的函数修改了 Fact 对象时，可能使规则再次被激活，从而导致死循环。将 no-loop 设置为 true 的目的是避免当前规则 then 部分被修改后的事实对象再次被激活，从而防止死循环的发生，即执行下面的规则。

创建规则文件 isNotLoop.drl，目录为 rules/testNoLoop，其内容为：

```
rule "testNoLoop1"
    //no-loop true
    when
```

```
        $p:Person(age==30);
    then
        $p.setAge(30);
        update($p);
        System.out.println("testNoLoop1 不设置 no-loop时的效果");
end
```

修改 kmodule.xml 配置文件，并添加如下配置：

```
<kbase name="isNoLoop" packages="rules.isNoLoop">
    <ksession name="isNoLoop"/>
</kbase>
```

创建 RulesNoLoop.java 文件，目录为 com.rulesAttributes，其内容为：

```
package com.rulesAttributes;

import com.pojo.Person;
import org.junit.Test;
import org.kie.api.KieServices;
import org.kie.api.runtime.KieContainer;
import org.kie.api.runtime.KieSession;

public class RulesNoLoop {
    @Test
    public void testNoLoop1() {
        KieServices kss = KieServices.Factory.get();
        KieContainer kc = kss.getKieClasspathContainer();
        KieSession ks = kc.newKieSession("isNoLoop");
        Person person = new Person();
        person.setName("张三");
        person.setAge(30);
        person.setClassName("一班");
        ks.insert(person);
        int count = ks.fireAllRules();
        System.out.println("总执行了" + count + "条规则");
        ks.dispose();
    }
}
```

执行 testNoLoop1() 方法，结果如图 4-2 所示。

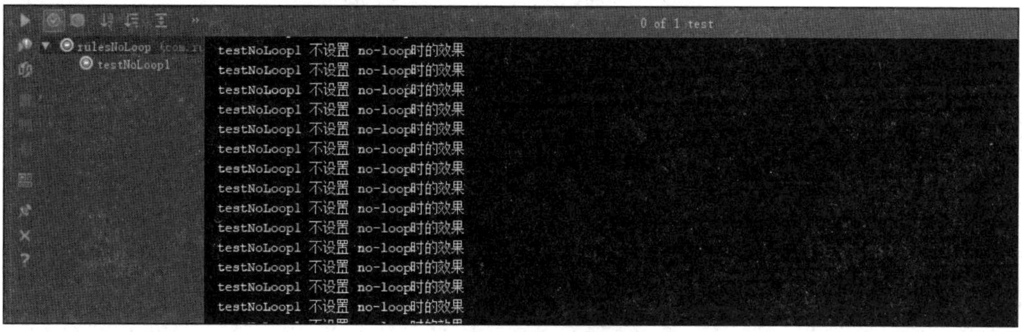

图4-2 死循环执行规则

防止结果为死循环的方法之一是修改 no-loop 属性。编辑 isNoLoop.drl 规则文件，为保证代码的完整性，注释 testNoLoop1 规则，添加 testNoLoop2 规则，其代码为：

```
package rules.isNoLoop
import com.pojo.Person;

/*rule "testNoLoop1"
    //no-loop true
    when
        $p:Person(age==30);
    then
        $p.setAge(30);
        update($p);
        System.out.println("testNoLoop1 不设置 no-loop时的效果");
end*/

rule "testNoLoop2"
    no-loop true
    when
        $p:Person(age==30);
    then
        $p.setAge(30);
        update($p);
        System.out.println("testNoLoop2 设置 no-loop时的效果");
end
```

执行 testNoLoop2() 方法，结果如图 4-3 所示。

图4-3　使用no-loop属性后的效果

设置规则体属性为 no-loop 并不是万无一失的，在某些情况下设置了 no-loop true，也会发生死循环。

编写 isNoLoop.drl 规则文件，其代码为：

```
package rules.isNoLoop
import com.pojo.Person;

/*rule "testNoLoop1"
```

```
    //no-loop true
    when
        $p:Person(age==30);
    then
        $p.setAge(30);
        update($p);
        System.out.println("testNoLoop1 不设置 no-loop时的效果");
end*/


/*rule "testNoLoop2"
    no-loop true
    when
        $p:Person(age==30);
    then
        $p.setAge(30);
        update($p);
        System.out.println("testNoLoop2 设置 no-loop时的效果");
end*/

rule "testNoLoop3"
    no-loop true
    when
        $p:Person(name=="张三");
    then
        $p.setAge(30);
        update($p);
        System.out.println("testNoLoop3 设置 no-loop时的效果");
end

rule "testNoLoop4"
    no-loop true
    when
        $p:Person(age==30);
    then
        $p.setName("张三");
        update($p);
        System.out.println("testNoLoop4 设置 no-loop时的效果");
end
```

执行 testNoLoop3() 和 testNoLoop4() 方法，结果如图 4-4 所示。

测试过程中，只有在 Fact 对象发生变化时才会出现死循环，如果在 LHS 部分的比较值并非 update 的修改值，那么会不会也出现这样的问题呢？为保证源码的完整性，现在将所有的规则体注释，编写一个 testNoLoop5 规则，代码为（省略其他被注释的规则）：

```
rule "testNoLoop5"
    //no-loop true
    when
        $p:Person(name=="张三");
```

```
    then
        $p.setAge(30);
        update($p);
        System.out.println("testNoLoop5 不设置 no-loop时的效果");
end
```

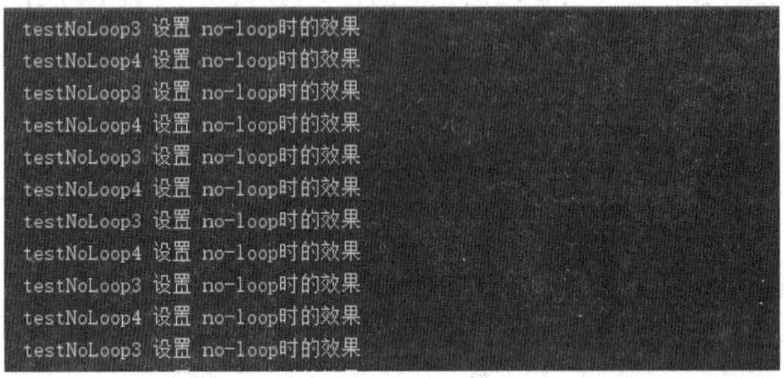

图4-4　设置no-loop发生死循环的效果

执行 testNoLoop5() 方法，结果如图 4-5 所示。

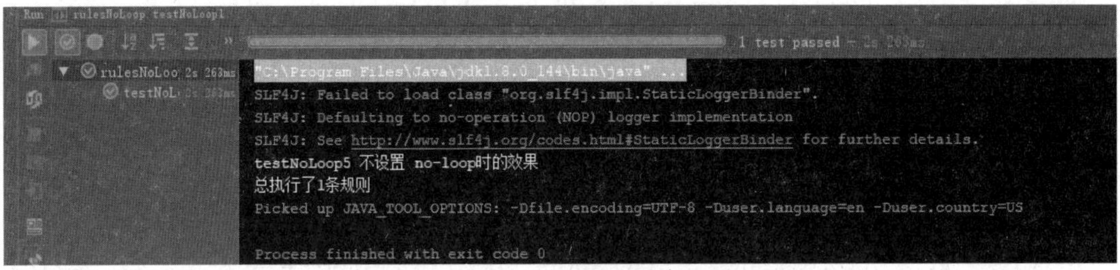

图4-5　未发生死循环

如图 4-5 所示，这个结果与预期的并不一样，出现这一结果的原因是什么？操作的 Fact 对象是同一个，难道每一个属性都是一个事实对象？带着这样的疑问，添加规则文件 testNoLoop6 并注释 rule testNoLoop5，其内容为：

```
rule "testNoLoop6"
    //no-loop true
    when
        $p:Person(name=="张三",age==30);
    then
        $p.setAge(30);
        update($p);
        System.out.println("testNoLoop6 不设置 no-loop时的效果");
end
```

执行 testNoLoop6() 方法，结果如图 4-6 所示。

图4-6　再次发生死循环

总结：当一个规则文件中，一个Fact（事实）对象通过Drools函数被修改，规则体将被再次激活。也就是说，在RHS部分使用了与update相类似的语法（insert同理），变更了Fact对象在规则中的内容，就会导致规则重新被激活和匹配。

再次激活的前提条件是被修改的事实对象与规则LHS部分的约束条件是包含关系。一个规则事实对象的变更会影响其他规则的结果，这一点在对象引用章节中有过简单说明。

4.2　属性ruleflow-group

默认值：N/A。

类型：String。

属性说明：ruleflow-group分为rule、flow和group3个部分，分别代表规则、流程、分组，即常说的规则流。详细内容请查阅规则流相关章节进行深度学习。

4.3　属性lock-on-active

默认值：false。

类型：Boolean。

属性说明：lock-on-active是指"锁定活跃"。既然它是规则体的属性，那一定是锁定规则的，而且是锁定活跃的规则。简单地说，当规则体设置该属性为true时，则当前只会被触发一次。当ruleflow-group或agenda-group再次被激活时，即使在规则体中设置了lock-on-active为true，该规则体也不能再次被激活，即无论如何更新规则事实对象，当前规则也只能被触发一次。这是no-loop的升级版，一个更强大的解决死循环的属性。下面通过例子结合no-loop说明中遇到的死循环问题提出解决方案。

创建规则文件 isLockNoActive.drl，目录为 rules/isLockNoActive，其内容为：

```
package rules.isLockNoActive
import com.pojo.Person;

rule "testLockNoActive1"
    lock-on-active true
    when
        $p:Person(name=="张三");
    then
        $p.setAge(30);
        update($p);
        System.out.println("testLockNoActive1 设置 no-loop时的效果");
end

rule "testLockNoActive2"
    lock-on-active true
    when
        $p:Person(age==30);
    then
        $p.setName("张三");
        update($p);
        System.out.println("testLockNoActive2 设置 no-loop时的效果");
end
```

修改 kmodule.xml 配置文件，并添加如下配置：

```
<kbase name="isLockNoActive" packages="rules.isLockNoActive">
    <ksession name="isLockNoActive"/>
</kbase>
```

创建 RulesLockOnActive.java 文件，目录为 com.rulesAttributes，其内容为：

```
package com.rulesAttributes;

import com.pojo.Person;
import org.junit.Test;
import org.kie.api.KieServices;
import org.kie.api.runtime.KieContainer;
import org.kie.api.runtime.KieSession;

public class RulesLockOnActive {
    @Test
    public void testLockOnActive() {
        KieServices kss = KieServices.Factory.get();
        KieContainer kc = kss.getKieClasspathContainer();
        KieSession ks = kc.newKieSession("isLockNoActive");
        Person person = new Person();
        person.setName("张三");
        person.setAge(30);
```

```
            person.setClassName("一班");
            ks.insert(person);
            int count = ks.fireAllRules();
            System.out.println("总执行了" + count + "条规则");
            ks.dispose();
        }
    }
```

执行 testLockOnActive () 方法，结果如图 4-7 所示。

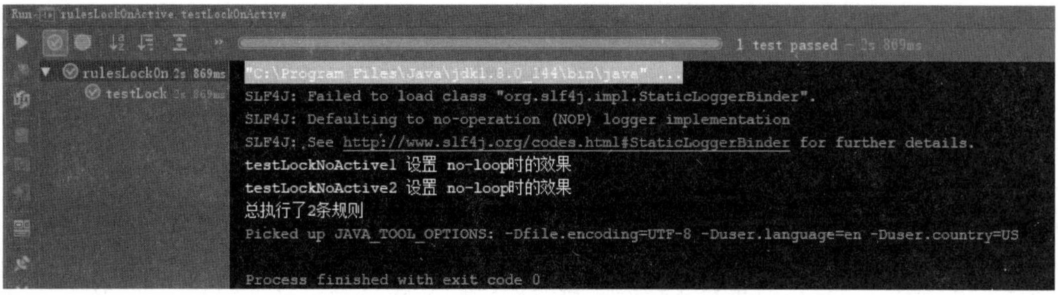

图4-7 lock-on-active属性的使用

4.4 属性salience

默认值：0。

类型：integer。

属性说明：规则体被执行的顺序，每一个规则都有一个默认的执行顺序，如果不设置 salience 属性，规则体的执行顺序为由上到下。salience 值可以是一个整数，但也可以是一个负数，其值越大，执行顺序越高，排名越靠前。Drools 还支持动态配置优先级。

创建规则文件 salience.drl，目录为 rules/isSalience，其内容为：

```
package rules.isSalience

rule "testSalience1"
salience 10
    when
    then
        System.out.println("hello testSalience1");
end

rule "testSalience2"
salience 20
    when
    then
```

```
        System.out.println("hello testSalience2");
end

rule "testSalience3"
salience 5
    when
    then
        System.out.println("hello testSalience3");
end
```

修改 kmodule.xml 配置文件，并添加如下配置：

```
<kbase name="isSalience" packages="rules.isSalience">
    <ksession name="isSalience"/>
</kbase>
```

创建 RulesSalience.java 文件，目录为 com.rulesAttributes，其内容为：

```
package com.rulesAttributes;

import org.junit.Test;
import org.kie.api.KieServices;
import org.kie.api.runtime.KieContainer;
import org.kie.api.runtime.KieSession;

public class RulesSalience {
    @Test
    public void testSalience () {
        KieServices kss = KieServices.Factory.get();
        KieContainer kc = kss.getKieClasspathContainer();
        KieSession ks = kc.newKieSession("isSalience");
        int count = ks.fireAllRules();
        System.out.println("总执行了" + count + "条规则");
        ks.dispose();
    }
}
```

执行 testSalience() 方法，结果如图 4-8 所示。

图4-8　salience属性的效果

属性说明中提到 Drools 支持动态 salience，因此编辑规则文件 salience.drl，并添加 testSalience4

规则，其内容为：

```
rule "testSalience4"
salience (Math.random() * 10 + 1)
    when
    then
        System.out.println("hello testSalience4");
end
```

再一次执行 testSalience() 方法，结果如图 4-9 所示。

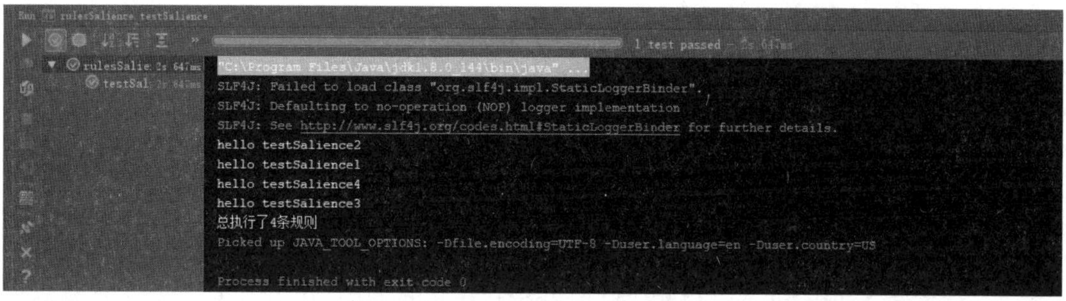

图4-9　salience属性的随机效果

4.5　属性enabled

默认值：true。

类型：Boolean。

属性说明：指规则是否可以被执行，若规则体设置为 enabled false，则规则体将视为永久不被激活。

创建规则文件 isEnabled.drl，目录为 rules/isEnabled，其内容为：

```
package rules.isEnabled

rule "testEnabled1"
    enabled  true
    when
    then
        System.out.println("testEnabled1 设置 enabled true");
end

rule "testEnabled2"
    enabled false
    when
    then
        System.out.println("testEnabled2 设置 enabled false");
end
```

修改 kmodule.xml 配置文件，并添加如下配置：

```xml
<kbase name="isEnabled" packages="rules.isEnabled">
    <ksession name="isEnabled"/>
</kbase>
```

创建 RulesEnabled.java 文件，目录为 com.rulesAttributes，其内容为：

```java
package com.rulesAttributes;

import org.junit.Test;
import org.kie.api.KieServices;
import org.kie.api.runtime.KieContainer;
import org.kie.api.runtime.KieSession;

public class RulesEnabled {
    @Test
    public void testEnabled() {
        KieServices kss = KieServices.Factory.get();
        KieContainer kc = kss.getKieClasspathContainer();
        KieSession ks = kc.newKieSession("isEnabled");
        int count = ks.fireAllRules();
        System.out.println("总执行了" + count + "条规则");
        ks.dispose();
    }
}
```

执行 testEnabled() 方法，结果如图 4-10 所示。

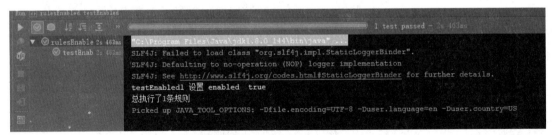

图4-10　enabled 属性效果

4.6　属性dialect

可能值： Java 或 Mvel。

类型： String。

属性说明： 用来定义规则中要使用的语言类型，支持 Mvel 和 Java 两种类型的语言，默认情况下是由包指定的。Java 语言在特殊情况下会用到，如 Accumulate[①]、引用 Java 中的语法等。

4.7 属性date-effective

默认值： N/A。

类型： String、日期、时间。

属性说明： 只有当前系统时间大于等于设置的时间或日期，规则才会被激活。在没有设置该属性的情况下，规则体不受时间限制。date-effective 的值是一个日期型的字符串，默认情况下，date-effective 可接受的日期格式为"dd-MMM-yyyy"，例如，设置 2018 年 08 月 07 日的 date-effective 为"date-effective "07-August-2018""。下面通过代码直接看效果。

创建规则文件 dateEffective.drl，目录为 rules/isDate，其内容为：

```
package rules.isDate

rule "testDateEffective1"
 date-effective "07-August-2018"
    when
        eval(true)
    then
        System.out.println("表示 当前系统日期 大于等于  date-effective 的参数");
end
```

修改 kmodule.xml 配置文件，并添加如下配置：

```
<kbase name="isDate" packages="rules.isDate">
    <ksession name="isDate"/>
</kbase>
```

创建 RulesDate.java 文件，目录为 com.rulesAttributes，其内容为：

```
package com.rulesAttributes;

import org.junit.Test;
import org.kie.api.KieServices;
import org.kie.api.runtime.KieContainer;
import org.kie.api.runtime.KieSession;

public class RulesDate {
    @Test
    public void testDate() {
        KieServices kss = KieServices.Factory.get();
```

[①] Accumulate 在规则体 LHS 部分中应用，是很有用的计算方法。

```
            KieContainer kc = kss.getKieClasspathContainer();
            KieSession ks = kc.newKieSession("isDate");
            int count = ks.fireAllRules();
            System.out.println("总执行了" + count + "条规则");
            ks.dispose();
    }
}
```

执行 testDate () 方法，结果如图 4-11 所示。

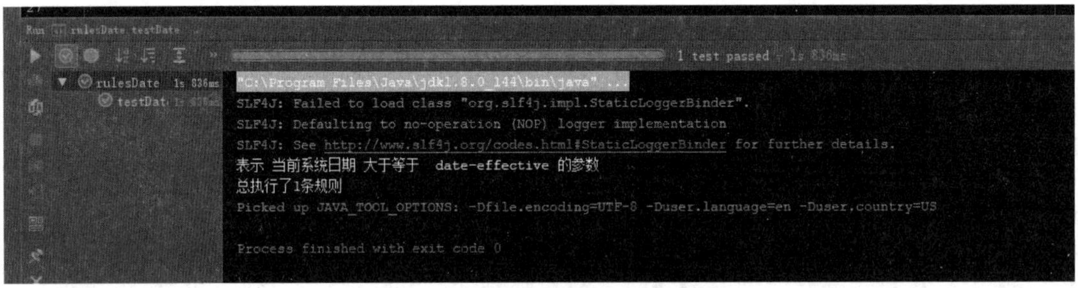

图4-11　date-effective属性效果

4.8　属性date-expires

默认值： N/A。

类型： String、日期、时间。

属性说明： date-expires 属性与 date-effective 属性是相反的，即只有当前系统时间小于设置的时间或日期，规则才会被激活。在没有设置该属性的情况下，规则体不受时间限制。date-expires 的值为一个日期型的字符串，默认情况下，date-expires 可接受的日期格式为"dd-MMM-yyyy"。例如，设置 2018 年 08 月 07 日的 date-effective 值为 "date-expires"07-August-2018""。

编辑规则文件 dateEffective.drl，目录为 rules/isDate，其内容为：

```
package rules.isDate

rule "testDateEffective1"
  date-effective "07-August-2018"
    when
        eval(true)
    then
        System.out.println("表示 当前系统日期 大于等于  date-effective 的参数");
end
```

```
rule "testDateExpires1"
 date-expires    "08-August-2018"
    when
        eval(true)
    then
        System.out.println("表示 当前系统日期 小于等于 date-expires 的参数
");
end
```

创建 RuleDate.java 文件，目录为 com.rulesAttributes，其内容为：

```
public class RulesDate {
    @Test
    public void testDate() {
        KieServices kss = KieServices.Factory.get();
        KieContainer kc = kss.getKieClasspathContainer();
        KieSession ks = kc.newKieSession("isDate");
        System.setProperty("drools.dateformat", "yyyy-MM-dd");
        int count = ks.fireAllRules();
        System.out.println("总执行了" + count + "条规则");
        ks.dispose();
    }
}
```

执行 testDate() 方法，结果如图 4-12 所示。

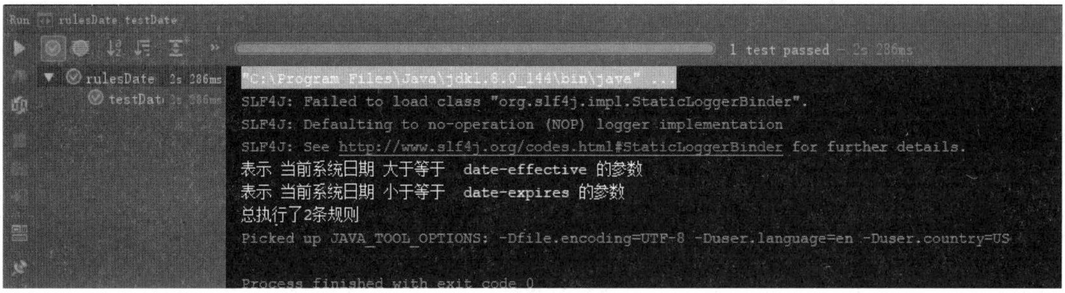

图4-12　date-expires属性效果

再次修改 dateEffective.drl 规则文件，并添加 testDateExpires2 规则，其代码为：

```
rule "testDateExpires2"
 date-expires "32-August-2018"
    when
        eval(true)
    then
        System.out.println("testDateExpires2 表示 当前系统日期 小于等于
date-expires 的参数");
end
```

执行 testDate() 方法，结果如图 4-13 所示。

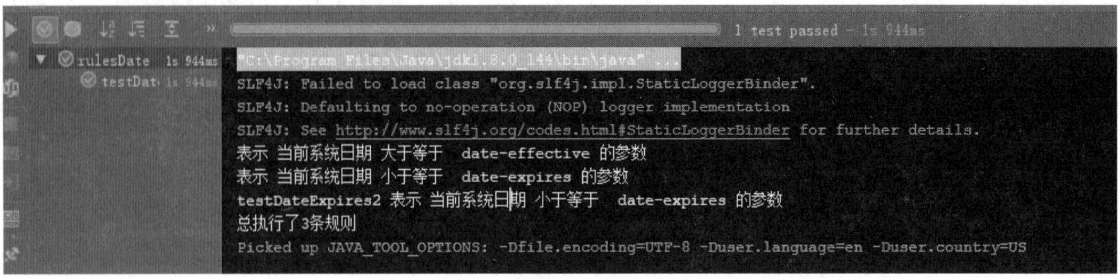

图4-13　date-expires属性的第二次测试效果

注意规则 testDateExpires2 中的属性，32 为具体日期，正常情况下每个月不可能超过 31 天，但这样的写法在规则中是可行的。经过多方面测试后，得出结论：当前设置的天数超过当前月份的最大天数后会自动累加至下个月；如果时间大于下个月的天数，则继续累加至下下月，以此类推。

Drools 规则引擎中日期属性格式为 dd-MMM-yyyy 是并不常用的格式，在开发过程中，一般都会写成 yyyy-MM-dd 形式。如果在规则中使用通俗的 Java 日期格式，需要在 Java 调用规则中添加代码"System.setProperty("drools.dateformat"，"yyyy-MM-dd HH:mm:ss");"。其中，时分秒为非必添项，一旦设置了该信息，规则中的日期格式就必须遵守当前设置的日期格式。需要注意，进行规则引擎格式化日期时，最好将其写在初始化 KIE 相关代码之前。

再次修改 dateEffective.drl 规则文件，添加 testDateExpires3 规则，并为保证可以正确输出，需要将其他规则体注释：

```
rule "testDateExpires3"
 date-expires "2018-08-08"
    when
        eval(true)
    then
        System.out.println("testDateExpires3 表示 当前系统日期 小于等于 date-expires 的参数");
end
```

修改执行调用规则代码（请注意加粗代码），其内容为：

```
@Test
public void testDate() {
    System.setProperty("drools.dateformat", "yyyy-MM-dd");
    KieServices kss = KieServices.Factory.get();
    KieContainer kc = kss.getKieClasspathContainer();
    KieSession ks = kc.newKieSession("isDate");
    int count = ks.fireAllRules();
    System.out.println("总执行了" + count + "条规则");
    ks.dispose();
}
```

执行 testDate () 方法，结果如图 4-14 所示。

图4-14　日期格式化后的效果

使用日期格式化时注意，需要在创建会话之前执行该代码，也就是在实例 KieSession 之前。如果是在整合 Spring 时，规则体中又要使用日期格式化，那么可以采用静态代码块的方式修改 Drools 规则引擎默认的时间格式，Java 的执行顺序为先静后动、先父后子。

规则体属性日期相关的两个参数，还有一些特殊的写法，但要看当前操作环境的输出日期格式。例如：

```
SimpleDateFormat sdf4 = new SimpleDateFormat("yyyy-MMM-dd");
Date currDate = new Date();
System.out.println(sdf4.format(currDate));
```

如果输出结果为 2018-Aug-07，就设置日志属性格式为 Aug，如果输出的结果为 2017- 八月 -08，就可以设置规则体的日期属性值为 08- 八月 -2018。

4.9　属性duration

默认值： 无。

类型： long。

属性说明： 表示定时器，如果当前规则 LHS 部分为 true，那么规则继续执行；如果该属性已经被弃用，那么通过新的属性 timer 来控制。

4.10　属性activation-group

默认值： N/A。

类型： String。

属性说明： activation-group 是指激活分组，通过字符串定义分组名称，具有相同组名称的规则体有且只有一个规则被激活，其他规则体的 LHS 部分仍然为 true 也不会再被执行。该属性受

salience 属性的影响，如当前规则文件中的其他规则未设计该属性，则视为规则处于被激活状态，并不受该属性的影响。

创建规则文件 dateEffective.drl，目录为 rules/isDate，其内容为：

```
package rules.isActivationGroup

rule "testActivationGroup1"
activation-group "testGroup"
    when
        eval(true)
    then
        System.out.println("规则 testActivationGroup1 设置属性 activation-group \"testGroup\" ");
end

rule "testActivationGroup2"
activation-group "testGroup"
    when
        eval(true)
    then
        System.out.println("规则 testActivationGroup2 设置属性 activation-group \"testGroup\" ");
end
```

修改 kmodule.xml 配置文件，并添加如下配置：

```
<kbase name="isActivationGroup" packages="rules.isActivationGroup">
    <ksession name="isActivationGroup"/>
</kbase>
```

创建 RulesActivationGroup.java 文件，目录为 com.rulesAttributes，其内容为：

```java
package com.rulesAttributes;

import org.junit.Test;
import org.kie.api.KieServices;
import org.kie.api.runtime.KieContainer;
import org.kie.api.runtime.KieSession;

public class RulesActivationGroup {
    @Test
    public void testActivationGroup() {
        KieServices kss = KieServices.Factory.get();
        KieContainer kc = kss.getKieClasspathContainer();
        KieSession ks = kc.newKieSession("isActivationGroup");
        int count = ks.fireAllRules();
        System.out.println("总执行了" + count + "条规则");
        ks.dispose();
    }
}
```

执行 testActivationGroup () 方法，结果如图 4-15 所示。

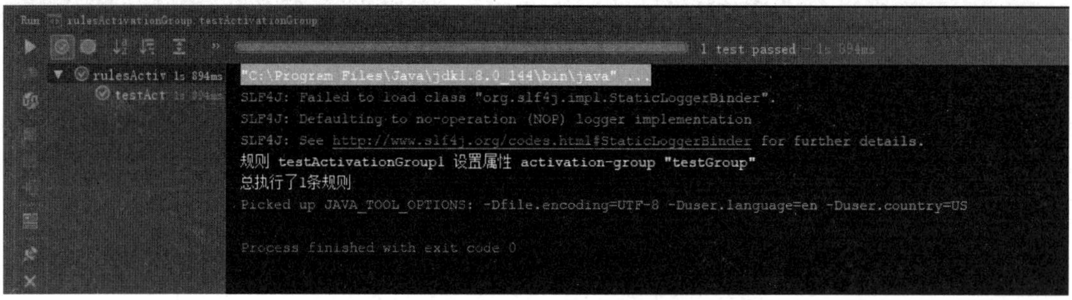

图4-15　activation-group属性的效果

activation-group 属性类似于规则流程中的 XOR 网关，当有规则体被执行完毕后，其他规则将不会被激活，即使其他规则中的 LHS 部分为 true。

图 4-15 中的输出结果一直都是 rule testActivationGroup1 的 RHS 部分，有两种方式可以使输出变为 testActivationGroup2：一种是根据规则匹配的特性，将 testActivationGroup1 规则的 LHS 部分设置为 false；另一种是通过设置 salience 属性。

修改 activationGroup.drl，添加 testAgs1 和 testAgs2 两个规则，其代码为：

```
rule "testAgs1"
salience 10
activation-group "testAgs"
    when
        eval(true)
    then
        System.out.println("规则 testAgs1 设置属性 activation-group \"testAgs\" ");
end

rule "testAgs2"
salience 11
activation-group "testAgs"
    when
        eval(true)
    then
        System.out.println("规则 testAgs2 设置属性 activation-group \"testAgs\" ");
end
```

执行 testActivationGroup () 方法，结果如图 4-16 所示。因为 testAgs2 的优先级高于 testAgs1，所以结果发生了变化。

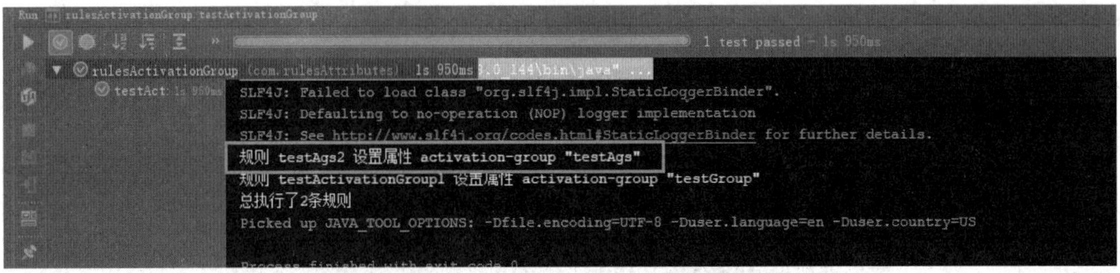

图4-16　activation-group设置的优先级效果

4.11　属性agenda-group

默认值： 无，需要通过 Java 设置。

类型： String。

属性说明： agenda-group 是议程分组，属于另一种可控的规则执行方式，是指用户可以通过配置 agenda-group 的参数来控制规则的执行，而且只有获取焦点的规则才会被激活。

创建规则文件 AgendaGroup.drl，目录为 rules/isAgendaGroup，其内容为：

```
package rules.isAgendaGroup

rule "testisAgendaGroup1"
agenda-group "ag1"
    when
        eval(true)
    then
        System.out.println("规则 testisAgendaGroup1 设置属性 agenda-group \"ag1\" ");
end

rule "testisAgendaGroup2"
agenda-group "ag2"
    when
        eval(true)
    then
        System.out.println("规则 testisAgendaGroup2 设置属性 agenda-group \"ag2\" ");
end
```

修改 kmodule.xml 配置文件，并添加如下配置：

```
<kbase name="isAgendaGroup" packages="rules.isAgendaGroup">
    <ksession name="isAgendaGroup"/>
</kbase>
```

创建 rulesAgendaGroup.java 文件，目录为 com.rulesAttributes，其内容为：

```java
package com.rulesAttributes;

import org.junit.Test;
import org.kie.api.KieServices;
import org.kie.api.runtime.KieContainer;
import org.kie.api.runtime.KieSession;
public class RulesAgendaGroup {
    @Test
    public void testAgendaGroup() {
        KieServices kss = KieServices.Factory.get();
        KieContainer kc = kss.getKieClasspathContainer();
        KieSession ks = kc.newKieSession("isAgendaGroup");
        ks.getAgenda().getAgendaGroup("ag1").setFocus();//让AgendaGroup
分组为ag1的获取焦点
        int count = ks.fireAllRules();
        System.out.println("总执行了" + count + "条规则");
        ks.dispose();
    }
}
```

执行 rulesAgendaGroup () 方法，结果如图 4-17 所示。

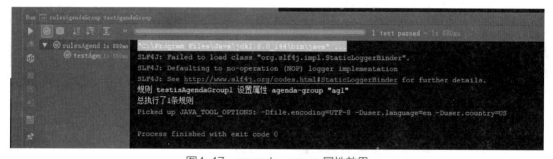

图4-17　agenda-group 属性效果

编辑 activationGroup.drl 规则文件，添加 testisAgendaGroup3 规则，其内容为：

```
rule "testisAgendaGroup3"
agenda-group "ag1"
    when
        eval(true)
    then
        System.out.println("规则 testisAgendaGroup3 设置属性 agenda-group \"ag1\" ");
end
```

执行 rulesAgendaGroup () 方法，结果如图 4-18 所示。

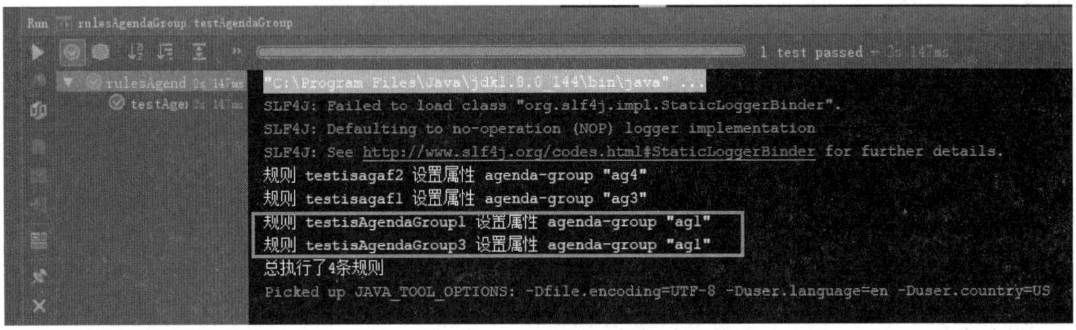

图4-18　设计两个agenda-group属性效果

如果有两个规则体的 agenda-group 属性相同，通过图 4-18 所示的输出结果可以得出结论，这样做是都可以被激活的。但有些情况下 Java 代码无法对所有的 agenda-group 进行管理，Java 代码中指定 getAgendaGroup() 的参数时，如果参数值不存在或未指定，则设置了 agenda-group 属性的规则体将无法再获取焦点，这是有风险的。

实际应用中的 agenda-group 可以与 auto-focus 属性一起使用，这样就不会出现上述问题了。如果将某个规则体的 auto-focus 属性设置为 true，那么即使该规则设置了 agenda-group 属性，也不需要在 Java 代码中设置。

编辑规则文件 agendaGroup.drl，并分别添加 testisagaf1、testisagaf2 规则，其代码为：

```
rule "testisagaf1"
agenda-group "ag3"
auto-focus true
    when
        eval(true)
    then
        System.out.println("规则 testisagaf1 设置属性 agenda-group \"ag3\"");
end

rule "testisagaf2"
agenda-group "ag4"
auto-focus true
    when
        eval(true)
    then
        System.out.println("规则 testisagaf2 设置属性 agenda-group \"ag4\"");
end
```

执行 rulesAgendaGroup () 方法，结果如图 4-19 所示。

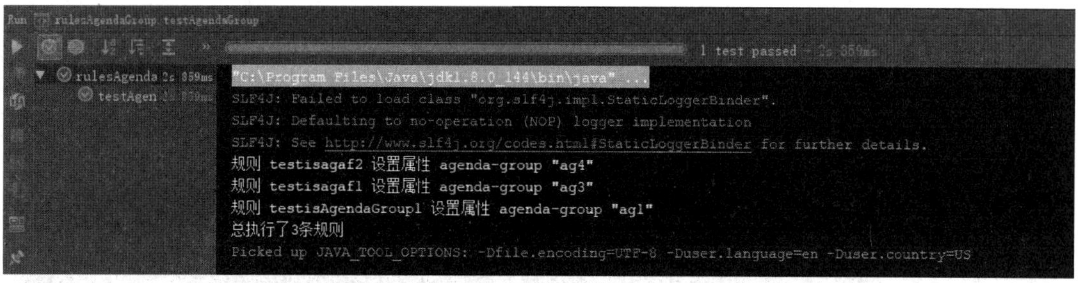

图4-19 agenda-group结合auto-focus属性效果

agenda-group 属性会受到其他属性的影响，如表 4-1 所示。

表4-1 agenda-group属性受其他属性影响

	agenda-group	activation-group	结果说明
是否同一Focus	是	是	只会执行其中的一个规则。执行顺序根据优化级控制，默认为从上到下
	否	是	只有获取焦点的规则才会被激活
	是	否	会执行多个规则，但只有获取Focus的规则才会被激活
	否	否	只有获取焦点的规则才会被激活

1. 第一种场景分析

编辑 agendaGroup.drl 规则文件，添加 testagag1、testagag2 规则，其代码为：

```
rule "testagag1"
agenda-group "ag1"
activation-group "agag"
    when
        eval(true)
    then
        System.out.println("规则 testagag1 设置属性 agenda-group \"ag1\" activation-group \"agag\"");
end

rule "testagag2"
activation-group "agag"
agenda-group "ag1"
    when
        eval(true)
    then
        System.out.println("规则 testagag2 设置属性 agenda-group \"ag1\" activation-group \"agag\"");
end
```

执行 rulesAgendaGroup() 方法，结果如图 4-20 所示。

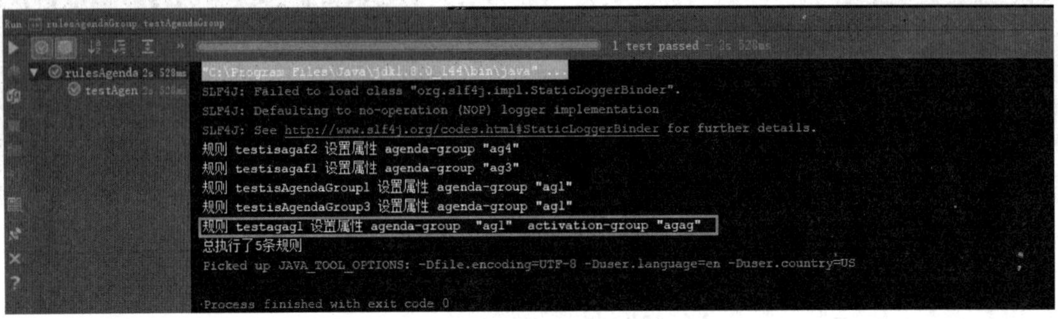

图4-20　第一种场景分析

2. 第三种场景分析

编辑 agendaGroup.drl 规则文件，添加 testagag3、testagag4 规则，其代码为：

```
rule "testagag3"
agenda-group "ag6"
activation-group "agag1"
    when
        eval(true)
    then
        System.out.println("规则 testagag3 设置属性 agenda-group \"ag6\" activation-group \"agag1\"");
end

rule "testagag4"
activation-group "agag2"
agenda-group "ag6"
    when
        eval(true)
    then
        System.out.println("规则 testagag4 设置属性 agenda-group \"ag6\" activation-group \"agag2\"");
end
```

编辑 RulesAgendaGroup.java 文件，添加 testAgendaGroup2() 方法并执行，其代码如下，输出结果如图 4-21 所示。

```
@Test
public void testAgendaGroup2() {
    KieServices kss = KieServices.Factory.get();
    KieContainer kc = kss.getKieClasspathContainer();
    KieSession ks = kc.newKieSession("isAgendaGroup");
    ks.getAgenda().getAgendaGroup("ag6").setFocus();//让AgendaGroup分组为ag6的获取焦点
    int count = ks.fireAllRules();
    System.out.println("总执行了" + count + "条规则");
```

```
        ks.dispose();
}
```

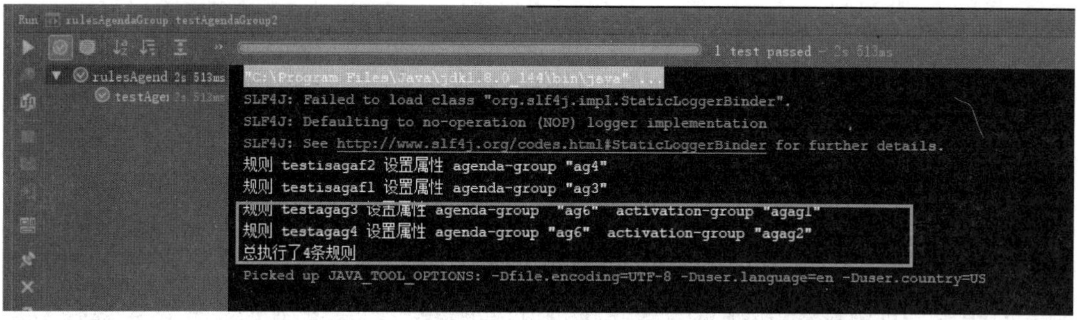

图4-21 第三种场景分析

3. 第四种场景分析

因为第二种场景与第四种场景情况相似，所以本节以第四种场景为例进行说明。

编辑 agendaGroup.drl 规则文件，添加 testagag5、testagag6 规则，代码为：

```
rule "testagag5"
agenda-group "ag8"
activation-group "agag1"
    when
        eval(true)
    then
        System.out.println("规则 testagag5 设置属性 agenda-group \"ag8\" activation-group \"agag1\"");
end

rule "testagag6"
activation-group "agag2"
agenda-group "ag9"
    when
        eval(true)
    then
        System.out.println("规则 testagag6 设置属性 agenda-group \"ag9\" activation-group \"agag2\"");
end
```

编辑 RulesAgendaGroup.java 文件，添加并执行 testAgendaGroup3() 方法，其代码如下，输出结果如图 4-22 所示。

```
@Test
public void testAgendaGroup3() {
    KieServices kss = KieServices.Factory.get();
    KieContainer kc = kss.getKieClasspathContainer();
    KieSession ks = kc.newKieSession("isAgendaGroup");
    ks.getAgenda().getAgendaGroup("ag8").setFocus();//让AgendaGroup分组为ag8的获取焦点
```

```
            int count = ks.fireAllRules();
            System.out.println("总执行了" + count + "条规则");
            ks.dispose();
}
```

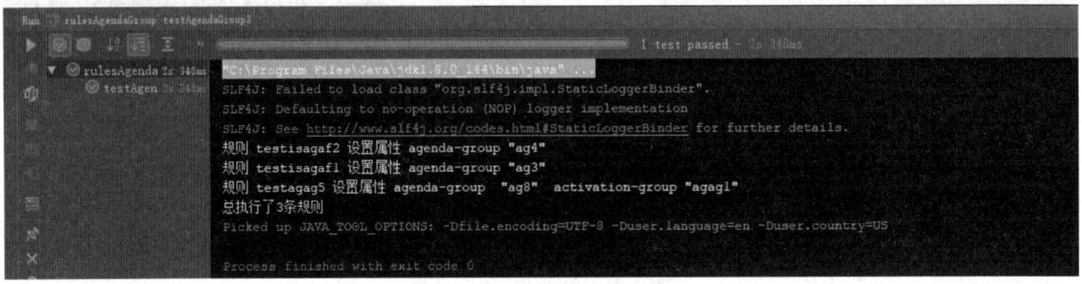

图4-22 第四种场景分析

4.12 属性auto-focus

默认值： false。

类型： Boolean。

属性说明： auto-focus 属性为自动获取焦点，即当前规则是否被激活。如果一个规则被执行，那么认为 auto-focus 为 true；如果单独设置，一般结合 agenda-group（在介绍 agenda-group 时有测试用例），当一个议程组未获取焦点时，可以设置 auto-focus 来控制。

4.13 属性timer

默认值： 无。

类型： 与 Java 定时器参数类型相似。

属性说明： timer 属性是一个定时器，用来控制规则的执行时间，主要有两种写法。

第一种写法的代码为：

```
timer ( int: <initial delay> <repeat interval>? )
timer ( int: 30s )
timer ( int: 30s 5m )
```

第二种写法的代码为：

```
timer ( cron: <cron expression> )
```

```
timer ( cron:* 0/15 * * * ? )
```

创建规则文件 timer.drl，目录为 rules/isTimer，其内容为：

```
package rules.isTimer

rule "testTimer1"
 timer (int: 3s)
    when
    then
        System.out.println("规则 testTimer1 3秒后执行");
end

rule "testTimer2"
 timer (cron:0/1 * * * * ?)
    when
    then
        System.out.println("规则 testTimer2 每一秒执行一次");
end
```

修改 kmodule.xml 配置文件，并添加如下代码：

```
<kbase name="isTimer" packages="rules.isTimer">
    <ksession name="isTimer"/>
</kbase>
```

创建 RulesATimer.java 文件，目录为 com.rulesAttributes，其内容为：

```
package com.rulesAttributes;

import org.kie.api.KieBase;
import org.kie.api.KieBaseConfiguration;
import org.kie.api.KieServices;
import org.kie.api.runtime.KieContainer;
import org.kie.api.runtime.KieSession;
import org.kie.api.runtime.KieSessionConfiguration;
import org.kie.api.runtime.conf.TimedRuleExecutionOption;

public class RulesATimer {
    public static void main(String[] args) throws InterruptedException {
        KieServices kss = KieServices.Factory.get();
        KieContainer kc = kss.getKieClasspathContainer();
        KieSessionConfiguration ksconf = KieServices.Factory.get().newKieSessionConfiguration();
        KieBaseConfiguration KieBaseConfiguration = KieServices.Factory.get().newKieBaseConfiguration();
        ksconf.setOption( TimedRuleExecutionOption.YES );
        KieBase KieBase = kc.newKieBase("isTimer", KieBaseConfiguration);
        KieSession kieSession = KieBase.newKieSession(ksconf, null);
        kieSession.fireAllRules();
        Thread.sleep(10000);
```

```
        kieSession.dispose();
    }
}
```

执行主函数方法，结果如图 4-23 所示。

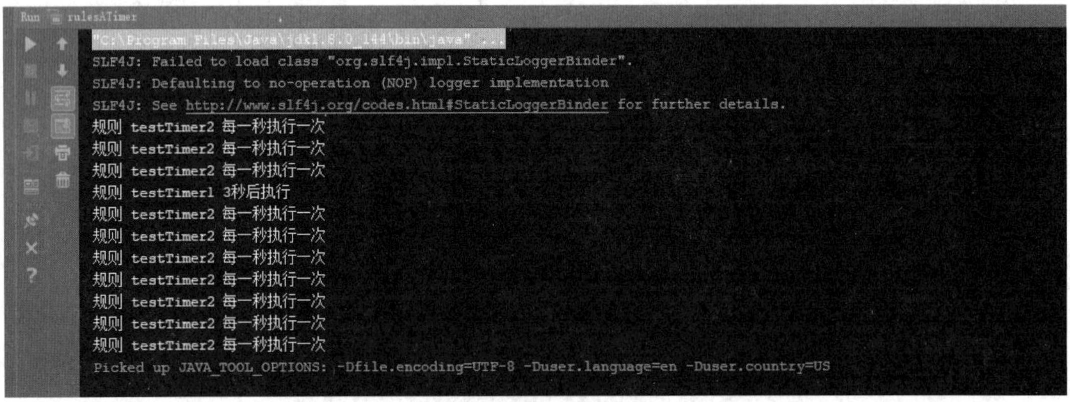

图4-23 timer属性效果

定时器的功能远不止如此，它还可以定义时间区间，即开始时间和结束时间。例如：

```
timer (int: 30s 10s; start=3-JAN-2018, end=5-JAN-2018)
```

定时器功能还可以通过变量进行赋值，但需要用到 declare 声明。例如：

```
declare Bean
    delay   : String = "30s"
    period  : long = 60000
end

rule "Expression timer"
    timer( expr: $d, $p )
when
    Bean( $d : delay, $p : period )
then
end
```

第5章 关键字及错误信息

5.1 关键字说明

Drools 规则引擎有硬关键字与软关键字之分。硬关键字为被保留，命名相关定义时，如对象、属性、方法、函数和应用于规则文本中的其他元素，编辑规则内容时不能使用硬关键字作为命名规范。硬关键字主要包括 true、false、null。编写规则时，一定要注意软关键字不像硬关键字那么强制，软关键字相比硬关键字要多，如果非要使用软关键字作为命名是没有问题的。软关键字包含 lock-on-active、date-effective、date-expires、no-loop、auto-focus、activation-group、agenda-group、ruleflow-group、entry-point、duration、package、import、dialec、salience、enabled、attributes、rule、extend、template、query、declare、function、global、eval、not、in、or、and、exists、forall、action、reverse、result、end、init 等。读者可以用驼峰格式使用这些（软或硬）单词作为方法的名称部分，如 notSomething() 或 accumulateSomething()。

DRL 语言的另一个改进是可以在规则文本中转义硬关键字。这个功能可以在编辑规则内容时减少使用关键字所带来的语法错误。编写规则内容时只需将当前关键字进行转义即可，如 Holiday('when' == "july")，只需用 "'" 符号括起来就可以解决语法错误的问题。

规则内容的任何地方都可以使用转义，但不包含 LHS 或 RHS 代码块中表达式参数的代码。

下面是正确使用关键字的示例，其内容为：

```
rule "validate holiday by eval"
    dialect "mvel"
    when
        h1 : Holiday( )
        eval( h1.when == "july" )
    then
        System.out.println(h1.name + ":" + h1.when);
End
rule "validate holiday"
    dialect "mvel"
    when
        h1 : Holiday( `when` == "july" )
    then
        System.out.println(h1.name + ":" + h1.when);
end
```

5.2 错误信息

Drools 5 以后引入了一个标准化的错误信息。标准化的目的在于更快、更容易地帮助用户发现和解决问题。下面将学习如何确定和解释这些错误信息，而且也会学到如何解决这些问题的一些相关提示。

异常信息格式如图 5-1 所示。

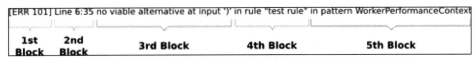

图5-1 异常信息格式

图 5-1 中分别说明了每一段信息所代表的含义。

```
1st Block：指错误代码。
2nd Block：指行列。
3rd Block：描述问题。
4th Block：指发生错误的规则名、函数、模板、查询等。
5th Block：指发生错误的pattern。
```

一般错误代码大多为 102 的语法问题，而官方给提供的错误例子，在测试过中只有代码 102 为正确的，其他的错误代码均不能直接体现，而且在 7.10 版本中的官方文档中的错误代码为 107，所以不能以官方文档为准。错误信息标准化是确定的，读者可根据在实践项目开发时遇到的错误进行定位。

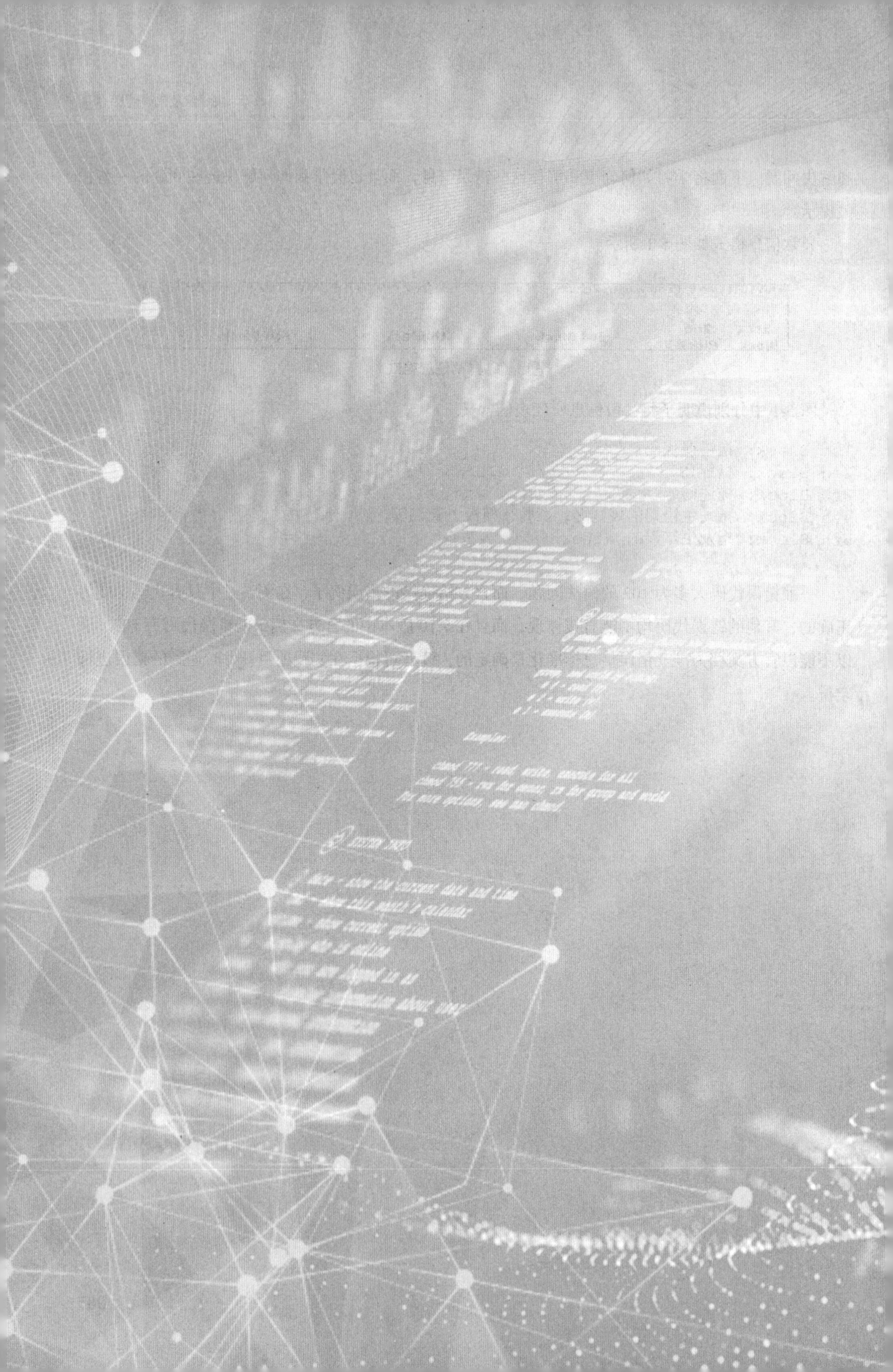

Drools 规则引擎技术指南

第三篇

中级篇

第6章 规则中级语法

6.1　package说明

介绍 package 之前需要先理解其组织结构图，如图 6-1 所示。

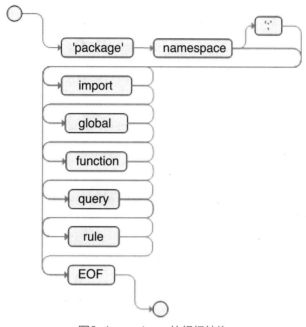

图6-1　package的组织结构

package 是定义在规则文件首行的，也是规则文件三大模块之一。介绍规则文件三大模块时就对 package 的作用做了部分说明，本节将对 package 的使用做详细阐述。package 为规则逻辑路径，定义时可不与物理目录相同，但为了更方便管理，强烈建议与物理目录同步。package 包含 import、global、funcation、query、rule、EOF。其中 import 的作用[1]已经做过详细说明，这里不再讲述。rule 是贯穿整个规则文件的核心，介绍的 pattern、运行符、约束连接、规则属性等都是 rule 中的内容。rule 的使用会在本节的 when、then 中详细说明。

介绍 package 时不得不再次说明 kmodule.xml 配置，当配置文件中 KieBase 设置了 packages 路径，则当前路径下的所有规则相关文件（规则文件、决策表、领域语言文件等）都会被加入规则库中，但当前路径下子文件夹的规则相关文件不包含在当前规则库中。package 参数本身是一个命名空间，并且不以任何方式关联文件或文件夹。因此，可以由多个规则目录为规则库构建源组合规则，有一个顶级的 package 配置，所有的规则都在其控制之下（在组合规则时）。虽然声明在不同名称下的资源不可能合并为同一个包，但是单个规则库可以用多个包来构建它。也就是说，packages 可以设置多个路径，并通过逗号分隔，但这是对配置文件 pakcages 而言的。如果强烈建议与物理目录同步，就不会遇到如下的问题。

---

[1]请参考对象引用章节。

创建规则文件 package.drl，目录为 rules/isPackage，其内容为：

```
package rules.isPackage

rule "testRuleNameOnly"
    when
    then
        System.out.println("testRuleNameOnly 设置 enabled  true");
end

rule "testRuleNameOnly"
    when
    then
        System.out.println("testRuleNameOnly 设置 enabled  true");
end
```

修改 kmodule.xml 配置文件，并添加如下配置：

```xml
<kbase name="isPackage" packages="rules.isPackage">
    <ksession name="isPackage"/>
</kbase>
```

创建 RulePackage.java 文件在 com.packages 目录，其内容为：

```java
package com.packages;

import org.junit.Test;
import org.kie.api.KieServices;
import org.kie.api.runtime.KieContainer;
import org.kie.api.runtime.KieSession;

public class RulePackage {
    @Test
    public void testPackage() {
        KieServices kss = KieServices.Factory.get();
        KieContainer kc = kss.getKieClasspathContainer();
        KieSession ks = kc.newKieSession("isPackage");
        int count = ks.fireAllRules();
        System.out.println("总执行了" + count + "条规则");
        ks.dispose();
    }
}
```

执行 testPackage() 方法，结果如图 6-2 所示。

图6-2　测试package 同包同规则名称的效果

结果一定是错的,规则文件代码 rule 名称有重复,出现汇报规则名称不唯一的错误。那么将第二个规则体的 name 参数改写一下就可以正常运行,这时再添加一个规则文件。

创建规则文件 package2.drl,目录为 rules/isPackage,它与 package.drl 的区别是 package 参数不同,其内容为:

```
package rules.isPackage2

rule "testRuleNameOnly"
    when
    then
        System.out.println("testRuleNameOnly 设置 enabled true");
end
```

注释 package.drl 文件中的第二个规则,执行 testPackage() 方法,结果如图 6-3 所示。

图6-3　不同package 同规则名称的效果

通过图 6-3 中的效果分析,同一物理目录下的规则相关文件都会被加载到规则库中,不同规则文件中不同的 package 会影响规则名称的定义。

再创建一个规则文件,测试子目录下的规则文件是否会被加载。创建规则文件 package3.drl,目录为 rules/isPackag/package2,其内容为:

```
package rules.isPackage.package2

rule "testRuleNameOnly"
    when
    then
        System.out.println("testRuleNameOnly 设置 enabled true");
end
```

执行 testPackage() 方法,结果没有变化,说明规则库不会加载子目录下的规则相关文件。

下面再次修改规则文件 package2.drl。编辑 package2.drl,内容为(为保证代码的完整性,请注意加粗部分):

```
/*package rules.isPackage2*/
package rules.isPackage
rule "testRuleNameOnly"
    when
    then
        System.out.println("testRuleNameOnly 设置 enabled true");
```

end

执行 testPackage () 方法，结果如图 6-4 所示。

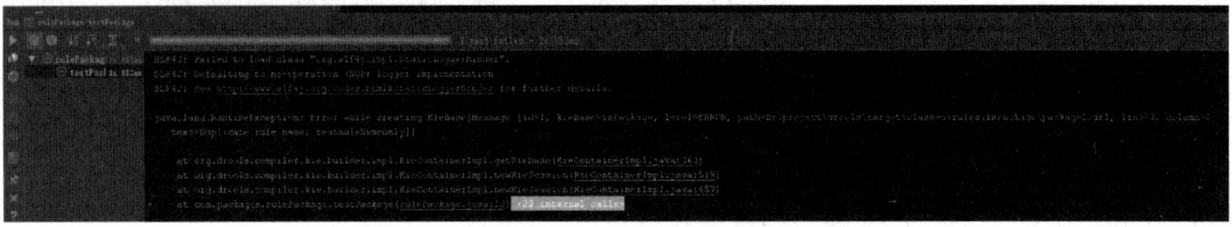

图6-4 不同规则文件中相同的包名称效果

结果分析，规则文件中 package 会命名并非没有用处，同一个规则库下逻辑目录中的规则名是唯一的。这也就是为什么在新版本（6.0 以后）中添加了 KIE 概念，引用了配置文件 kmodule.xml，并非像 5.x 版本中直接指定规则文件的方式加载规则库。package 名称不像规则命名那样允许有空格。在规则文件中，其他元素节点顺序是任意的，但要主要规则体中不能嵌套规则体，保证语法开始与结束。package 必须在文件顶部，除了规则模板文件这一特例。这只是对规则模板文件来说的，模板生成规则内容后，package 照样还是规则内容中的首行。

6.2　global全局变量

介绍 global 之前需要先理解其组织结构，如图 6-5 所示。

图6-5 global 组织结构

全局变量是一个很有用的元素节点，它是通过关键字 global class（对象类型） name 组成的。class 可以是任意的，能为规则提供操作数据或服务等功能，特别是在规则 RHS 部分中使用程序提供的服务功能，如在 RHS 部分添加日志功能、修改数据表、发送 E-mail 等。

global 全局变量与 Fact（事实）对象不同，不会因为值变化而影响到规则的再次激活。

创建规则文件 global1.drl，目录为 rules/isGlobal，内容为（注意加粗部分）：

```
package rules.isGlobal
import com.pojo.Person;
global   java.lang.Integer count;

rule "global的使用update1"
    when
    then
```

```
            System.out.println("globalupdate1------------>count"+count);
            count=10;
            System.out.println("globalupdate1------------>count"+count);
end
rule "global的使用update2"
    when
    then
        System.out.println("globalupdate2------------>count"+count);
end
rule "global的使用update3"
    when
        $total : Double() from
           accumulate(Person(),
                  init(Double totls = 0.0),
action(count=20;totls+=1;System.out.println(totls+">>>>>>>>"+count);),
                  reverse( totls-=1;
System.out.println(totls+"<<<<<<<");),//,
                  result( totls )
            )
    then
        count = 90;
        System.out.println($total+"globalupdate3reverse>>>>>>>>
count"+count);//+$s.count+$s.name);
end
```

修改 kmodule.xml 配置文件，并添加如下配置：

```
<kbase name="isGlobal" packages="rules.isGlobal">
    <ksession name="isGlobal"/>
</kbase>
```

创建 RulesGlobal.java 文件，目录为 com.ruleGlobal，其内容为：

```
package com.ruleGlobal;

import com.pojo.Person;
import org.junit.Test;
import org.kie.api.KieServices;
import org.kie.api.runtime.KieContainer;
import org.kie.api.runtime.KieSession;

public class RulesGlobal {
    @Test
    public void testGlobal() {
        KieServices kss = KieServices.Factory.get();
        KieContainer kc = kss.getKieClasspathContainer();
        KieSession ks =kc.newKieSession("isGlobal");
        Person person=new Person();
        person.setAge(2);
        person.setName("zs");
        ks.insert(person);
```

```
            ks.setGlobal("count",2017);
            int count = ks.fireAllRules();
            System.out.println(ks.getGlobal("count"));
            System.out.println("总共执行了"+count+"条规则");
            ks.dispose();
        }
}
```

执行 testGlobal() 方法，结果如图 6-6 所示。

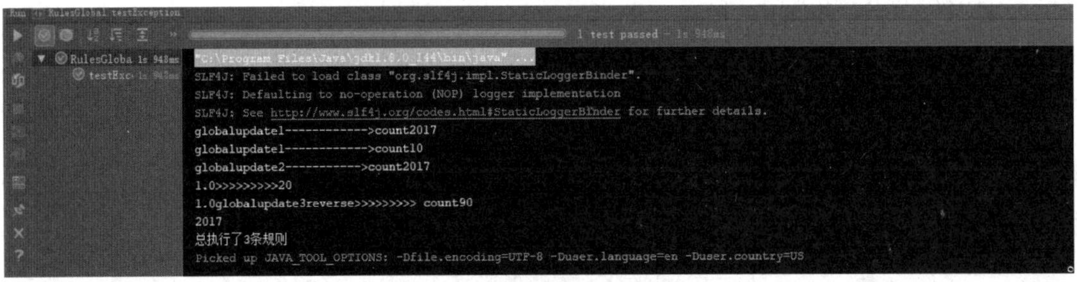

图6-6　global 测试用例的效果

通过控制台的输出结果可以看出，当前规则体中修改的值包装类的全局变量只会影响到当前规则体。

如果多个 package 使用相同标识声明的全局变量，那么它们的类型必须是相同的，并且它们所有引用都是相同的全局变量。

创建规则文件 global2.drl，目录为 rules/isGlobal，内容为（注意加粗部分）：

```
package rules.isGlobal
import com.pojo.Person;
global  java.lang.Boolean count;

rule "global2的使用update1"
    when
    then
        System.out.println("global2update1------------->count"+count);
end
```

执行 testGlobal() 方法，结果如图 6-7 所示。

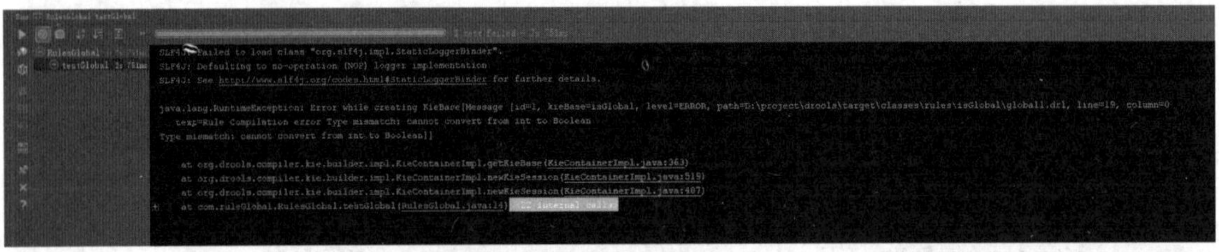

图6-7　异常使用global的效果

重新编辑 global2.drl，其内容为（为保证代码完整性，请注意加粗部分）：

```
package rules.isGlobal
import com.pojo.Person;
//global    java.lang.Boolean count;
global    java.lang.Integer count;

rule "global2的使用update1"
    when
    then
System.out.println("global2update1------------>count"+count);
end
```

执行 testGlobal () 方法，结果如图 6-8 所示。

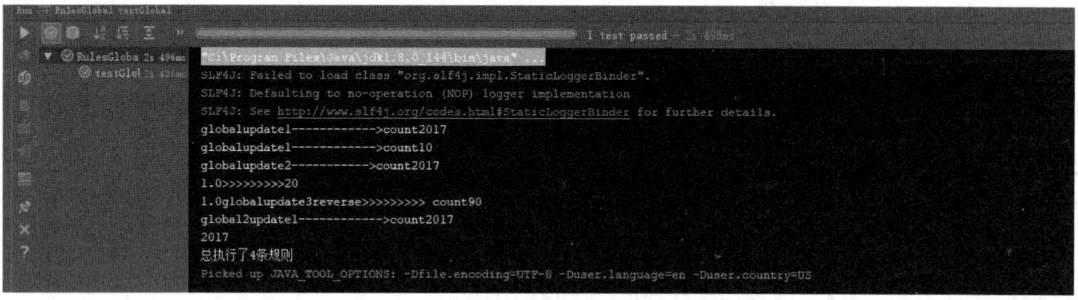

图6-8　修改后使用global关键字的效果

1. 分析总结global的功能一

全局变量定义成常量或包装类型时，该值对整个规则而言是不变的。但是如果在同一个段规则代码中改变了 global 值，那么只针对这段规则代码而言，使用的是被修改后的 global 值，对其他规则代码或元素节点中的 global 不会有影响。可以理解为它是当前规则代码或其他元素节点中的 global 副本。规则内部修改不会影响全局的使用。

global 作为全局变量方便存放在规则中，虽然说全局变量的值是不可以变化的，但并不提倡用于数据共享。原因是针对不同类型全局变量中的内容也可能会发生变化。

上述的测试用例中展示了全局变量的部分功能，并且在最后提到：全局变量定义为常量或包装类型值时是不变的，但并不是所有的类型都是不变的。下面通过测试用例来证明这一点。

创建 global3.drl，其代码为：

```
package rules.isGlobal
import com.pojo.Person;
//global    java.lang.Boolean count;
global    java.util.List    list;
global    com.pojo.Person    person;

rule "global3的使用update1"
    when
```

```
        then
System.out.println("global3update1------------>list.size"+list.size());
        list.add("aaa");
        list.add("aaa");
        list.add("aaa");
        person.setName("张小三");
end

rule "global3的使用update2"
    when
    then
System.out.println("global3update1------------>list.size"+list.size());
System.out.println("global3update1------------>person"+person.
getName());
end
```

编辑 RulesGlobal.java 文件，目录为 com.ruleGlobal，并重写 Person 的 toString() 方法，其内容为：

```java
package com.ruleGlobal;

import com.pojo.Person;
import org.junit.Test;
import org.kie.api.KieServices;
import org.kie.api.runtime.KieContainer;
import org.kie.api.runtime.KieSession;

import java.util.ArrayList;

public class RulesGlobal {
    @Test
    public void testGlobal() {
        KieServices kss = KieServices.Factory.get();
        KieContainer kc = kss.getKieClasspathContainer();
        KieSession ks =kc.newKieSession("isGlobal");
        Person person=new Person();
        person.setAge(2);
        person.setName("zs");
        ks.insert(person);
        ks.setGlobal("count",2017);
        ks.setGlobal("list",new ArrayList<>());
        ks.setGlobal("person",person);
        int count = ks.fireAllRules();
        System.out.println(ks.getGlobal("count"));
        System.out.println(ks.getGlobal("person").toString());
        System.out.println("总执行了"+count+"条规则");
        ks.dispose();
    }
}
```

执行 testGlobal() 方法，结果如图 6-9 所示。

```
2018-37-29 10:37:53 [DEBUG] org.drools...DefaultAgenda - State was INACTIVE is now FIRING_ALL_RULES
global2update1------------>count50
globalupdate1------------>count2017
globalupdate1------------>count10
globalupdate2------------>count2017
1.0>>>>>>>>>20
1.0globalupdate3reverse>>>>>>>>> count90
global3update1------------>list.size0
global3update1------------>list.size3
global3update1------------>person张小三
2018-37-29 10:37:53 [DEBUG] org.drools...DefaultAgenda - State was FIRING_ALL_RULES is now HALTING
2018-37-29 10:37:53 [DEBUG] org.drools...DefaultAgenda - State was HALTING is now INACTIVE
2017
Person[name='张小三', age=2, className='null']
总执行了6条规则
2018-37-29 10:37:53 [DEBUG] org.drools...DefaultAgenda - State was INACTIVE is now DISPOSED
```

图6-9　变化的全局变量

2. 分析总结global的功能二

全局变量如果定义成集合类或JavaBean时，在规则体RHS部分中进行修改，则规则库或Java代码中的值都会发生变化。这证明了全局变量并非不可变的值，但也正是这一点会引发出一个问题，如果在多个地方使用并修改了全局变量，就可能会导致最终结果并不是设计师所想的。

根据源码进行说明，代码执行到"ks.setGlobal("count",2017);"时会调用StatefulKnowledgeSessionImpl的setGlobal()方法，其内容为：

```java
public void setGlobal(final String identifier,
                      final Object value) {
    // Cannot set null values
    if ( value == null ) {
        return;
    }
    try {
        this.kBase.readLock();
        startOperation();
        // Make sure the global has been declared in the RuleBase
        Class type = this.kBase.getGlobals().get( identifier );
        if ( (type == null) ) {
            throw new RuntimeException( "Unexpected global [" + identifier + "]" );
        } else if ( !type.isInstance( value ) ) {
            throw new RuntimeException( "Illegal class for global. " + "Expected [" + type.getName() + "], " + "found [" + value.getClass().getName() + "]." );
        } else {
            this.globalResolver.setGlobal( identifier,
                                           value );
        }
    } finally {
        endOperation();
        this.kBase.readUnlock();
```

        }
}
```

代码执行到该方法后又调用了 MapGlobalResolver 类的 setGlobal() 方法,其内容为:

```
public void setGlobal(String identifier, Object value) {
    this.map.put( identifier,
                  value );
}
```

this.map 是一个 MapGlobalResolver 类的私有成员变量,而且是一个线程安全的 map,其内容为:

```
public MapGlobalResolver() {
    this.map = new ConcurrentHashMap<String, Object>();
}
```

通过部分源码分析,可以看出 global 的 value 是一个 object 类型,这就证明 global 的类型是任意的,并且全局变量是可以通过服务方式进行使用的。既然什么类型都可以设置,那将操作 Service 的服务通过全局变量的形式注入规则中是否可行呢?答案是一定的,只要 Java 代码中设置了全局变量的值,并且在规则文件中指定该值的类型就可以使用了。

全局变量比较难理解的一点,应该就是集合元素内容的类型了。关于集合内容的类型说明,对于集合中放的 value 可以通过源码得知,setGlobal 其实放的是 Object,所以在规则文件中,读取的也是 Object,所以定义的值,或者是泛型都要进行类型强制才能将值取出。

定义 global 时要注意,global 是不会放到工作内存中的,但是如果在设计规则时要定义,有两个规则文件中都用到了同一个全局变量,这两个 global 的内容不会因为其他调用的改变而改变。得出的结论是,global 不是用来做数据共享的,KieSession 会影响到 global 的用法。也就是说,在 "KieSession.setGlobal("perosn",p);" 这个值时,该 global 只针对于这一次创建的 KieSession 生效,如果该 global 不属于此 KieSession,则无法正常获取。

使用全局变量时应注意点以下几点。

(1) 常量值是不能改变的。

(2) 包装类是不能改变的。

(3) 类似 JavaBean、List 类的操作是可以改变内容的,但内存地址不会变。

上述内容中都是通过有状态的 KieSession 进行操作的。存在有状态的 KieSession 必然存在无状态[①] 的 KieSession。下面通过无状态 KieSession 操作全局变量。

设置全局变量,其内容为:

```
...
StatelessKieSession ksession = kbase.newStatelessKieSession();
ExecutionResults bresults =
    ksession.execute( CommandFactory.newSetGlobal( "stilton", new Cheese( "stilton" ), true));
```

---

① 无状态的 KieSession 是指 StatelessKieSession。
```

```
Cheese stilton = bresults.getValue( "stilton" );
...
```

获取全局变量，其内容为：

```
...
StatelessKieSession ksession = kbase.newStatelessKieSession();
ExecutionResults bresults =
    ksession.execute( CommandFactory.getGlobal( "stilton" );
Cheese stilton = bresults.getValue( "stilton" );
...
```

特别声明：使用 global 时，其实是可以通过传递参数的方式将操作数据库传入规则中的，但并不建议这样做。原因很简单，在事务的处理中规则可能会存在问题，而且明明已经获取了服务，又何必通过 global 的方式传入规则中再进行一步操作，这无形中给规则与业务数据库处理增加了耦合。

6.3 query查询

在介绍 query 之前需要先理解其组织结构，如图 6-10 所示。

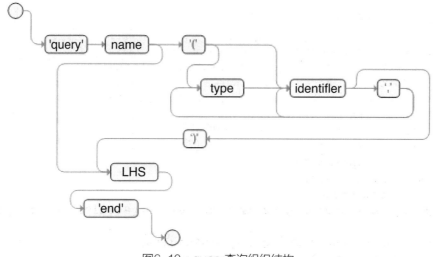

图6-10　query查询组织结构

元素 query 指查询，从结构图上看，它以 query 开头、以 end 结束，其中包含 query name，查询参数是可以选择的，多个参数以逗号为分隔符，查询是一种条件匹配的方式，因为它只包含 LHS 部分，因此不需要指定 when 或 then。如何判断条件匹配是否正确，可以通过 Java 代码进行获取。查询的 name 在当前规则库的逻辑路径下是唯一的，这点与规则名称约束相似。

创建规则文件 query.drl，目录为 rules/isQuery，其内容为（注意加粗部分）：

```
package rules.isQuery

import   com.pojo.Person;

query "person age is 30"
    person:Person(age==30)
end
```

修改 kmodule.xml 配置文件，并添加如下配置：

```xml
<kbase name="isQuery" packages="rules.isQuery">
    <ksession name="isQuery"/>
</kbase>
```

创建 RulesQuery.java 文件，目录为 com.ruleQuery，其内容为：

```java
package com.ruleQuery;

import com.pojo.Person;
import org.junit.Test;
import org.kie.api.KieServices;
import org.kie.api.runtime.KieContainer;
import org.kie.api.runtime.KieSession;
import org.kie.api.runtime.rule.QueryResults;
import org.kie.api.runtime.rule.QueryResultsRow;

public class RulesQuery {
    @Test
    public void testQuery() {
        KieServices kss = KieServices.Factory.get();
        KieContainer kc = kss.getKieClasspathContainer();
        KieSession ks =kc.newKieSession("isQuery");
        Person person1 = new Person("张三", 35);
        Person person2 = new Person("李四", 30);
        Person person3 = new Person("王五", 50);
        ks.insert(person1);
        ks.insert(person2);
        ks.insert(person3);
        QueryResults queryResults = ks.getQueryResults("person age is 30");
        for (QueryResultsRow q : queryResults) {
            Person p= (Person) q.get("person");
            System.out.println("输出符合查询条件的实体对象name为"+p.getName());
        }
        ks.dispose();
    }
}
```

执行 testQuery() 方法，结果如图 6-11 所示。

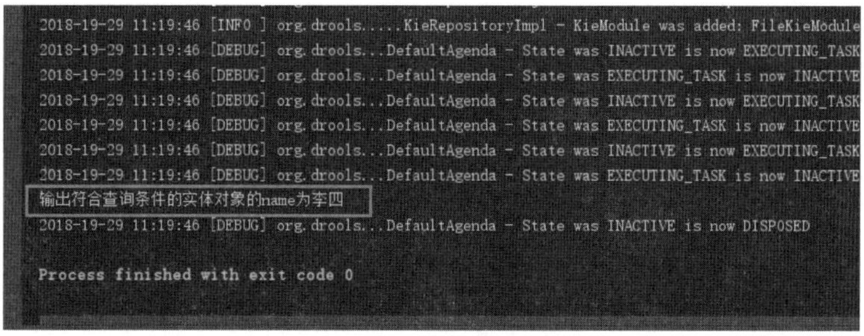

图6-11 query查询效果

输出内容与期待结果是一致的，规则文件中并没有使用任何的规则体，而 query 的目的只是为了判断 insert 到规则中的 Fact 对象是否满足条件。query 很简单，判断 Person 属性是否满足 age==30，如果满足，则返回一个集合。

query 是可以添加参数的，下面通过例子来进行说明。

编辑规则文件 query.drl，目录为 rules/isQuery，其内容为（注意加粗部分）：

```
query "person age is 30 and name is 张小三"(String $name)
    person:Person(name==$name,age==30)
end package rules.isQuery

import   com.pojo.Person;

query "person age is 30"
    person:Person(age==30)
end

query "person age is 30 and name is 张小三"(String $name)
    person:Person(name==$name,age==30)
end
```

编辑 RulesQuery.java 文件，并添加 testQuery2() 方法，其代码为：

```
@Test
public void testQuery2() {
    KieServices kss = KieServices.Factory.get();
    KieContainer kc = kss.getKieClasspathContainer();
    KieSession ks =kc.newKieSession("isQuery");
    Person person1 = new Person("张三", 35);
    Person person2 = new Person("李四", 30);
    Person person3 = new Person("王五", 50);
    Person person4 = new Person("张小三", 30);
    ks.insert(person1);
    ks.insert(person2);
    ks.insert(person3);
    ks.insert(person4);
    Object[] objects=new Object[]{"张小三"};
    QueryResults queryResults = ks.getQueryResults("person age is 30
```

```
and name is 张小三",objects);
    for (QueryResultsRow q : queryResults) {
        Person p= (Person) q.get("person");
        System.out.println("输出符合查询条件的实体对象的name为"+p.getName());
    }
    ks.dispose();
}
```

执行 testQuery2() 方法，结果如图 6-12 所示。

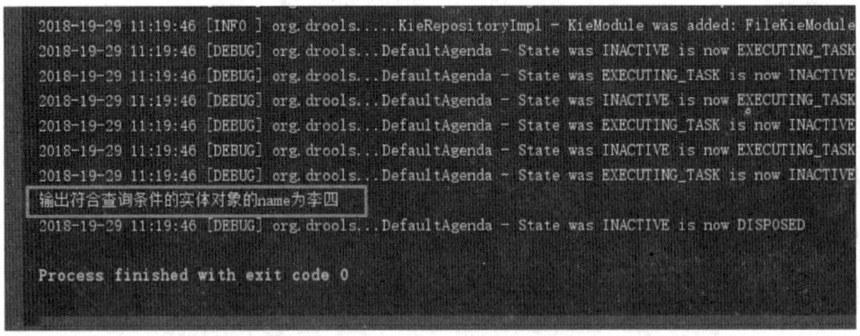

图6-12　query的参数效果

将所要传入的值按 DRL 文件定义好的类型顺序依次写入。$name 主要为了区别于 Person 属性的名称。

6.4　function函数

在介绍 function 函数之前需要先理解其组织结构，如图 6-13 所示。

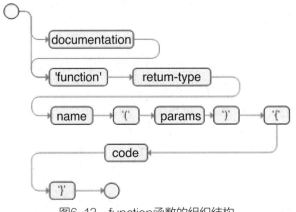

图6-13　function函数的组织结构

function 函数写法与 JavaScript 中的函数相似，规则函数是 Java 类方法的一种变形，它的参数是非必填信息，返回值也是非必填的。一个规则使用函数的好处是可以在同一个地方保持所有的逻辑，同一个逻辑路径下的函数是一个全局的函数。当然规则函数发生变化，则意味着所有调用该函数的规则体都将发生变化。

规则中的函数有两种形式：一种是 import，可以引用 Java 静态方法；另一种需要在规则中添加关键字 function。

使用第二种规则自定义函数方式，创建规则文件 function1.drl，目录为 rules/isFunction，其内容为（注意加粗部分）：

```
package rules.isFunction

rule "function1"
    when
    then
        function01();
        System.out.println("函数function02()的返回值"+function02());
        function03("张小三");
        System.out.println("函数function04()的返回值"+function04("李小四"));
end
function void function01(){
    System.out.println("输出一个无参无返回值的函数");
}
function String function02(){
    System.out.println("输出一个无参有返回值的函数");
    return "Hello";
}
function void function03(String name){
    System.out.println("输出一个有参无返回值的函数，输出参数是"+name);
}
function String function04(String name){
    System.out.println("输出一个有参有返回值的函数，输出参数是"+name);
    return name;
}
```

修改 kmodule.xml 配置文件，并添加如下配置：

```
<kbase name="isFunction" packages="rules.isFunction">
    <ksession name="isFunction"/>
</kbase>
```

创建 RulesFunction.java 文件，目录为 com.rulesFunction，其内容为：

```
package com.rulesFunction;

import org.junit.Test;
import org.kie.api.KieServices;
import org.kie.api.runtime.KieContainer;
import org.kie.api.runtime.KieSession;
```

```
public class RulesFunction {
    @Test
    public void testFunction() {
        KieServices kss = KieServices.Factory.get();
        KieContainer kc = kss.getKieClasspathContainer();
        KieSession ks = kc.newKieSession("isFunction");
        int count = ks.fireAllRules();
        System.out.println("总执行了"+count+"条规则");
        ks.dispose();
    }
}
```

执行 testFunction () 方法，结果如图 6-14 所示。

图6-14 引用function函数的第一种方式

通过上述简单例子可以得出结论，function 函数与一个 Java 方法没有什么区别。

验证同逻辑路径下函数的全局性。创建规则文件 function2.drl，目录为 rules/isFunction，其内容为（注意加粗部分）：

```
package rules.isFunction

rule "function2"
    when
    then
        function01();
        System.out.println("函数function02()的返回值"+function02());
        function03("张小小");
        System.out.println("函数function04()的返回值"+function04("李小小"));
end
```

执行 testFunction () 方法，结果如图 6-15 所示。

通过图 6-15 的输出结果，证明 function 函数在同逻辑路径下是全局性的，与全局变量相似。

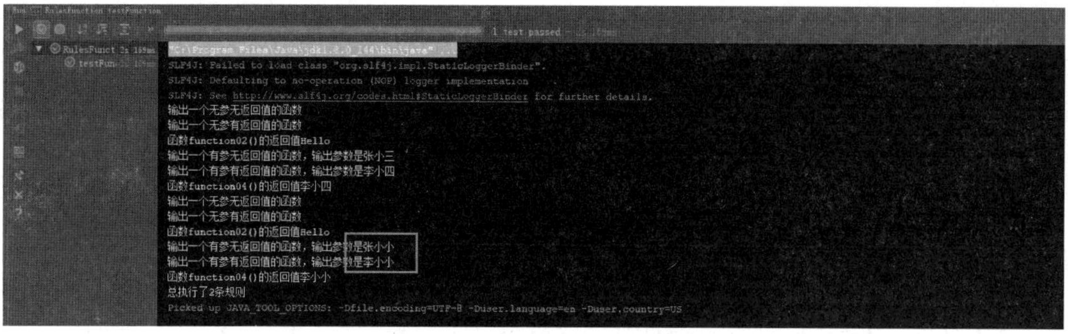

图6-15 function函数的全局性效果

验证不同逻辑路径下函数的全局性，创建规则文件 function1.drl，目录为 rules/isFunction，其内容为（注意加粗部分）：

```
package rules.isFunction2

rule "function1"
    when
    then
        function01();
        System.out.println("函数function02()的返回值"+function02());
        function03("张不三");
        System.out.println("函数function04()的返回值"+function04("李不四"));
end
```

执行 testFunction() 方法，结果如图 6-16 所示。

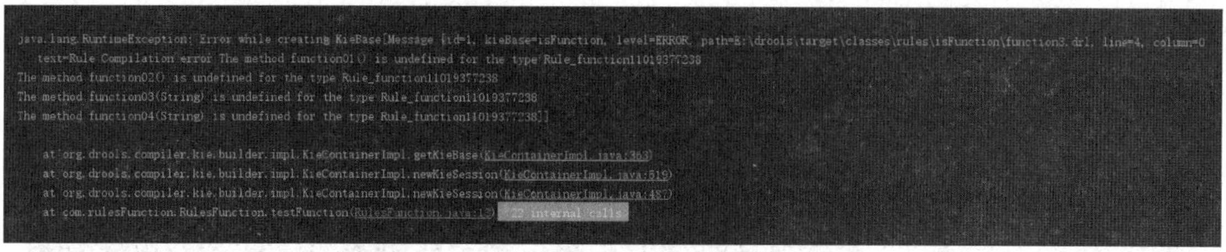

图6-16 不同逻辑路径下的函数全局性效果

通过图 6-16 中的效果，证明不同逻辑路径下的函数不可使用。

function 函数的第一种形式，是通过 Java 静态方法的方式。创建 FunctionStatic.java 文件，目录为 com.rulesFunction，其内容为：

```
package com.rulesFunction;

public class FunctionStatic {
    public static void testStatic1(){
        System.out.println("输出一个无参无返回值的静态方法");
    }
```

```
    public static String testStatic2() {
        System.out.println("输出一个无参有返回值的静态方法");
        return "Hello";
    }
    public static void testStatic3(String name){
        System.out.println("输出一个有参无返回值的静态方法,输出参数是"+name);
    }
    public static String testStatic4(String name){
        System.out.println("输出一个有参有返回值的静态方法,输出参数是"+name);
        return name;
    }
}
```

创建规则文件 function4.drl,目录为 rules/isFunction,其内容为(注意加粗部分):

```
package rules.isFunction

import function com.rulesFunction.FunctionStatic.testStatic1;
import function com.rulesFunction.FunctionStatic.testStatic2;
import function com.rulesFunction.FunctionStatic.testStatic3;
import function com.rulesFunction.FunctionStatic.testStatic4;

rule "function4"
    when
    then
        testStatic1();
        System.out.println("函数function02()的返回值"+testStatic2());
        testStatic3("张小三");
        System.out.println("函数function04()的返回值"+testStatic4("李小四"));
end
```

执行 testFunction() 方法,结果如图 6-17 所示,这是为证明函数可以正常执行注释 function3.drl 规则文件。

图6-17　function函数的静态方法效果

有些场景下,LHS 部分无法将所有的约束条件陈列其中,使用 function 函数多数也是在 RHS 部分,用来做其他的约束或处理事件,至于 when 中是否可以使用 function 函数,答案是肯定的。when 作为规则体的比较部分存在,其效果与 Java 语法中的 if 功能大相径庭。在上述的测试中,函数是可以有返回值的,直接在 when 中使用 function,需要的返回值是 Boolean 类型。

创建规则文件 function5.drl,目录为 rules/isFunction,其内容为(注意加粗部分):

```
package rules.isFunction

rule "function5"
    when
        eval(function05());
    then
        System.out.println("测试when中使用函数");
end
function Boolean function05(){
    return true;
}
```

执行 testFunction () 方法,结果如图 6-18 所示。

图6-18　when中使用function函数的效果

规则体 when 部分并不能单独使用函数,只有通过返回值是 Boolean 类型的"eval();"进行引用才可以,但如果函数作为返回值与 Fact 对象的属性进行比较也是可行的,使其 when 结果为 Boolean 即可。强调 function 不能在 when 部分单独使用。

6.5　declare声明

在介绍 declare 声明之前需要先理解其组织结构,如图 6-19 所示。

declare 声明在规则引擎中的功能主要有两个:一是声明新类型,二是声明元数据类型。

声明新类型,与 JavaBean 功能一样,但方式却比 JavaBean 简单。在之前的讲述过程中,规则中操作事实对象都通过 Java 代码 insert 到规则中进行处理。然而有些时候,并非所有的情况都要编辑 JavaBean。如果既想使用 JavaBean 中的特点,又不想多创建 JavaBean 文件,那么使用声明是再好不过的。

声明元数据类型,fact 对象包含了一些特性,这些特性称为类元信息,如需要当前属性长度是固定的,那就在属性声明前添加元数据。一般元数据用于之前讲的查询,复杂事件处理和属性字段约束居多。

declare 声明的语法以 declare 开头，以 end 结束。name 参数与 JavaBean 定义类名一样。

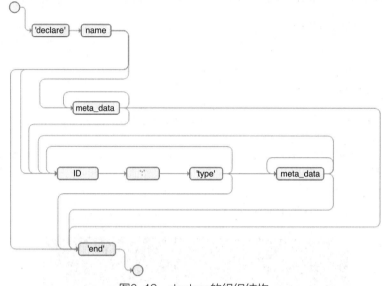

图6-19　declare的组织结构

创建规则文件 declare.drl，目录为 rules/isDeclare，其内容为（注意加粗部分）：

```
package rules.isDeclare

declare Person
    name:String
    age:int
end

rule "declareInsert"
    when

    then
        insert(new Person("张三",20));
end

rule "declareTest"
    when
        $p:Person(name=="张三")
    then
        System.out.println("使用通过declare来实现insert 后进行操作");
end
```

修改 kmodule.xml 配置文件，并添加如下配置：

```
<kbase name="isDeclare" packages="rules.isDeclare">
    <ksession name="isDeclare"/>
</kbase>
```

创建 RulesDeclare.java 文件，目录为 com.ruleDeclare，其内容为：

```java
package com.ruleDeaclare;

import org.junit.Test;
import org.kie.api.KieServices;
import org.kie.api.runtime.KieContainer;
import org.kie.api.runtime.KieSession;

public class RulesDeclare {
    @Test
    public void testDeclare() {
        KieServices kss = KieServices.Factory.get();
        KieContainer kc = kss.getKieClasspathContainer();
        KieSession ks =kc.newKieSession("isDeclare");
        int count = ks.fireAllRules();
        System.out.println("总执行了"+count+"条规则");
        ks.dispose();
    }
}
```

执行 testDeclare () 方法，结果如图 6-20 所示。

图6-20　declare声明的使用效果

上述例子中声明一个名称为 Person 的新事实类型。该事实类型将会有两个属性：name 和 age。定义声明属性类型可以是任意的 Java 有效类型，并且包括用户创建的任何其他类，甚至可以是当前规则中已有的声明类型。

声明时，还可以通过指定属性类型来进行使用，其内容为：

```
...
declare Person
    name : String
    dateOfBirth : java.util.Date
    age: int
end
...
```

上述例子中，dateOfBirth 是 java.util.Date 类型，来自 Java API。或者也可以为如下形式：

```
...
```

```
import java.util.Date
declare Person
    name : String
    dateOfBirth : Date
    age: int
end
...
```

通过使用 import 子句，可以避免每次编写时必须书写完全合格类名，如上述代码，使用 import 子句，就可避免必须使用类的全名（类路径＋类名）。

声明一个新事实类型时，Drools 会在编译时生成实现表示事实类型的一个 Java 类的字节码。生成的 Java 类，将是一个一对一 Java Bean 类型定义的映射，所以对于前面的例子，生成 Java 类后，其内容为：

```
public class Person implements Serializable {
    private String name;
    private java.util.Date dateOfBirth;
    private int age;
    // getters and setters      //此处省略这些方法
    // equals/hashCode
    // toString
}
```

因为生成的类是一个简单的 Java 类，相当于在当前规则中 import 一个对象，所以在同逻辑路径下的规则体中都是可以使用的，与全局变量、函数查询的共享范围是一致的。

创建规则文件 declare2.drl，目录为 rules/isDeclare，其内容为（注意加粗部分）：

```
package rules.isDeclare

rule "declareTest2"
    when
        $p:Person(age==20)
    then
        System.out.println("declareTest2 使用通过declare来实现insert 后进行操作");
end
```

执行 testDeclare() 方法，结果如图 6-21 所示。

对于声明并不太好使的问题在于 Java 代码中难以获取这个声明，并不是不可以获取，只是相对麻烦一些。当然声明的定义原本就是为了对 JavaBean 的扩展或并不希望对 JavaBean 进行修改。在使用常用的声明时，则必须要对它进行 insert 操作。

第 6 章 规则中级语法

图6-21 declare场景的测试效果

若规则文件中声明的类型与 import 引用的类名相同时，则系统会报错。创建规则文件 declare3.drl，目录为 rules/isDeclare2，其内容为（注意加粗部分）：

```
package rules.isDeclare2
import com.pojo.Person;

declare Person
    name:String
    age:int
end
/*declare Person
end*/

rule "declare类型名相同测试"
    when
    then
        System.out.println("declare类型名相同测试");
end
```

执行 testDeclare () 方法，结果如图 6-22 所示，执行直接报错。

图6-22 使用declare场景异常

但如果将声明类型中的属性删除，执行 testDeclare () 方法，结果如图 6-23 所示。

原因是规则文件编译冲突，这是正常的。因为出现这类情况时，规则并不知道自己需要用哪一种引用类型，所以只能是系统报错。

声明功能有一个强大的继承功能。其中，关键字 extends 的继承方式与 Java 相似，除了上述讲到的基本语法不变外，在声明 name 后添加 extends 关键字，并指明要继承的类就可以了。

113

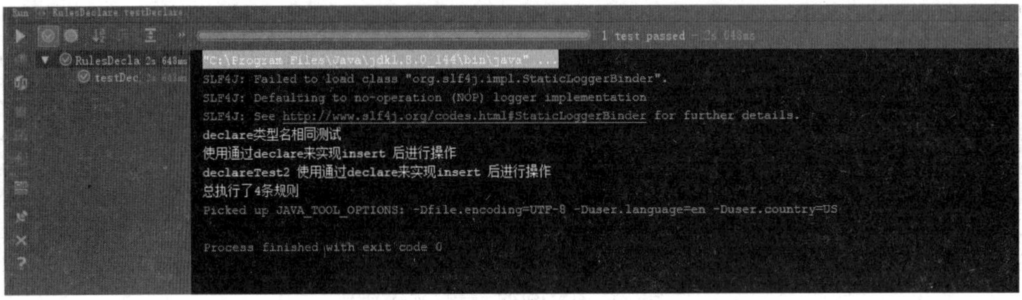

图6-23 删除声明类属性

创建规则文件declare4.drl，目录为rules/isDeclare，其内容为：

```
package rules.isDeclare

declare PersonEx extends com.pojo.Person
    type:String
end

rule "declareInsertExt"
    when
    then
        PersonEx p=new PersonEx();
        p.setType("1");
        p.setAge(20);
        p.setName("张三");
        insert(p);
end

rule "declareTestExtends"
    when
        $p:PersonEx(name=="张三")
    then
        System.out.println("输出PersonEx的name"+$p.getName());
        System.out.println("输出PersonEx的age"+$p.getAge());
        System.out.println("输出PersonEx的type"+$p.getType());
end
```

执行testDeclare()方法，结果如图6-24所示。

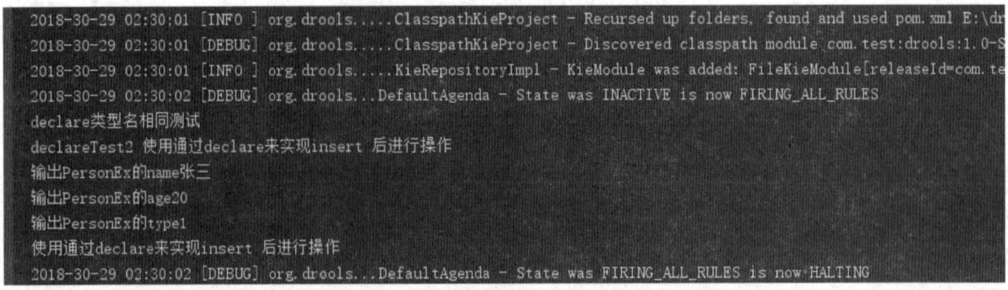

图6-24 使用extends关键字的效果

6.6　规则when

规则文件中处理业务的核心在于规则体的使用，简称规则。前面已经详细介绍了规则的语法结构，本节将对 when 中所用到的功能做进一步说明。其介绍方式还是由简单逐步到复杂。

书写规范，与 Java 语法相似，不能使用全角输入法，LHS 部分为 when 与 then 中间语法，规则属性判断过程，只有规则在 LHS 中匹配结果为 true 时，才视为规则成立，才会转到 RHS 部分。如果 LHS 为空，则表示该规则永为 true。当 WorkingMemory（Fact 事实对象）发生变化时，该规则可能会被再次激活，解决方案请查看规则属性 no-loop 的用法。

pattern 匹配模式中讲到，pattern 是一种约束条件，了解匹配模式要先了解其组织结构，如图 6-25 所示。

图6-25　pattern组织结构

匹配模式由零个或多个约束组成，并且有一个可选择的模式绑定，在它的最简单格式中，没有内部约束条件，通过一个给定类型的事实就可以进行模式匹配了。注意类型不必是一些事实对象的实际类，模式可以引用子类、接口、函数等。

引用匹配的对象，可以为一个模式绑定变量，如 $c，前缀符号 ($) 是可选的；复杂的规则中非常有用，它有助于编码者区别变量与字段，如 $p:Person(age==30);。模式中的括号内部是所有匹配的地方，约束可以是一个字段，内嵌计算（Inline Eval），或者一个约束组。约束可以使用 ","、"&&" 或 "||" 符号分隔，如图 6-26 所示。

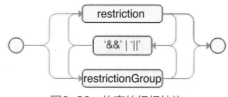

图6-26　约束的组织结构

比较运算符是有优先级的，与 Java 一致。进行"与，或"运算时要注意短路机制。

对象属性有 3 种类型的限制：单值限制、复合值限制和多限制。单值限制就是经常提到的类似上述说明中的例子，复合限制类似于数据库查询语句中的 in 或 not in。

1. 复合值限制in/not in

in/not in 复合限制组织结构如图 6-27 所示。

复合值限制是指超过一种匹配值的限制条件，如 Sql 语句中的 in，语法格式与 Sql 多匹配相似。以括号为第二参数，括号内比较值以逗号分隔，比较值可以是变量、文字、返回值或标识符等，其

内部功能与 "!=" "==" 运算符的多限制列表类似。

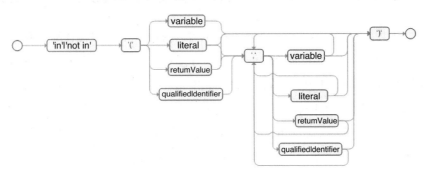

图6-27 复合限制的组织结构

创建规则文件 in.drl，目录为 rules/isIn，其内容为：

```
package rules.isIn
import com.pojo.Person;
import com.pojo.School;

rule "in的复合规则"
    when
        School($cn:className)
        Person(className in("五班","六班",$cn) )
    then
        System.out.println("验证in的复合限制规则");
end

rule "not in的复合规则"
    when
        School($cn:className)
        Person(className not in("五班","六班",$cn) )
    then
        System.out.println("验证not in的复合限制规则");
end
```

修改 kmodule.xml 配置文件，并添加如下配置：

```
<kbase name="isIn" packages="rules.isIn">
    <ksession name="isIn"/>
</kbase>
```

创建 RuleIn.java 文件，目录为 com.ruleIn，其内容为：

```
package com.ruleIn

import com.pojo.Person;
import com.pojo.School;
import org.junit.Test;
import org.kie.api.KieServices;
import org.kie.api.runtime.KieContainer;
```

```
import org.kie.api.runtime.KieSession;

public class RuleIn {
    @Test
    public void testIn() {
        KieServices kss = KieServices.Factory.get();
        KieContainer kc = kss.getKieClasspathContainer();
        KieSession ks =kc.newKieSession("isIn");
        Person person=new Person();
        person.setClassName("三班");
        ks.insert(person);
        School school=new School();
        school.setClassName("三班");
        ks.insert(school);
        Person person2=new Person();
        person2.setClassName("四班");
        ks.insert(person2);
        int count = ks.fireAllRules();
        System.out.println("总执行了"+count+"条规则");
        ks.dispose();
    }
}
```

执行 testIn () 方法，结果如图 6-28 所示。

图6-28　复合限制条件效果

多限制还有另外的展现方式，简单的多限制使用 &&、Person(age > 30 && < 40)，也可用"，"替换 &&。复杂多约束使用分组多限制 Person(age ((> 30 && < 40) || (> 20 && < 25)))，约束符 and/or 用法比较简单，使用法则与 Java 语法的约束是一致的，因为有优先级的问题，所以在设计过程中需要注意短路机制。

2. 条件元素 eval

eval 组织结构如图 6-29 所示。

图6-29　eval组织结构

条件元素 eval 是在最开始的测试用例中就使用过的。它可以是任何语义代码，并返回一个 boolean 类型，也可以绑定规则 LHS 中的变量、函数或直接写常量进行比较。但在实际编码过程中，要尽可能地少用 eval 作为比较符，因为会导致引擎的性能问题。读者可以参考书中经典 Hello World 章节中的用例，或者在 funcation 函数中进行测试。

3. 条件元素 not

not 组织结构如图 6-30 所示。

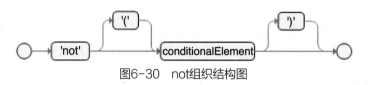

图6-30　not组织结构图

条件元素 not 是判断在工作内存中是否还存在某个值，当 not EC 成立时就代表当前工作内存中不存在 EC，可以看成"一定没有这个值"。

创建规则文件 not.drl，目录为 rules/isNot，其内容为（注意加粗部分）：

```
package rules.isNot
import com.pojo.Person;

rule "测试Not"
    when
        not Person();
    then
        System.out.println("测试Person 一定不在工作内存中");
end
rule "测试not Not"
    when
        not(not Person());
    then
        System.out.println("测试Person 一定在工作内存中");
end
```

修改 kmodule.xml 配置文件，并添加如下配置：

```
<kbase name="isNot" packages="rules.isNot">
    <ksession name="isNot"/>
</kbase>
```

创建 RuleNot.java 文件，目录为 com.ruleNot，其内容为：

```
package com.ruleNot;

import com.pojo.Person;
import com.pojo.School;
import org.junit.Test;
import org.kie.api.KieServices;
import org.kie.api.runtime.KieContainer;
import org.kie.api.runtime.KieSession;
```

```
public class RuleNot {
    @Test
    public void testNot() {
        KieServices kss = KieServices.Factory.get();
        KieContainer kc = kss.getKieClasspathContainer();
        KieSession ks =kc.newKieSession("isNot");
        Person person2=new Person();
        person2.setClassName("四班");
        ks.insert(person2);
        int count = ks.fireAllRules();
        System.out.println("总执行了"+count+"条规则");
        ks.dispose();
    }
}
```

执行 testNot () 方法，注释 insert 代码，结果如图 6-31 所示，不注释 insert 代码，结果如图 6-32 所示。

图6-31　测试not约束效果（1）

图6-32　测试not约束效果（2）

4. 条件元素 exists

exists 组织结构如图 6-33 所示。

图6-33　exists组织结构

条件元素 exists 的功能与 not 的功能是相反的，指在工作内存中是否存在某个东西，可以看作"至少有一个"。

创建规则文件 exists.drl，目录为 rules/isExists，其内容为（注意加粗部分）：

```
package rules.isExists
import com.pojo.Person;

rule "测试Exists"
    when
        exists Person();
    then
        System.out.println("测试Person 一定在工作内存中");
end

rule "测试not Exists"
    when
     not (exists Person());
    then
        System.out.println("测试Person 一定在不工作内存中");
end
```

修改 kmodule.xml 配置文件，并添加如下配置：

```
<kbase name="isExists" packages="rules.isExists">
    <ksession name="isExists"/>
</kbase>
```

创建 RuleExists.java 文件，目录为 com.ruleExists，其内容为：

```
package com.ruleExists;

import com.pojo.Person;
import com.pojo.School;
import org.junit.Test;
import org.kie.api.KieServices;
import org.kie.api.runtime.KieContainer;
import org.kie.api.runtime.KieSession;

public class RuleExists {
    @Test
    public void testExists() {
        KieServices kss = KieServices.Factory.get();
        KieContainer kc = kss.getKieClasspathContainer();
        KieSession ks =kc.newKieSession("isExists");
        Person person=new Person();
        person.setClassName("三班");
        person.setAge(35);
        //ks.insert(person);
        int count = ks.fireAllRules();
        System.out.println("总执行了"+count+"条规则");
        ks.dispose();
    }
}
```

执行 testExists () 方法，不注释 insert 代码的结果如图 6-34 所示，注释 insert 代码的结果如图 6-35 所示。

图6-34　exists测试效果（1）

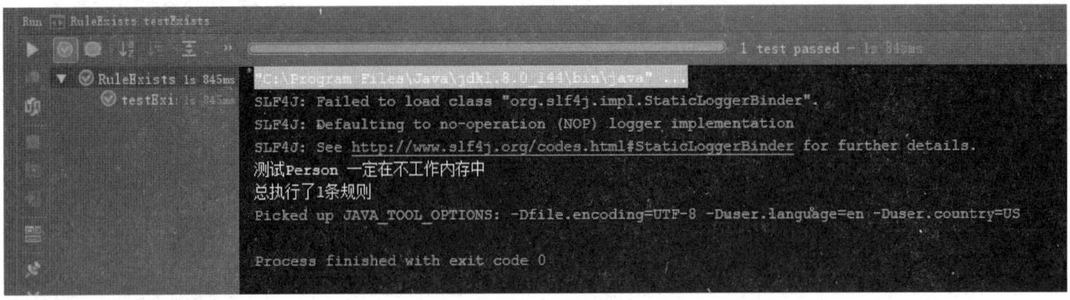

图6-35　exists测试效果（2）

5．条件元素 forall

forall 组织结构如图 6-36 所示。

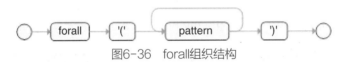

图6-36　forall组织结构

条件元素 forall 的功能与 eval() 的功能是相似的，通过模式匹配对 forall 进行一个判断，当完全匹配时，forall 为 true。

创建规则文件 forall.drl，目录为 rules/isForall，其内容为（注意加粗部分）：

```
package rules.isForall
import com.pojo.Person;

rule "测试forall"
    when
        forall($p : Person(name =="张三")
               Person( this == $p,age==30))
    then
        System.out.println("测试forall");
end
```

修改 kmodule.xml 配置文件，并添加如下配置：

```xml
<kbase name="isForall" packages="rules.isForall">
    <ksession name="isForall"/>
</kbase>
```

创建 RuleForAll.java 文件，目录为 com.ruleForall，其内容为：

```java
package com.ruleForall;

import com.pojo.Person;
import org.junit.Test;
import org.kie.api.KieServices;
import org.kie.api.runtime.KieContainer;
import org.kie.api.runtime.KieSession;

public class RuleForAll {
    @Test
    public void testForAll() {
        KieServices kss = KieServices.Factory.get();
        KieContainer kc = kss.getKieClasspathContainer();
        KieSession ks =kc.newKieSession("isForall");
        Person person=new Person();
        person.setAge(30);
        person.setName("张三");
        ks.insert(person);
        int count = ks.fireAllRules();
        System.out.println("总执行了"+count+"条规则");
        ks.dispose();
    }
}
```

执行 testForAll () 方法，如图 6-37 所示。

图6-37　forall测试效果

6. 条件元素 from

from 组织结构如图 6-38 所示。

图6-38　from组织结构

条件元素 from 是一个很有意思的约束，它可以让用户指定任意的资源，用于 LHS 部分的数据匹配，也可以用来对集合进行遍历，还可以用来对 Java 服务进行访问并就结果进行遍历（在 global 全局变量提到 golobal 可以提供 Java 服务）。

修改 Person.java 文件，添加私有属性 School school，并实现 get set 方法。创建规则文件 from.drl，目录为 rules/isFrom，其内容为（注意加粗部分）：

```
package rules.isFrom
import com.pojo.Person;
import com.pojo.School;

rule "测试from"
    when
        $p:Person($ps:school)
        $s:School(className=="一班") from $ps;
    then
        System.out.println("测试from");
end
```

修改 kmodule.xml 配置文件，并添加如下配置：

```xml
<kbase name="isFrom" packages="rules.isFrom">
    <ksession name="isFrom"/>
</kbase>
```

创建 RuleFrom.java 文件，目录为 com.ruleFrom，其内容为：

```java
package com.ruleFrom;

import com.pojo.Person;
import com.pojo.School;
import org.junit.Test;
import org.kie.api.KieServices;
import org.kie.api.runtime.KieContainer;
import org.kie.api.runtime.KieSession;

public class RuleFrom {
    @Test
    public void testFrom() {
        KieServices kss = KieServices.Factory.get();
        KieContainer kc = kss.getKieClasspathContainer();
        KieSession ks =kc.newKieSession("isFrom");
        Person person=new Person();
        person.setAge(30);
        person.setName("张三");
        person.setSchool(new School("一班"));
        ks.insert(person);
        ks.insert(new School("二班"));
        int count = ks.fireAllRules();
        System.out.println("总执行了"+count+"条规则");
```

```
        ks.dispose();
    }
}
```

执行 testFrom () 方法，如图 6-39 所示。

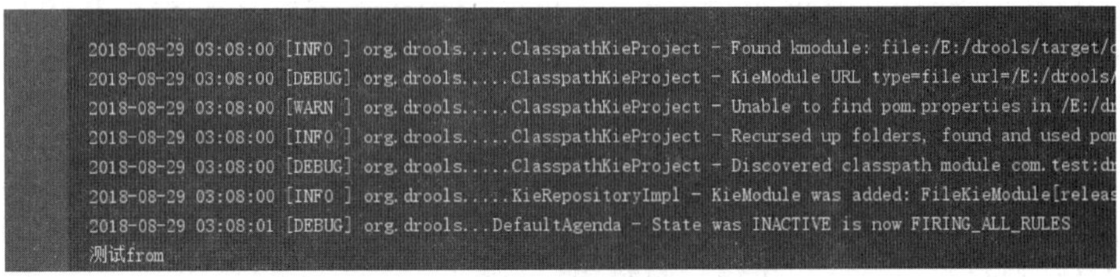

图6-39 from第一次测试效果

上述规则文件内容也可以换成另外一种方式，其内容为：

```
rule "测试from2"
    when
        $p:Person()
        $s:School(className=="一班") from $p.school;
    then
        System.out.println("测试from2");
end
```

条件元素 from 支持对象源，返回一个对象集合，在这种情况下，from 将会遍历集合中的所有对象，并分别匹配它们每一个对象值。

例如，输出所有一班的学生名字：编辑规则文件 from.drl 并添加测试 from3List 规则，其内容为（注意加粗部分）：

```
rule "测试from3List"
    when
        $s:School()
        $p:Person(className=="一班") from $s.classNameList
    then
        System.out.println("测试from3List"+$p.getName());
end
```

编辑 RuleFrom.java 文件，并添加 testFromList() 方法，其内容为：

```
@Test
public void testFromList() {
    KieServices kss = KieServices.Factory.get();
    KieContainer kc = kss.getKieClasspathContainer();
    KieSession ks =kc.newKieSession("isFrom");
    Person person=new Person();person.setName("张三");person.setClassName("一班");
    Person person2=new Person();person2.setName("李四");person2.setClassName("一班");
    Person person3=new Person();person3.setName("王五");person3.
```

```
setClassName("二班");
    Person person4=new Person();person4.setName("赵六");person4.
setClassName("一班");
    School school=new School();
    List classnameList=new ArrayList();
    classnameList.add(person);
    classnameList.add(person2);
    classnameList.add(person3);
    classnameList.add(person4);
    school.setClassNameList(classnameList);
    ks.insert(school);
    int count = ks.fireAllRules();
    System.out.println("总执行了"+count+"条规则");
    ks.dispose();
}
```

执行 testFromList () 方法，如图 6-40 所示。

图6-40　from遍历对象效果

使用 form 时，设计者必须要注意使用属性功能，特别是与 lock-on-active 规则属性联合使用时，因为该属性可能产生不一样的结果。

from 不仅可以在对象、集合中应用，还可以在函数中使用。编辑规则文件 from.drl 并添加测试 from 的使用 function 规则，其内容为（注意加粗部分）：

```
import function com.ruleFrom.RuleFrom.listfrom;
rule "from的使用function"
    when
        Person($name:name);
        $ps:Person(name==$name) from listfrom("张三",10)
    then
        System.out.println("from的使用function,通过function 传值的方法"+$ps.getName()+$ps.getAge());
end
```

编辑 RuleFrom.java 文件，并添加 listfrom () 方法，其内容为：

```
/**
 * 用来测试from   function的方法
 * @param name
 * @param integer
```

```
 * @return
 */
public   static List<Object> listfrom(String name,Integer integer) {
    List list = new ArrayList();
    for (int i = 0; i < 5; i++) {
        Person person = new Person();
        person.setName(name);
        person.setAge(integer+i);
        list.add(person);
    }
    return  list;
}
```

执行 testFrom () 方法，如图 6-41 所示。

图6-41　from测试函数效果

通过上述的例子，证明规则 from 是可以调用 function 函数的，在某些情况下，global 全局变量是可以替代 function 的，而且在介绍 global 时已经强调过，使用 global 可以调用服务，在使用 from 时，可以通过 global 来调用 Java 提供的服务功能。例如，调用 Spring 提供的查询功能时，将依赖注入的 JavaBean 以全局变量的方式放在规则中，通过 from 进行集合遍历。

7. 条件元素 collect

collect 组织结构如图 6-42 所示。

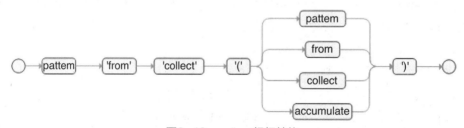

图6-42　collect组织结构

条件元素 collect 从组织结构上看，它需要结合 from 来使用，之前对 from 已进行了说明，相信读者对 from 的遍历功能并不陌生。collect 从字面意思来看是一个收集的功能，也就是说 from 是使用遍历的，而 from collect 是用来汇总的。而且在 collect 后的参数中还可以是匹配、遍历、收集、统计的功能，因此 collect 是一个十分强大的功能。

创建规则文件 collect.drl，目录为 rules/isCollect，其内容为（注意加粗部分）：

```
package rules.isCollect
import com.pojo.Person;
import com.pojo.School;
import java.util.ArrayList;


rule "测试from collect"
    when
        $al : ArrayList() from collect( $p:Person(className=="一班"))
    then
        System.out.println("测试from collect 测试结果为 ArrayList 大小为->"+$al.size());
end


rule "测试from collect pattern"
    when
        $al : ArrayList(size>=3) from collect( $p:Person(className=="一班"))
    then
        System.out.println("测试from collect pattern 测试结果为 ArrayList 大小为->"+$al.size());
end
```

修改 kmodule.xml 配置文件,并添加如下配置:

```
<kbase name="isCollect" packages="rules.isCollect">
    <ksession name="isCollect"/>
</kbase>
```

创建 RuleFromCollect.java 文件,目录为 com.ruleCollect,其内容为:

```
package com.ruleCollect;

import com.pojo.Person;
import com.pojo.School;
import org.junit.Test;
import org.kie.api.KieServices;
import org.kie.api.runtime.KieContainer;
import org.kie.api.runtime.KieSession;

import java.util.ArrayList;
import java.util.List;

public class RuleFromCollect {

    @Test
    public void testFromCollect(){
        KieServices kss = KieServices.Factory.get();
        KieContainer kc = kss.getKieClasspathContainer();
        KieSession ks =kc.newKieSession("isCollect");
```

```
            Person person=new Person();
            person.setName("张三");
            person.setClassName("一班");
            ks.insert(person);
            Person person2=new Person();
            person2.setName("李四");
            person2.setClassName("一班");
            ks.insert(person2);
            Person person3=new Person();
            person3.setName("王五");
            person3.setClassName("二班");
            ks.insert(person3);
            Person person4=new Person();
            person4.setName("赵六");
            person4.setClassName("一班");
            ks.insert(person4);
            int count = ks.fireAllRules();
            System.out.println("总执行了"+count+"条规则");
            ks.dispose();
        }
}
```

执行 testFromCollect() 方法，如图 6-43 所示。

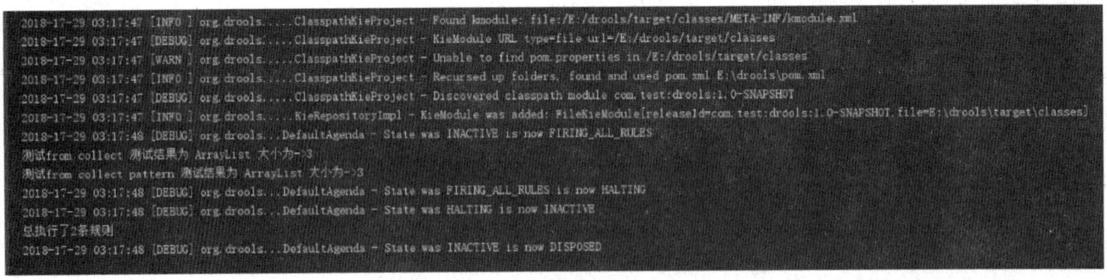

图6-43 collect测试效果

在 collect 参数中使用 from。编辑规则文件 collect.drl 并添加测试 from 的 collect from 规则，其内容为（注意加粗部分）：

```
rule "测试from collect from "
    when
        $s:School()
        $al : ArrayList(size>=3) from collect( $p:Person(className=="一班") from $s.classNameList )
    then
        System.out.println("测试from collect from  测试结果为 ArrayList 大小为->"+$al.size());
end
```

编辑 RuleFrom.java 文件并添加 testFromCollectFrom() 方法，其内容为：

```
@Test
```

```java
public void testFromCollectFrom(){
    KieServices kss = KieServices.Factory.get();
    KieContainer kc = kss.getKieClasspathContainer();
    KieSession ks =kc.newKieSession("isCollect");
    Person person=new Person();
    person.setName("张三");
    person.setClassName("一班");
    Person person2=new Person();
    person2.setName("李四");
    person2.setClassName("一班");
    Person person3=new Person();
    person3.setName("王五");
    person3.setClassName("二班");
    Person person4=new Person();
    person4.setName("赵六");
    person4.setClassName("一班");
    School school=new School();
    List classnameList=new ArrayList();
    classnameList.add(person);
    classnameList.add(person2);
    classnameList.add(person3);
    classnameList.add(person4);
    school.setClassNameList(classnameList);
    ks.insert(school);
    int count = ks.fireAllRules();
    System.out.println("总执行了"+count+"条规则");
    ks.dispose();
}
```

执行 testFromCollectFrom () 方法，如图 6-44 所示。

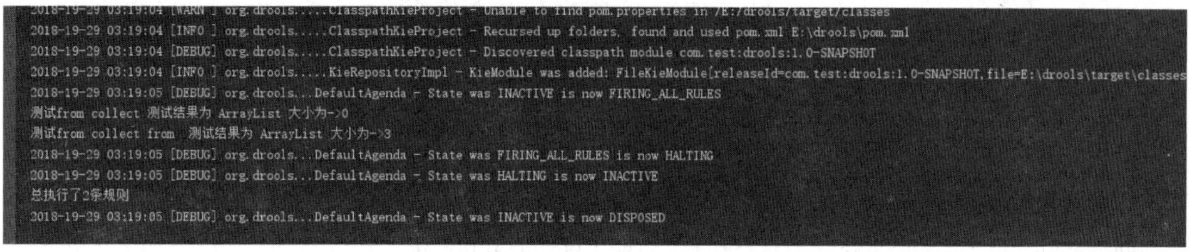

图6-44　collect使用From效果

8. 条件元素 accumulate

accumulate 组织结构如图 6-45 所示。

条件元素 accumulate 是一个更为灵活的 collect，它可以实现 collect 做不了的事。如条件元素 accumulate 的参数可以求一些不同的值，这是 collect 做不到的。

它主要做的事是允许规则迭代整个对象的集合，为每个元素定制执行动作，并在结束时返回一个结果对象，accumulate 不仅支持预定义的累积函数的使用，而且可以使用其他内置函数，重要的是可使用自定义函数进行特殊化操作。常用的 accumulate 函数有求最大值、求最小值和求和等。

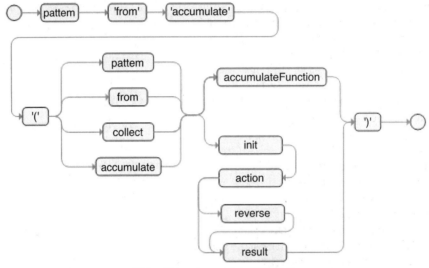

图6-45　accumulate组织结构

创建规则文件 accumulate.drl，目录为 rules/isAccumulate，其内容为：

```
package rules.isAccumulate
import com.pojo.Person;

rule "测试 Accumulate 第一种 取对象中的最大值,和最小值"
    when
            accumulate(Person($age:age),$min:min($age),$max:max($age))
    then
        System.out.println("传入的对象最小值为"+$min+"最大值为"+$max);
end
```

修改 kmodule.xml 配置文件，并添加如下配置：

```
<kbase name="isAccumulate" packages="rules.isAccumulate">
    <ksession name="isAccumulate"/>
</kbase>
```

创建 RuleAccumulate.java 文件，目录为 com.rulesAccumulate，其内容为：

```
package com.rulesAccumulate;

import com.pojo.Person;
import com.pojo.School;
import org.junit.Test;
import org.kie.api.KieServices;
import org.kie.api.runtime.KieContainer;
import org.kie.api.runtime.KieSession;

import java.util.ArrayList;
```

```java
import java.util.List;
public class RuleAccumulate {
    @Test
    public void testAccumulate(){
        KieServices kss = KieServices.Factory.get();
        KieContainer kc = kss.getKieClasspathContainer();
        KieSession ks =kc.newKieSession("isAccumulate");
        Person person=new Person();
        person.setName("张三");
        person.setAge(50);
        ks.insert(person);
        Person person2=new Person();
        person2.setName("李四");
        person2.setAge(20);
        ks.insert(person2);
        Person person3=new Person();
        person3.setName("王五");
        person3.setAge(24);
        ks.insert(person3);
        Person person4=new Person();
        person4.setName("赵六");
        person4.setAge(15);
        ks.insert(person4);
        int count = ks.fireAllRules();
        System.out.println("总执行了"+count+"条规则");
        ks.dispose();
    }
}
```

执行 testAccumulate () 方法，如图 6-46 所示。

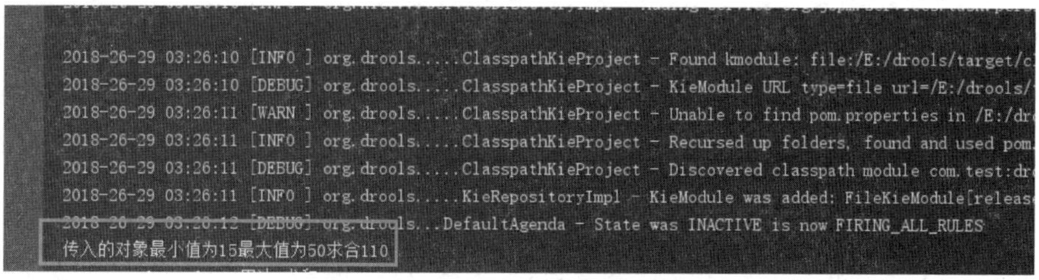

图6-46 accumulate测试效果

Drools 附带有内置的 accumulate 功能，包括 average（平均值）、min（最小值）、max（最大值）、count（统计）、sum（求和）、collectList（返回 List）和 collectSet（返回 HastSet）。

第二种形式的 accumulate 只支持身后兼容，下面是 inline 的语法结构：

```
<result pattern>from accumulate(<source pattern>,init(<init code>),
action(<action code>),
```

```
reverse(<reverse code>),
result(<result expression>))
```

语法分析如下。

<init code>：init 是做初始化用的，简单地说，在 source pattern 遍历完之后，就已经触发，类似 for 的开头。

<action code>：action 会执行所有满足条件的源对象进行操作，类似 for 的方法体。在里面可写 Java 脚本。

<reverse code>：这是一个可选的被选方言的语义代码块，如果存在，将为不再匹配资源模式的每个资源对象的执行。这个代码块的目的是不在 <action code> 块中做任何计算。所以，当一个资源对象被修改或删除时，引擎可能做递减计算，极大地提升了这些操作的性能。

<result expression>：返回值，是根据 action 上面两个遍历出来的结果进行一个返回，这个返回值中也可以进行计算。

<result pattern>：返回值类型，在 <result expression> 返回值的类型中再一次进行匹配，如果匹配不成功则返回 false。

编辑规则文件 accumulate.drl，并添加"测试 accumulatefrom 第二种用法"规则，其内容为（注意加粗部分）：

```
package rules.isAccumulate
import com.pojo.Person;


rule "测试 Accumulate 第一种 取对象中的最大值和最小值"
    when
accumulate(Person($age:age),$min:min($age),$max:max($age),$sum:sum($age))
    then
        System.out.println("传入的对象最小值为"+$min+"最大值为"+$max+"求合"+$sum);
end


rule "测试accumulatefrom第二种用法"
when
$total : Integer() from
        accumulate(Person($value:age),
                init( Integer total =0;),
                action( total += $value; ),
                result( total )
        )
then
System.out.println($total+"accumulate from 用法 求和");
end
```

执行 testAccumulate () 方法，如图 6-47 所示。

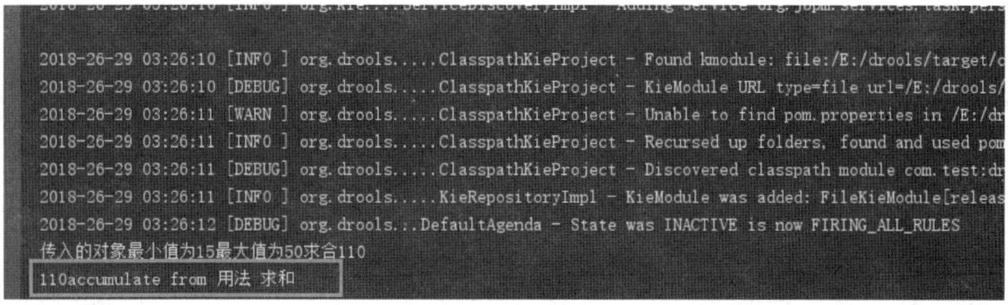
2018-26-29 03:26:10 [INFO] org.drools.....ClasspathKieProject - Found kmodule: file:/E:/drools/target/c
2018-26-29 03:26:10 [DEBUG] org.drools.....ClasspathKieProject - KieModule URL type=file url=/E:/drools/
2018-26-29 03:26:11 [WARN] org.drools.....ClasspathKieProject - Unable to find pom.properties in /E:/dr
2018-26-29 03:26:11 [INFO] org.drools.....ClasspathKieProject - Recursed up folders, found and used pom
2018-26-29 03:26:11 [DEBUG] org.drools.....ClasspathKieProject - Discovered classpath module com.test:dr
2018-26-29 03:26:11 [INFO] org.drools.....KieRepositoryImpl - KieModule was added: FileKieModule[releas
2018-26-29 03:26:12 [DEBUG] org.drools.:.DefaultAgenda - State was INACTIVE is now FIRING_ALL_RULES
传入的对象最小值为15最大值为50求合110
110accumulate from 用法 求和

图6-47 accumulate的第二种用法效果

init 是初始化，action 是遍历并计算，result 是返回结果，这个规则可以返回类型，而且是在 from 前面定义的类型。上述例子中的返回值是 total，它是一个 Integer 包装类，所以在 from 前面只能用 Intger() 来接收，当然要看返回值的具体类型，也可以返回 String。但返回值类型必须有 toString() 才能正常返回。这里需要提醒设计者，设计的方言必须按语法进行规则的编码。

reverse 是可选项，编辑 Person.java 文件添加 “private Double dous;” 并实现 get set 方法。编辑规则文件 accumulate.drl 并添加如下两个规则，其内容为（注意加粗部分）：

```
rule "测试accumulatefrom用法reverse"
dialect "mvel"
when
$total : Double() from
       accumulate(Person($age:age),
                  init(Double totls = 0.0),
                  action(totls+=$age;
System.out.println(totls+">>>>>>>>>");),
reverse(totls-=$age;System.out.println(totls+"<<<<<<<<");),
                  result( totls )
         )
then
        System.out.println($total+"");
end


rule "测试accumulatefrom3用法reverse"
dialect "mvel"
when
$ps:Person(dous>=3)
then
$ps.setDous(1.2);
   update($ps);
   System.out.println($ps.dous);
end
```

编辑 RuleAccumulate.java 并添加 testAccumulate2 方法，其内容为：

```
@Test
public void testAccumulate2(){
```

```
    KieServices kss = KieServices.Factory.get();
    KieContainer kc = kss.getKieClasspathContainer();
    KieSession ks =kc.newKieSession("isAccumulate");
    Person person=new Person();person.setName("张三");person.
setAge(50);person.setDous(5.0);    ks.insert(person);
    Person person2=new Person();person2.setName("李四");person2.
setAge(20);   ks.insert(person2);
    Person person3=new Person();person3.setName("王五");person3.
setAge(25);   ks.insert(person3);
    Person person4=new Person();person4.setName("赵六");person4.
setAge(15);   ks.insert(person4);
    int count = ks.fireAllRules();
    System.out.println("总执行了"+count+"条规则");
    ks.dispose();
}
```

执行 testAccumulate2() 方法，如图 6-48 所示。

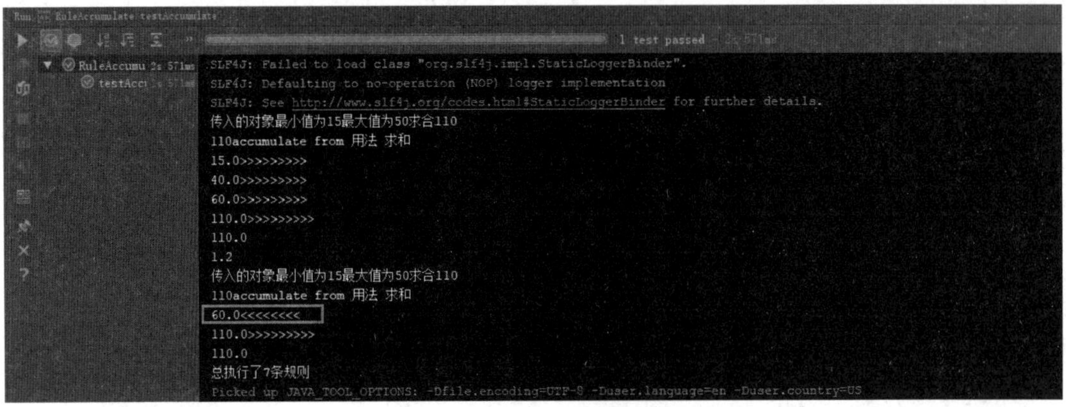

图6-48　reverse的使用效果

分析上述结果，规则首先会执行 from 用法的 reverse 规则，因为有满足的条件，但执行完成后，会再执行 from3 用法的 reverse 规则，这个规则用 update 方法将值改了，所以规则会再次被激活。

针对上述的例子做以下总结。

（1）accumulate 的使用，有一个很重要的函数 action。这个函数提供了匹配源模式的执行动作。

（2）将 action 看成两个状态，当源对象匹配源模式时，定会触发 action, 将触发过 action 的源对象称为有状态的（等同于标记），反之为无状态的。

（3）当传入的源对象在 RHS 中或其他规则的 RHS 中发生改变 (update,insert…) 时，所有满足条件规则的将会再次被激活。

（4）当规则再次被执行且遇到 accumulate 时，则有状态的源对象会先执行 reverse 函数进行"回滚"操作，并将修改后的源对象再次与 accumulate 条件进行比较，若比较为 true，则该对象(被修改过)会再次触发 action 函数；若比较为 false，则不会执行 action 函数。

（5）当规则再次被执行且遇到 accumulate 时，则无状态的源对象会与 accumulate 条件进行比较。若比较为 true，则该对象（被修改过的）会第一次触发 action 函数；若比较为 false，则不会执行 action 函数。

根据上述所有测试的结果得出以下结论。

① accumulate 有 3 种形式，分别是 inline、$min:min(XXX) 和 min()。

② inline 形式虽然官方不推荐，但不代表不能使用。

③ $min:min(XXX) 只是其中的一种写法，如果在函数前面加了 $XXX 变量，则 accumulate 不是使用 from 方式，至于如何取 $XXX 变量的值，与模式方式是一样的。根据上面的介绍，是可以通过在函数的结束符 ";" 后面加条件约束，引用的方法也同模式方式一样。例如，在规则代码 accumulate 的第一种写法中加一个条件，其内容为：

```
rule "测试 Accumulate 第一种 取对象中的最大值和最小值"
    when
accumulate(Person($age:age),$min:min($age),$max:max($age),$sum:sum($age);
$max>=5)
    then
        System.out.println("传入的对象最小值为"+$min+"最大值为"+$max+"求合"+$sum);
end
```

④ min() 写法，前面没有加任何的变量引用说明，当然也不可能在函数的结束符 ";" 后面再加条件约束，并且想要使用函数中返回的结果，在这种没有引用变量的前提下，则必须要引用 from 关键字，但 from 关键字有一个问题，即返回的结果。只有在使用函数时不能有变量引用，才能正常使用 from。

⑤ 使用 accumulate 语法时，如果 source pattern 结束符后面函数的结果有变量，那么不能使用 from，且使用 from 返回结果必须是一个。

上述说的其实都是 accumulate 中自带的一些函数和语法，accumulate 的强大功能远不止如此。

accumulate 函数可以自定义，实现一个新的 accumulate 函数，所以需要创建一个 Java 类，该类实现 AccumulateFunction 接口，在 7.10 版本中与官方说明不一样。规则使用自定义 accumulate 时，需要加关键字，如 import accumulate com.rulesAccumulate.TestAccunmulateneeded factorial。

getResult 在自定义的类中可以写多个，同一个类可以写多个函数，并且可以对原来的方法进行修改。继承了 AccumulateFunction 接口相当于要重写里面的内置函数。

内置的函数（总和、平均值等）是由引擎自动导入的，只有用户自定义定制的累积函数需要显式导入。新建 TestAccunmulateneeded 并实现 AccumulateFunction 目录为 com/rulesAccumulate，其内容为：

```
package com.rulesAccumulate;
```

```java
import org.kie.api.runtime.rule.AccumulateFunction;

import java.io.*;

public class TestAccunmulateneeded implements AccumulateFunction{

    public static class Factorial implements Externalizable {
        public Factorial(){}

        public double total = 1;
        @Override
        public void writeExternal(ObjectOutput out) throws IOException {
            out.writeDouble(total);
        }

        @Override
        public void readExternal(ObjectInput in) throws IOException, ClassNotFoundException {
            total=in.readDouble();
        }
    }

    @Override
    public Class<?> getResultType() {
        System.out.println("getResultType");
        return Number.class;

    }

    @Override
    public Serializable createContext() {
        System.out.println("createContext");
        return new Factorial();
    }

    @Override
    public void init(Serializable serializable) throws Exception {
        Factorial factorial= (Factorial) serializable;
        factorial.total=1;
        System.out.println("init");
    }

    @Override
    public void accumulate(Serializable serializable, Object o) {
```

```
        Factorial factorial= (Factorial) serializable;
        factorial.total *= ((Number)o).doubleValue();
        System.out.println("accumulate");
    }

    @Override
    public void reverse(Serializable serializable, Object o) throws Exception {
        Factorial factorial= (Factorial) serializable;
        factorial.total /= ((Number)o).doubleValue();
        System.out.println("reverse");
    }

    @Override
    public Object getResult(Serializable serializable) throws Exception {
        Factorial factorial= (Factorial) serializable;
        Double d =new Double(((Factorial) serializable).total ==1?1:((Factorial) serializable).total);
        return d;
    }

    @Override
    public boolean supportsReverse() {
        System.out.println("supportsReverse");
        return true;
    }

    @Override
    public void writeExternal(ObjectOutput out) throws IOException {
        System.out.println("writeExternal");
    }

    @Override
    public void readExternal(ObjectInput in) throws IOException, ClassNotFoundException {
        System.out.println("readExternal");
    }
}
```

为了更好地体现自定义函数的使用，在所有的重写方法中都添加了输出，新建 accumulate2.drl 文件目录为 rules/isAccumulate/customize，其内容为：

```
package rules.isAccumulate
import com.pojo.Person;
import accumulate com.rulesAccumulate.TestAccunmulateneeded factorial

rule "自定义函数阶乘"
dialect "mvel"
when
```

```
    accumulate(Person($value:age !=null ),$factorial:factorial($value))
then
System.out.println("自定义函数>>>>>>>>>>>>"+$factorial);
end
```

修改 kmodule.xml 配置文件，并添加如下配置：

```
<kbase name="customize" packages="rules.isAccumulate.customize">
    <ksession name="customize"/>
</kbase>
```

编辑 RuleAccumulate.java 并添加 testAccumulate3 方法，其内容为：

```
@Test
public void testAccumulate3(){
    KieServices kss = KieServices.Factory.get();
    KieContainer kc = kss.getKieClasspathContainer();
    KieSession ks =kc.newKieSession("customize");
    Person person=new Person();person.setName("张三");person.setAge(1);
ks.insert(person);
    Person person2=new Person();person2.setName("李四");person2.setAge(2);   ks.insert(person2);
    Person person3=new Person();person3.setName("王五");person3.setAge(3);   ks.insert(person3);
    Person person4=new Person();person4.setName("赵六");person4.setAge(4);   ks.insert(person4);
    int count = ks.fireAllRules();
    System.out.println("总执行了"+count+"条规则");
    ks.dispose();
}
```

执行 testAccumulate3 () 方法，如图 6-49 所示。

图6-49　自定义accumulate效果

通过上述操作，就可以实现自定义函数来操作特殊业务了。它的用法和 sum、min 内置函数相似，但是自定义的名称如果是 sum、min 等，是不会对原方法进行重写的。

9. 条件元素规则继承

条件元素规则继承与 Java 相似。在学习 Drools 规则引擎的语法过程中,大部分操作都离不开使用 Java 脚本对一类 fact 对象进行操作,因为 Drools 规则引擎是基于 Java 开发的开源技术。编写 Java 代码时,如果发现有重复代码,一般的做法是写一些公共方法或通过继承的方式进行重构。Drools 规则引擎的语法也存在这类条件继承的逻辑关系,其目的是为了减少相同代码的出现,从而方便代码的管理与优化。

下面为一般规则文件的写法:

```
package rules.isExtends
import   com.pojo.Person
import   com.pojo.School

rule "test    No1"
    when
        $p:Person(name=="张小三")
    then
        System.out.println("test    No1");
end

rule "test    No2"
    when
        $p:Person(name=="张小三")
        $s:School(className=="一班")
    then
        System.out.println("test    No2");
end
```

上述代码中,两个规则都有相同的约束条件,如 $p:Person(name=="张小三"),这类写法是没有问题的,但对于维护规则的程序员来说是件比较头疼的事。因此,解决方案可以通过条件继承的方式进行约束。新建 extends.drl 文件,目录为 rules/isExtends,其内容为:

```
package rules.isExtends
import   com.pojo.Person
import   com.pojo.School

rule "test extends No1"
    when
        $p:Person(name=="张小三")
    then
        System.out.println("***********************");
    end

rule "test extends No2" extends "test extends No1"
    when
        $s:School(className=="一班")
    then
```

```
        System.out.println("-----------------------");
end
```

修改 kmodule.xml 配置文件，并添加如下配置：

```xml
<kbase name="isExtends" packages="rules.isExtends">
    <ksession name="isExtends"/>
</kbase>
```

创建 RuleExtends.java 文件，目录为 com.ruleExtends，其内容为：

```java
package com.ruleExtends;

import com.pojo.Person;
import com.pojo.School;
import org.junit.Test;
import org.kie.api.KieServices;
import org.kie.api.runtime.KieContainer;
import org.kie.api.runtime.KieSession;

public class RuleExtends {
    @Test
    public void testExtends() {
        KieServices kss = KieServices.Factory.get();
        KieContainer kc = kss.getKieClasspathContainer();
        KieSession ks =kc.newKieSession("isExtends");
        Person person=new Person();
        person.setName("张小三");
        ks.insert(person);
        School school=new School();
        school.setClassName("一班");
        ks.insert(school);
        int count = ks.fireAllRules();
        System.out.println("总执行了"+count+"条规则");
        ks.dispose();
    }
}
```

执行 testExtends() 方法，结果如图 6-50 所示。

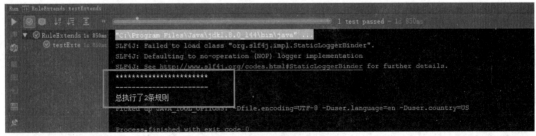

图6-50　条件继承效果

10. 条件元素do对应多then同条件

条件元素 do 对应多 then 同条件，实际项目中用到的地方并不多。规则继承允许避免重写条件设置多个规则，如使用 do 关键词可以有效地避免规则 then 部分出现多次相同的条件。它们基本上是用一个标识符来标记额外的子句，以制订一个规则体现为几个。确定规则应该指向定义的 then 部分。例如，如果想要写出规则继承中看到的两条规则作为一个单一规则，可以新建 do.drl 文件，目录为 rules/isDoThen，其内容为：

```
package rules.isDoThen
import com.pojo.Person

rule "testDoNo1"
    when
        $p:Person(name=="张小三")
        do[then01]
    then
        System.out.println("------------------------");
    then[then01]
        System.out.println("************************");
end
```

修改 kmodule.xml 配置文件，并添加如下配置：

```
<kbase name="isDoThen" packages="rules.isDoThen">
    <ksession name="isDoThen"/>
</kbase>
```

新建 RulesDoThen.java 文件，目录为 com.ruleDoThen，其内容为：

```
package com.ruleDoThen;

import com.pojo.Person;
import org.junit.Test;
import org.kie.api.KieServices;
import org.kie.api.runtime.KieContainer;
import org.kie.api.runtime.KieSession;

public class RulesDoThen{
    @Test
    public void testDoThen () {
        KieServices kss = KieServices.Factory.get();
        KieContainer kc = kss.getKieClasspathContainer();
        KieSession ks = kc.newKieSession("isDoThen");
        Person person=new Person();
        person.setName("张小三");
        person.setAge(35);
        ks.insert(person);
        int count = ks.fireAllRules();
        System.out.println("总执行了" + count + "条规则");
        ks.dispose();
```

```
        }
}
```

执行 testDoThen() 方法,结果如图 6-51 所示。

两个 then 部分的值都被输出了,因为在 LHS 部分中让其先执行 then[then001] 的 RHS 部分。do 在 RHS 部分多了一个带 "name" 的 then,这里就先称之为 then 的名称,然后正常 insert 实体 Person 到 fact 中,调用规则后,当条件满足时,都会被输出,只是先输出了 "*****…"。

使用 do 关键字标记可以去一个具体的后果,即 then。在所有的条件规则都是 true 时,两个 then 都会执行,图 6-51 就是一个很好的证明。如果定义了两个不同的规则和规则继承一样,必须非常小心地使用这个特性。虽然小规模的业务使用 do 关键字可以节省很多重写的工作,但从长远来看,并不是非常合适,不仅无形中提高了规则的可读性,而且结果不可预测。

do 语法使用注意的点很多,do 可以是多个,then 也可以是多个,但要注意的是 do 后面的参数必须是 RHS 部分中所包含的值,否则就会显示异常,如下面的例子。

图6-51　do测试效果

编辑 do.drl 并新添加 testDoNo2 规则,其内容为:

```
rule "testDoNo2"
    when
        $p:Person(name=="张小三")
        do[then01]
        do[then02]
    then
        System.out.println("------------------------");
    then[then01]
        System.out.println("**********************");
    then[then03]
        System.out.println("//////////////////////");
end
```

执行 testDoThen() 方法,结果如图 6-52 所示。

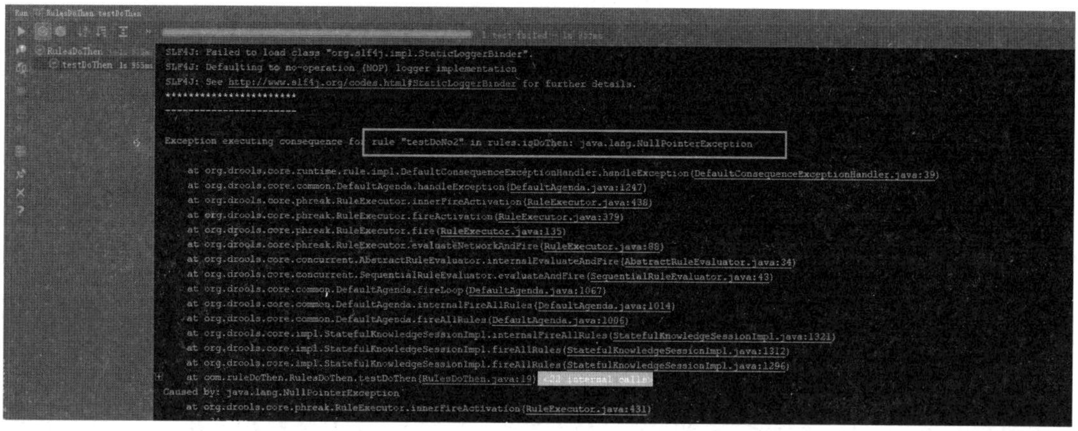

第 6 章 规则中级语法

图6-52 错误的使用do关键字

LHS 部分使用两个 do，且指向同一个 then。编辑 do.drl 并新添加 testDoNo3，其内容为（并注释 testDoNo2 规则）：

```
rule "testDoNo3"
    when
        $p:Person(name=="张小三")
        do[then03]
        do[then03]
    then
        System.out.println("-----------------------");
    then[then01]
        System.out.println("**********************");
    then[then03]
        System.out.println("///////////////////////");
end
```

执行 testDoThen () 方法，结果如图 6-53 所示。

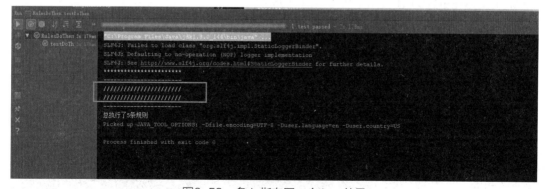

图6-53 多do指向同一个then效果

do 可以指定 then 的一个操作，而没有 then 参数的结果部分是一个默认值。then 部分中的 fact 对象如果发生了变化，可能是灾难性的。

143

下面通过例子来进行说明。编辑 do.drl 并新添加 testDoNo4，其内容为：

```
rule "testDoNo4"
    when
        $p:Person(name=="张小三")
        do[then01]
        do[then03]
    then
        System.out.println("-----------------------");
    then[then01]
        System.out.println("***********************");
        $p.setName("王五");
    then[then03]
        System.out.println("///////////////////////");
        $p.setName("李四");
end
```

编辑 RulesDoThen.java 并添加 testDoThen2() 方法，其内容为：

```
@Test
public void testDoThen2() {
    KieServices kss = KieServices.Factory.get();
    KieContainer kc = kss.getKieClasspathContainer();
    KieSession ks = kc.newKieSession("isDoThen");
    Person person=new Person();
    person.setName("张小三");
    person.setAge(35);
    ks.insert(person);
    int count = ks.fireAllRules();
    System.out.println("总执行了" + count + "条规则");
    System.out.println("输出Person的name"+person.getName());
    ks.dispose();
}
```

执行 testDoThen2() 方法，结果如图 6-54 所示。

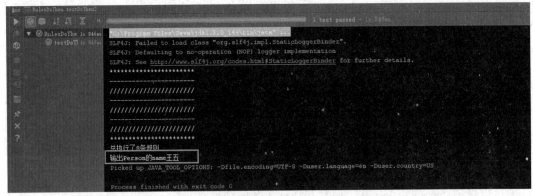

图6-54　then修改后使用do的效果

then 部分都会执行，但 do 的最后也会影响到 then 最终的结果。测试没有使用任何修改代码，得出了这样的结果。那么通过使用 update 的方式进行操作是否会对结果造成影响呢？

编辑 do.drl 并新添加 testDoNo5，其内容为（并注释 testDoNo4 规则）：

```
rule "testDoNo5"
 when
      $p:Person(name=="张小三")
      do[then01]
      do[then03]
   then
      System.out.println("-----------------------");
      $p.setName("赵六");
      update($p);
   then[then01]
      System.out.println("***********************");
      $p.setName("王五");
      update($p);
   then[then03]
      System.out.println("/////////////////////");
      $p.setName("李四");
      update($p);
end
```

执行 testDoThen2 () 方法，结果如图 6-55 所示。

图6-55　then修改后使用do效果（1）

结果很意外，只执行了 then 的结果部分，那么将 do[then01] 与 do[then02] 换一下位置并注释 then 中的 update 函数。再次执行 testDoThen2 () 方法，结果如图 6-56 所示。

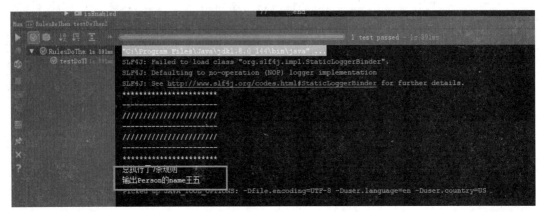

图6-56　then修改后使用do效果（2）

结果只执行了then02，fact被正确修改后，会影响到最终结果。通过上述的论证，得出如下结论。

（1）使用do时，do的参数与then的参数必须有对应关系。

（2）then可以不与do成对出现，如规则testDoNo3。

（3）使用do时，do后的参数在RHS中是必须要存在的，参数是一个指针。

（4）使用修改时，在非update情况下，then部分都会被执行。

（5）使用修改时，在update情况下，then部分是有优先级的，是根据do的顺序控制的，但没有参数的then可能会改变其他的RHS部分。

（6）使用do时，顺序控制是至关重要的。

（7）无参数的then，优先级可能会低于指定的do。

（8）do只能在其他条件元素之后使用。

6.7 规则then

rule是then与end之间的部分，是规则体中的重要部分之一，主要用来处理相关的业务。RHS是真正做事的部分，业务操作时，基本都是通过操作Fact事实对象，并将修改过或已经得到的结果返回Java代码，并进行处理。

RHS部分是规则的结果或行动部分的通用名称，该部分应包含要执行操作的列表。RHS部分如果再次进行条件判断或执行其他语法显然是不明智的，LHS部分已经提供了相关的判断，这样的写法与"硬编码"又有什么区别呢？

一般来说，RHS部分是规则中的最小操作。只有规则体LHS部分全部满足条件时才会执行这部分。规则体的RHS部分也应保持较小，从而提高规则的可读性和可维护性，如果发现RHS部分中需要调用其他语法或添加条件，那么应该尝试将该规则分解成多个规则。RHS部分的主要目的是插入、删除或修改工作内存数据（Fact 事实对象）。下面罗列一下 RHS 部分经常操作的一些命名。

update(Fact 事实对象)：它会告诉引擎一个对象已经改变了（一个绑定到 LHS 部分上的引用，即 $p:Person 中的 $p），修改成功后在工作内存中会发生变化，可能会导致规则再次被激活。在经典的 Hello World 章节中有例子说明。只有真正将工作内存中的值改变时，其他规则体才能正常对变化后的 Fact 事实对象进行判断操作。

insert(new Object())：将一个新的 Fact 事实对象放入工作内存中。它不仅可以操作引用的 Fact 事实对象，还可以操作 declare 声明的 Fact 事实对象。这一点在 declare 声明章节有过说明。

insertLogical(new Object())：与 insert 类似，但是当没有更多的事实来支持当前触发规则的 LHS 部分时，该 Fact 事实对象将被自动删除。使用的并不是很多。

delete(handle)：从工作内存中删除一个 Fact 事实对象，其语法与 update 相似，都是通过引用 LHS 部分上绑定的值。

Update 是修改 Fact 事实对象的一种方式，Drools 还提供了另一种方式，即 modfiy，modfiy 的功能与 update 是一样的，但写法却有很大的不同。

modfiy 的语言扩展为事实更新提供了结构化的方法，它将更新操作与多个设置器调用相结合，以更改对象的字段，其代码结构为：

```
modify ( <fact-expression> ) {
    <expression> [ , <expression> ]*
}
```

括号中的 <fact-expression> 必须产生一个事实对象引用。{} 中的表达式列表应该由给定对象的 setter 调用组成，不需要通常的对象引用来编写，这是由编译器自动添加的，下面通过一个例子来具体说明。

新建规则文件 modify.drl 文件，目录为 rules/isModify，其内容为（注意加粗部分）：

```
package rules.isModify
import com.pojo.Person;

rule "测试modfiy"
no-loop true
salience 20
    when
        $p:Person(name=="张三")
    then
        modify($p){
            setAge(50),
            setName("李四")
        }
        System.out.println("将Person为张三的人修改为"+$p.getName()+"年龄修改为"+$p.getAge());
end
```

```
rule "测试update"
no-loop true
salience 10
    when
        $p:Person(name=="李四")
    then
        $p.setName("王五");
        $p.setAge(60);
        update($p);
        System.out.println("将Person为李四的人修改为"+$p.getName()+"年龄修改为 "+$p.getAge());
end
```

修改 kmodule.xml 配置文件，并添加如下配置：

```
<kbase name="isModify" packages="rules.isModify">
    <ksession name="isModify"/>
</kbase>
```

创建 RulesModify.java 文件，目录为 com.ruleModify，其内容为：

```
package com.ruleModify;

import com.pojo.Person;
import org.junit.Test;
import org.kie.api.KieServices;
import org.kie.api.runtime.KieContainer;
import org.kie.api.runtime.KieSession;

public class RulesModify {
    @Test
    public void testModify(){
        KieServices kss = KieServices.Factory.get();
        KieContainer kc = kss.getKieClasspathContainer();
        KieSession ks = kc.newKieSession("isModify");
        Person person = new Person();
        person.setName("张三");
        person.setAge(30);
        ks.insert(person);
        int count = ks.fireAllRules();
        System.out.println("总执行了" + count + "条规则");
        ks.dispose();
    }
}
```

执行 testModify() 方法，结果如图 6-57 所示。

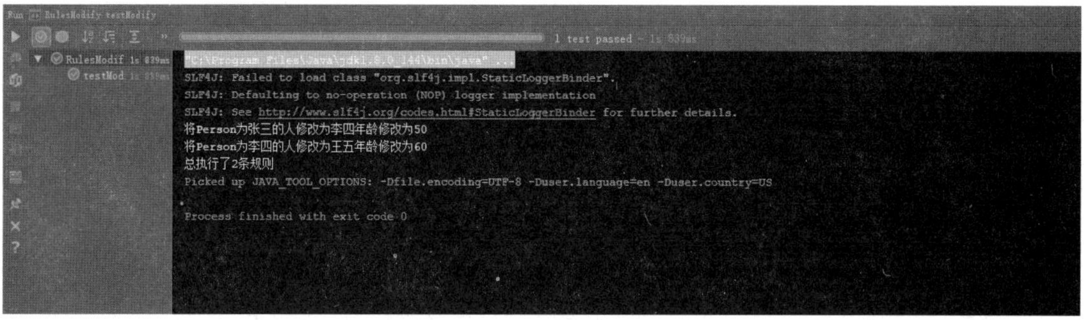

图6-57 modify测试效果

通过上述例子可以看出，modify 与 update 是不一样的，update 必须先对 Fact 事实对象的引用值进行 setter 操作，再执行 update 命名。而 modify 需要在参数中先对 Fact 事实对象进行引用，在块中进行 setter 调用。

Drools 为 RHS 部分提供了几个操作命名，下面分别进行介绍。

drools.halt()：立即终止规则执行。只要有一条规则执行了 drools.halt()，则其他规则将不再进行判断，直接返回结束。这也是常用的一种手段。

新建规则文件 isHalt.drl 文件，目录为 rules/isRHSCommand，其内容为（注意加粗部分）：

```
package rules.isRHSCommand
import com.pojo.Person;

rule "测试drools.halt()1"
no-loop true
salience 20
    when
    then
        System.out.println("输出测试drools.halt()1");
        drools.halt();
end

rule "测试drools.halt()2"
no-loop true
salience 10
    when
    then
        System.out.println("输出测试drools.halt()2");
        drools.halt();
end
```

修改 kmodule.xml 配置文件，并添加如下配置：

```
<kbase name="isRHSCommand" packages="rules.isRHSCommand">
    <ksession name="isRHSCommand"/>
</kbase>
```

创建 RuleRHSCommand.java 文件，目录为 com.ruleRHSCommand，其内容为：

```
package com.ruleRHSCommand;

import org.junit.Test;
import org.kie.api.KieServices;
import org.kie.api.runtime.KieContainer;
import org.kie.api.runtime.KieSession;

public class RuleRHSCommand {

    @Test
    public void testHalt() {
        KieServices kss = KieServices.Factory.get();
        KieContainer kc = kss.getKieClasspathContainer();
        KieSession ks =kc.newKieSession("isRHSCommand");
        int count = ks.fireAllRules();
        System.out.println("总执行了"+count+"条规则");
        ks.dispose();
    }
}
```

执行 testHalt () 方法，结果如图 6-58 所示。

图6-58　drools.halt的测试效果

（1）drools.getWorkingMemory()：返回 WorkingMemory 对象。

（2）drools.setFocus(String s)：将焦点放在指定的议程组上。

（3）drools.getRule().getName()：返回规则的名称。

（4）drools.getTuple()：返回与当前正在执行的规则匹配的数组，并且为 drools.getActivation() 提供相应的激活（这些调用对于日志记录和调试很有用）。

6.8　kmodule配置说明

kmodule.xml 是重要的配置文件，它关系到规则库是否可以成功地创建及 KieSession 的正确使用，将 kmodule.xml 文件放到 src/main/resources/META-INF/ 文件夹下，其基本内容为：

```
<?xml version="1.0" encoding="UTF-8"?>
```

```
<kmodule xmlns="http://www.drools.org/xsd/kmodule">
    <kbase name="rules" packages="rules.rulesHello">
        <ksession name="testhelloworld"/>
    </kbase>
</kmodule>
```

1. KieBase 属性说明

KieBase 属性说明如表 6-1 所示。

表6-1　kBase属性说明

属性名称	是否必填	值	说　明
name	是	string	用于从KieContainer检索该KieBase的名称。既是规则库的名称，又是唯一的强制属性。最好是规范命名，见名知意
packages	否	string	默认情况下，加载资源文件夹中所有Drools相关文件，此属性目的是将当前路径下的规则文件在KieBase中编译。仅限于指定的软件包路径下的所有文件。当然这个值可以是多个，中间用逗号分隔，如果路径是子文件，则必须用小数点进行区别，就像Java包一样
includes	否	string	它的应用与它的意思是一样，既然是在KieBase中出现，那其功能一定是包括其他的KieBase，指一个KieBase被另一个KieBase2的配置中设置了includes='KieBase1'，那KieBase2就有了KieBase1所有资源了，当然这个值是可以多个，中间用逗号分隔开，有点类似继承的关系
default	否	True/false	用来表示当前KieBase是否是此模块的默认值，因此可以从KieContainer中创建，而不会传递任何名称。每个模块中最多只能有一个默认的KieBase，默认是false
equalsBehavior	否	Identity/equality	将新Fact事实对象插入工作内存中时，定义了Drools的操作。使用了身份属性，那么它会创建一个新的FactHandle，除非相同的对象不存在于工作内存中，而只有新插入的对象与已存在的事实对象不相等（根据其等同的方法）时才会相同
eventProcessingMode	否	Cloud/stream	当以云模式编译时，KieBase将事件视为正常事实，而在流模式下可以对其进行时间推理。多在Workbench中使用
declarativeAgenda	否	Disabled/enabled	定义声明议程是否启用

2. KieSession属性说明

KieSession 属性说明如表 6-2 所示。

表6-2 KieSession属性说明

属性名称	是否必填	值	说明
name	是	string	这是KieSession的名称。用于从KieContainer中获取KieSession。这是唯一的强制属性，用于指定操作规则的会话
type	否	stateful/stateless	指当前KieSession的类型是有状态的还是无状态的（默认不指定为有状态的）
default	否	true/false	定义此KieSession是否为该模块的默认值，如果该值为true，则可以从KieContainer中创建，而不会传递任何名称。在每个模块中，每个类型最多可以有一个默认的KieSession，且是有状态的
clockType	否	realtime/seudo	定义事件时间戳是由系统时钟还是由应用程序控制的seudo时钟确定。该时钟对于单元测试时间规则特别有用
beliefSystem	否	simple/tms/defeasible	定义KieSession使用的belief System的类型

第7章
指定规则名调用

规则文件放在同一个目录下,或者将好多的规则体放在同一个规则文件内,执行调用规则代码时,满足条件的规则都会被执行,介绍 Drools 配置文件时,解决这一类问题的手段之一是指定要执行的规则名称。

创建规则文件 ruleName.drl,目录为 rulesTwo/isRuleName,其内容为(注意加粗部分):

```
package rulesTwo.isRuleName
import com.pojo.Person;
import com.pojo.School;

rule "指定规则名一"
    when
    then
        System.out.println("调用规则"+drools.getRule().getName());
end

rule "指定规则名二"
    when
    then
        System.out.println("调用规则"+drools.getRule().getName());
end
```

修改 kmodule.xml 配置文件,并添加如下配置:

```
<kbase name="isRuleName" packages="rulesTwo.isRuleName">
    <ksession name="isRuleName"/>
</kbase>
```

创建 RuleName.java 文件,目录为 comTwo.ruleName,其内容为:

```
package comTwo.ruleName;

import org.junit.Test;
import org.kie.api.KieServices;
import org.kie.api.runtime.KieContainer;
import org.kie.api.runtime.KieSession;

public class RuleName {

    @Test
    public void testRuleName(){
        KieServices kss = KieServices.Factory.get();
        KieContainer kc = kss.getKieClasspathContainer();
        KieSession ks =kc.newKieSession("isRuleName");
        int count = ks.fireAllRules();
        System.out.println("总执行了"+count+"条规则");
        ks.dispose();
    }

}
```

执行 testRuleName() 方法，结果如图 7-1 所示。

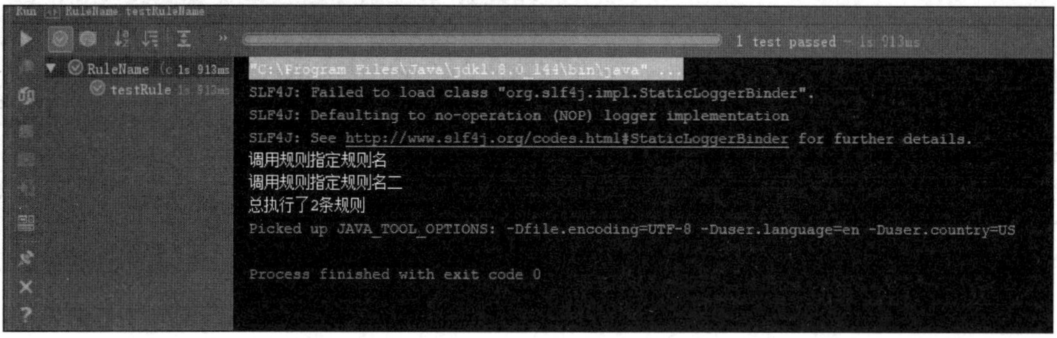

图7-1 未指定规则名称的效果

毫无疑问，输出的是两个规则的规则名，那就可以稍修改一下执行调用规则的方法。编辑 RuleName.java 文件，新建 testRuleName2() 方法，其内容为：

```
@Test
public void testRuleName2(){
    KieServices kss = KieServices.Factory.get();
    KieContainer kc = kss.getKieClasspathContainer();
    KieSession ks =kc.newKieSession("isRuleName");
    int count = ks.fireAllRules(new RuleNameEqualsAgendaFilter("指定规则名"));
    System.out.println("总执行了"+count+"条规则");
    ks.dispose();
}
```

执行 testRuleName2() 方法，结果如图 7-2 所示。

图7-2 指定规则名称的效果

结果只输出了代码中指定的规则名称，解释效果之前先来看 fireAllRules 方法的参数 new RuleNameEqualsAgendaFilter(" 指定规则名 ")，即规则名称等于 AgendaFilter。进入 fireAllRules 中查看一下它的结构：StatefulRuleSession 是一个接口，其中包括 fireAllRules 与 fireUntilHalt 两个接口，其结构如图 7-3 所示。

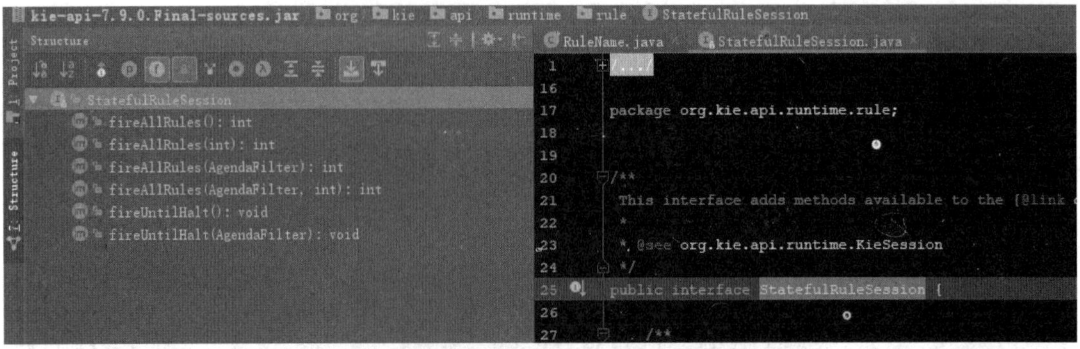

图7-3 StatefulRuleSession结构

与fireAllRules相关的方法主要是用来操作调用规则的，从图7-3中可以看出，它有很多参数返回值都是Int，只成功执行了几条规则。直接找到刚调用的代码接口，即"int fireAllRules(AgendaFilter agendaFilter);"中的AgendaFilter，发现也是一个接口。下面看一下它的实现类，如图7-4所示。

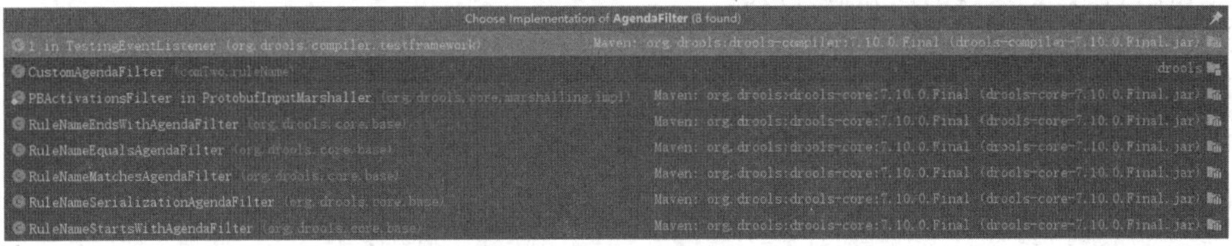

图7-4 AgendaFilter结构

通过结构图，找到了上述例子中的使用类RuleNameEqualsAgendaFilter，如图7-5所示。

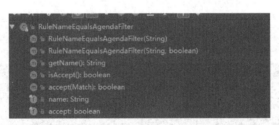

图7-5 RuleNameEqualsAgendaFilter结构

RuleNameEqualsAgendaFilter类有两个构成方法，其功能是精确匹配指定的执行规则名。默认可以是执行规则，也可以是精通匹配指定不激活的规则名。

创建规则文件ruleName2.drl，目录为rulesTwo/isRuleName，其内容为（注意加粗部分）：

```
package rulesTwo.isRuleName
import com.pojo.Person;
import com.pojo.School;

rule "指定规则名三"
```

```
        when
        then
            System.out.println("调用规则"+drools.getRule().getName());
end
rule "指定规则名四"
        when
        then
            System.out.println("调用规则"+drools.getRule().getName());
end
```

当规则文件被加载到规则库时，同逻辑路径下的规则名唯一，这里新建了两个规则，即指定规则名三和指定规则名四。执行 testRuleName2() 方法，结果与图 7-2 中是一样的。

图 7-4 中 AgendaFilter 的实现类不止一个，下面分别对它们进行说明。其中，"RuleNameEndsWithAgendaFilter"是根据指定的规则名称后缀过滤。

编辑 RuleName.java 文件，新建 testRuleName3() 方法，其内容为：

```
@Test
public void testRuleName3(){
    KieServices kss = KieServices.Factory.get();
    KieContainer kc = kss.getKieClasspathContainer();
    KieSession ks =kc.newKieSession("isRuleName");
    int count = ks.fireAllRules(new RuleNameEndsWithAgendaFilter("三"));
    System.out.println("总执行了"+count+"条规则");
    ks.dispose();
}
```

执行 testRuleName3() 方法，结果如图 7-6 所示。

图7-6　RuleNameEndsWithAgendaFilter测试效果

"RuleNameStartsWithAgendaFilter"是根据指定的规则名称前缀过滤。与 RuleNameEndsWithAgendaFilter 例子相反，编辑 RuleName.java 文件，新建 testRuleName4() 方法，其内容为：

```
@Test
public void testRuleName4(){
    KieServices kss = KieServices.Factory.get();
    KieContainer kc = kss.getKieClasspathContainer();
    KieSession ks =kc.newKieSession("isRuleName");
```

```
    int count = ks.fireAllRules(new RuleNameStartsWithAgendaFilter("指定"));
    System.out.println("总执行了"+count+"条规则");
    ks.dispose();
}
```

执行 testRuleName4() 方法，结果如图 7-7 所示。

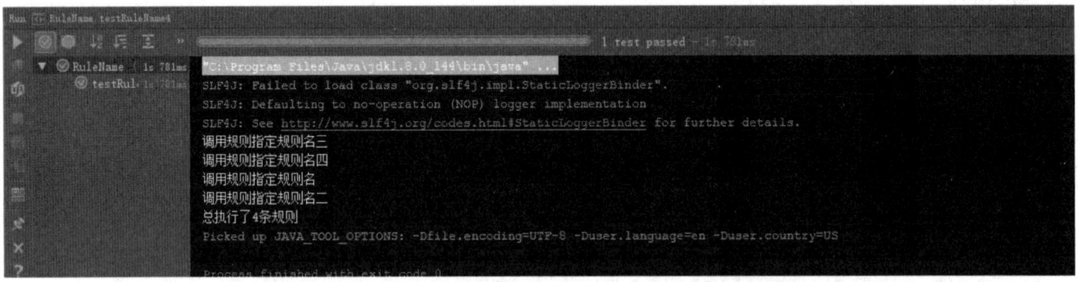

图7-7 RuleNameStartsWithAgendaFilter测试效果

"RuleNameMatchesAgendaFilter" 是根据指定的规则名称正则匹配过滤。创建规则文件 ruleName3.drl，目录为 rulesTwo/isRuleName，其内容为：

```
package rulesTwo.isRuleName
import com.pojo.Person;
import com.pojo.School;


rule "abc"
    when
    then
        System.out.println("调用规则"+drools.getRule().getName());
end

rule "123"
    when
    then
        System.out.println("调用规则"+drools.getRule().getName());
end
```

编辑 RuleName.java 文件，新建 testRuleName5() 方法，其内容为：

```
@Test
public void testRuleName5(){
    KieServices kss = KieServices.Factory.get();
    KieContainer kc = kss.getKieClasspathContainer();
    KieSession ks =kc.newKieSession("isRuleName");
    int count = ks.fireAllRules(new RuleNameMatchesAgendaFilter("\\w{0,10}"));
    System.out.println("总执行了"+count+"条规则");
    ks.dispose();
}
```

执行 testRuleName5() 方法，结果如图 7-8 所示。

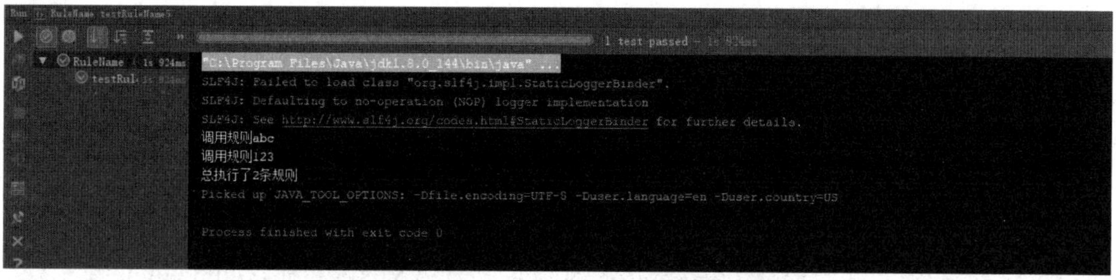

图7-8　RuleNameMatchesAgendaFilter测试效果

使用正规匹配过滤时要指定正则匹配的长度，相当于指定规则名的长度。前面讲述了 Drools 自带的匹配过滤规则名，下面介绍自定义的方式。AgendaFilter 是一个接口，需要写一个类并实现 AgendaFilter 接口。创建 CustomAgendaFilter.java 文件，目录为 comTwo/ruleName，其内容为：

```java
package comTwo.ruleName;

import org.kie.api.runtime.rule.AgendaFilter;
import org.kie.api.runtime.rule.Match;

import java.util.Set;

public class CustomAgendaFilter  implements AgendaFilter {
    private final Set<String> ruleNamesThatAreAllowedToFire;//传入的rule name
    public CustomAgendaFilter(Set<String> ruleNamesThatAreAllowedToFire) {
        this.ruleNamesThatAreAllowedToFire = ruleNamesThatAreAllowedToFire;
    }
    @Override
    public boolean accept(Match match) {
        return ruleNamesThatAreAllowedToFire.contains(match.getRule().getName());
    }
}
```

编辑 RuleName.java 文件，新建 testRuleName6() 方法，其内容为：

```java
@Test
public void testRuleName6() {
    KieServices kss = KieServices.Factory.get();
    KieContainer kc = kss.getKieClasspathContainer();
    KieSession ks = kc.newKieSession("isRuleName");
    Set set=new HashSet();
    set.add("指定规则名");
    set.add("指定规则名四");
    set.add("123");
    int count = ks.fireAllRules(new CustomAgendaFilter(set));
```

```
System.out.println("总执行了" + count + "条规则");
ks.dispose();
}
```

执行 testRuleName6() 方法，结果如图 7-9 所示。

图7-9　CustomAgendaFilter测试效果

通过自定义指定激活规则也是很有趣的事情，这样可为开发扩展提供良好的定义环境。

对指定规则名调用方法做以下总结。

（1）"int fireAllRules()"执行所有满足条件的规则。

（2）"int fireAllRules(int max)"执行规则的最大数量，简单地说，即执行多少条规则。

（3）"int fireAllRules(AgendaFilter agendaFilter)" 指定规则名的方式。

（4）"int fireAllRules(AgendaFilter agendaFilter, int max);"指定规则名，并设置执行规则的最大数量，在精确匹配中是不适用的。

第8章
Spring整合Drools

8.1 Spring+Drools简单配置

整合 Spring 一直都是 Java 程序员必须掌握的，Drools 与 Spring 整合也是学习 Drools 必须要掌握的，整合过程共分为 Spring+Drools 和 Spring+Drools+web 的方式两个部分，这两个项目均为 Maven 管理。

创建一个 Maven 项目，并引用相关联的 jar 包。配置 pom.xml 文本，其内容为：

```xml
<?xml version="1.0" encoding="UTF-8"?>
<project xmlns="http://maven.apache.org/POM/4.0.0"
         xmlns:xsi="http://www.w3.org/2001/XMLSchema-instance"
         xsi:schemaLocation="http://maven.apache.org/POM/4.0.0 http://maven.apache.org/xsd/maven-4.0.0.xsd">
    <modelVersion>4.0.0</modelVersion>

    <groupId>com.droolsSpring</groupId>
    <artifactId>droolsSpring</artifactId>
    <version>1.0-SNAPSHOT</version>

    <properties>
        <drools.version>7.10.0.Final</drools.version>
        <log4j2.version>2.5</log4j2.version>
        <spring.version>4.2.6.RELEASE</spring.version>
    </properties>
    <dependencies>
        <dependency>
            <groupId>org.drools</groupId>
            <artifactId>drools-compiler</artifactId>
            <version>${drools.version}</version>
        </dependency>
        <dependency>
            <groupId>org.kie</groupId>
            <artifactId>kie-spring</artifactId>
            <version>${drools.version}</version>
        </dependency>
        <dependency>
            <groupId>junit</groupId>
            <artifactId>junit</artifactId>
            <version>4.11</version>
        </dependency>
        <dependency>
            <groupId>org.springframework</groupId>
            <artifactId>spring-core</artifactId>
            <version>${spring.version}</version>
        </dependency>
        <dependency>
            <groupId>org.springframework</groupId>
            <artifactId>spring-beans</artifactId>
```

```xml
            <version>${spring.version}</version>
        </dependency>
        <dependency>
            <groupId>org.springframework</groupId>
            <artifactId>spring-context</artifactId>
            <version>${spring.version}</version>
        </dependency>
        <dependency>
            <groupId>org.springframework</groupId>
            <artifactId>spring-test</artifactId>
            <version>${spring.version}</version>
        </dependency>
        <dependency>
            <groupId>org.springframework</groupId>
            <artifactId>spring-context-support</artifactId>
            <version>${spring.version}</version>
        </dependency>
        <dependency>
            <groupId>org.springframework</groupId>
            <artifactId>spring-tx</artifactId>
            <version>${spring.version}</version>
        </dependency>
        <dependency>
            <groupId>commons-logging</groupId>
            <artifactId>commons-logging</artifactId>
            <version>1.1.1</version>
        </dependency>
    </dependencies>
</project>
```

创建规则文件 ruleName3.drl，目录为 rules/isString，其内容为：

```
package rules.isString

rule "测试Spring+Drools规则"
    when
    then
        System.out.println("调用规则"+drools.getRule().getName());
end
```

创建 Spring.xml 文件，目录为 resources，其内容为：

```xml
<?xml version="1.0" encoding="UTF-8"?>
<beans xmlns="http://www.springframework.org/schema/beans"
       xmlns:xsi="http://www.w3.org/2001/XMLSchema-instance"
       xmlns:kie="http://drools.org/schema/kie-spring"
       xsi:schemaLocation="
       http://drools.org/schema/kie-spring
       http://drools.org/schema/kie-spring.xsd
       http://www.springframework.org/schema/beans
       http://www.springframework.org/schema/beans/spring-beans.xsd">
```

```xml
<!-- 与kmodule.xml的配置是相似的 -->
<kie:kmodule id="kmodule">
    <kie:kbase name="kbase" packages="rules.isString">
        <kie:ksession name="ksession"  />
    </kie:kbase>
</kie:kmodule>
<bean id="kiePostProcessor"
class="org.kie.spring.annotations.KModuleAnnotationPostProcessor"/>
</beans>
```

创建 TestSpring.java 文件，目录为 com/ruleString，其内容为：

```java
package com.ruleString;

import org.junit.Test;
import org.junit.runner.RunWith;
import org.kie.api.cdi.KSession;
import org.kie.api.runtime.KieSession;
import org.springframework.test.context.ContextConfiguration;
import org.springframework.test.context.junit4.SpringJUnit4ClassRunner;

@RunWith(SpringJUnit4ClassRunner.class)
@ContextConfiguration({"classpath:Spring.xml"})
public class TestSpring {
    @KSession("ksession")//注：这里的值与配置文件中的值是一样的
    KieSession ksession;

    @Test
    public void runRules() {
        int count = ksession.fireAllRules();
        System.out.println("总执行了" + count + "条规则");
        ksession.dispose();
    }
}
```

创建完成后，执行 runRules() 方法，如图 8-1 所示。

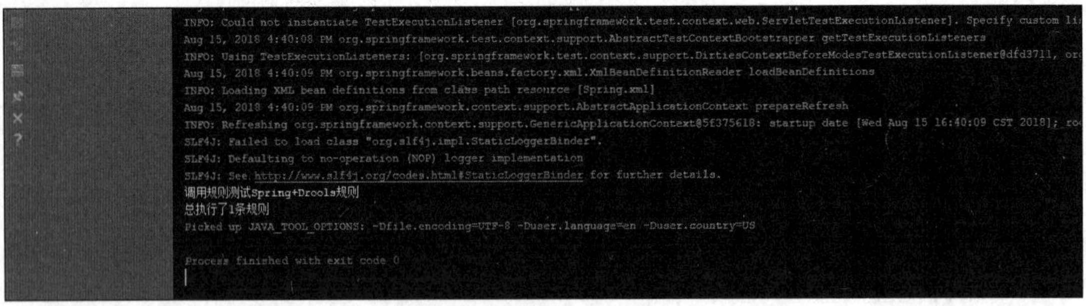

图8-1 整合Spring效果

分析 Spring.xml 与 TestSpring.java 两个文件。Spring.xml 与之前所讲到的 kmodule.xml 配置文件很相似，相比 kmodule.xml 在标签上多了 "kie:" 的前缀，这也是整合 Spring 必须用到的标签。

kiePostProcessor bean 中的 class 是指调用规则通过注释的方式，而 TestSpring.java 文件中的调用规则更是简单，规则的调用离不开 KieSession 会话，而注解 @KieSession 正是根据在 Spring.xml 配置文件中的 kiePostProcessor bean 而来的。其中，@KieSession 的参数正是 Spring.xml 中配置 KieSession 的 name 值。既可以通过 KieSession 注解，又可以通过 KieBase 进行注入。编辑 TestSpring.java 文件，其内容为：

```java
package com.ruleString;

import org.junit.Test;
import org.junit.runner.RunWith;
import org.kie.api.KieBase;
import org.kie.api.cdi.KBase;
import org.kie.api.cdi.KSession;
import org.kie.api.runtime.KieSession;
import org.springframework.test.context.ContextConfiguration;
import org.springframework.test.context.junit4.SpringJUnit4ClassRunner;


@RunWith(SpringJUnit4ClassRunner.class)
@ContextConfiguration({"classpath:Spring.xml"})
public class TestSpring {
    @KSession("ksession")//注：这里的值与配置文件中的值是一样的
    KieSession ksession;

    @KBase("kbase")//注：这里的值与配置文件中的值是一样的
    KieBase kBase;

    @Test
    public void runRules() {
        int count = ksession.fireAllRules();
        System.out.println("总执行了" + count + "条规则");
        ksession.dispose();
    }

    @Test
    public void runRulesKbase() {
        KieSession kieSession = kBase.newKieSession();
        int count = kieSession.fireAllRules();
        System.out.println("总执行了" + count + "条规则");
        ksession.dispose();
    }
}
```

执行 runRulesKieBase () 方法，结果与图 8-1 是一样的。但这里并没有指定 session 的名称，只是默认值为 KieSession。

官方提供了 Drools+Spring 配置文件使用的两种方式，如图 8-2 所示。

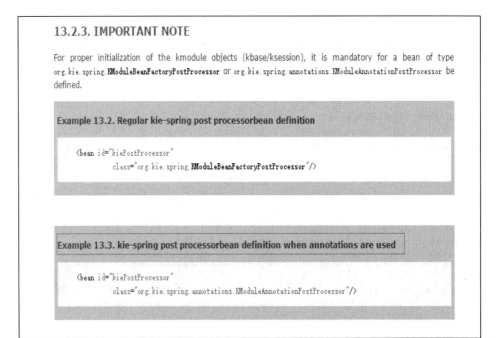

图8-2 官方配置文件说明

通过 Maven 整合 Drools+Spring 时要注意：用 @Test 运行时，有时 Spring 是找不到规则文件的，就是规则文件中的规则体恒为 true，但一直报执行 0 条规则。解决方案有以下两种。

方法一：可以找到项目 target/test-classes 下的规则文件，将其全部都放到里面。

方法二：在 pom.xml 配置文件中既可以写下面这段代码，又可以解决 Spring 找不到规则文件的问题，其内容为：

```
<build>
    <testResources>
        <testResource>
            <directory>
                ${project.basedir}/src/main/resources
            </directory>
        </testResource>
    </testResources>
</build>
```

还有一些很有趣的问题，Drools 的版本是 6.4，在配置 Spring 时会遇到 <kie:XXX> 标签找不到的情况，这时可以将 http://drools.org/schema/kie-spring.xsd 替换为 https://raw.github.com/droolsjbpm/droolsjbpm-integration/master/kie-spring/src/main/resources/org/kie/spring/kie-spring-6.0.0.xsd。如果还有问题，就要检查网络是否可以访问到这两个地址，或者 pom.xml 是否正确。

8.2　Drools整合Spring+Web

整合 Spring+Web 也是 Java 程序员必须掌握的,上节已将 Drools+Spring 整合成功,本节介绍如何在 Web 项目中使用 Drools。编辑 pom.xml,添加 Web 的引用包,其内容为:

```xml
<dependency>
    <groupId>org.springframework</groupId>
    <artifactId>spring-web</artifactId>
    <version>${spring.version}</version>
</dependency>
<dependency>
    <groupId>org.springframework</groupId>
    <artifactId>spring-webmvc</artifactId>
    <version>${spring.version}</version>
</dependency>
```

新建 web.xml,目录为 webapp/WEB-INF,其内容为:

```xml
<?xml version="1.0" encoding="UTF-8"?>
<web-app xmlns:xsi="http://www.w3.org/2001/XMLSchema-instance"
        xmlns="http://java.sun.com/xml/ns/javaee"
        xsi:schemaLocation="http://java.sun.com/xml/ns/javaee http://java.sun.com/xml/ns/javaee/web-app_3_0.xsd" version="3.0"
metadata-complete="true">
  <display-name>Archetype Created Web Application</display-name>

    <context-param>
        <param-name>contextConfigLocation</param-name>
        <param-value>classpath*:applicationContext.xml</param-value>
    </context-param>

    <!-- Spring MVC servlet -->
    <servlet>
        <servlet-name>SpringMVC</servlet-name>
<servlet-class>org.springframework.web.servlet.DispatcherServlet</servlet-class>
        <init-param>
            <param-name>contextConfigLocation</param-name>
            <param-value>classpath:applicationContext.xml</param-value>
        </init-param>
        <load-on-startup>1</load-on-startup>
        <async-supported>true</async-supported>
    </servlet>
    <servlet-mapping>
        <servlet-name>SpringMVC</servlet-name>
        <!-- 此处可以配置成*.do,对应struts的后缀习惯 -->
        <url-pattern>/</url-pattern>
    </servlet-mapping>
    <welcome-file-list>
```

```xml
        <welcome-file>/index.jsp</welcome-file>
    </welcome-file-list>
</web-app>
```

新建 indea.jsp，目录为 webapp/WEB-IN，目的是测试 tomcat 是否启动成功，其内容为：

```jsp
<%@ page language="java" contentType="text/html; charset=UTF-8"
pageEncoding="UTF-8"%>
<!DOCTYPE html>
<html>
<head>
    <meta http-equiv="Content-Type" content="text/html; charset=UTF-8">
    <title>aaaaaaaaaaaaaaaaaaa</title>
</head>
<body>
aaaaaaaaaaaaaaaaaaaaaa
</body>
</html>
```

新建 showUser.jsp 文件，目录为 webapp/WEB-INF/view，其内容为：

```jsp
<%@ page language="java" import="java.util.*" pageEncoding="utf-8" %>
<!DOCTYPE HTML PUBLIC "-//W3C//DTD HTML 4.01 Transitional//EN">
<html>
<head>
    <title>测试</title>
</head>
<body>
KSession方式——${ruleKSession} <br>
KBase方式——${ruleKBase}<br>
Autowired方式——${ruleAutowired} <br>
</body>
</html>
```

新建 SpringMVC 配置文件 applicationContext.xml，目录为 resources，其内容为：

```xml
<?xml version="1.0" encoding="UTF-8"?>
<beans xmlns="http://www.springframework.org/schema/beans"
       xmlns:xsi="http://www.w3.org/2001/XMLSchema-instance"
       xmlns:context="http://www.springframework.org/schema/context"
       xmlns:mvc="http://www.springframework.org/schema/mvc"
       xsi:schemaLocation="http://www.springframework.org/schema/beans
   http://www.springframework.org/schema/beans/spring-beans-3.1.xsd
   http://www.springframework.org/schema/context
   http://www.springframework.org/schema/context/spring-context-3.1.xsd
http://www.springframework.org/schema/mvc http://www.springframework.org/schema/mvc/spring-mvc.xsd">

    <!-- 开启注解-->
    <mvc:annotation-driven />
    <context:component-scan base-package="com.web.controller"/>
    <bean id="ruleService" class="com.web.service.RuleService"/>
```

```xml
<!-- 配置视图解析器
作用:在controller中指定页面路径的时候就不用写页面的完整路径名称了,可以直接写页面去掉扩展名的名称
-->
<bean class="org.springframework.web.servlet.view.InternalResourceViewResolver">
    <!-- 真正的页面路径 =  前缀 + 去掉后缀名的页面名称 + 后缀 -->
    <!-- 前缀 -->
    <property name="prefix" value="/WEB-INF/view/"></property>
    <!-- 后缀 -->
    <property name="suffix" value=".jsp"></property>
</bean>
<import resource="Spring.xml"/>
</beans>
```

新建 pakcage com.Web 包，Web 包下新建两个子包，分别是 controller 和 service。新建 RuleName Controller.java 文件，目录为 com.web.controller，其内容为：

```java
package com.web.controller;


import com.web.service.RuleService;
import org.kie.api.cdi.KSession;
import org.kie.api.runtime.KieSession;
import org.springframework.beans.factory.annotation.Autowired;
import org.springframework.stereotype.Controller;
import org.springframework.ui.Model;
import org.springframework.web.bind.annotation.RequestMapping;

@Controller
@RequestMapping(value = "/rule")
public class RuleNameController {

    @Autowired
    private RuleService ruleService;

    @KSession("ksession")//注：这里的值与配置文件中的值是一样的
    KieSession ksession;

    @RequestMapping(value = "/name")
    public String name(Model model) {
        String ruleKSession = ruleService.ruleKSession();
        String ruleKBase = ruleService.ruleKBase();
        String ruleAutowired = ruleService.ruleAutowired();
        model.addAttribute("ruleKSession", ruleKSession);
        model.addAttribute("ruleKBase", ruleKBase);
        model.addAttribute("ruleAutowired", ruleAutowired);
        return "showUser";
    }
```

```java
    @RequestMapping(value = "/name2")
    public String name2(Model model) {
        model.addAttribute("ruleKSession", "在Controller层使用 @KSession
执行规则"+ksession.fireAllRules());
        return "showUser";
    }
}
```

新建 RuleService.java 文件，目录为 com/web/service，其内容为：

```java
package com.web.service;

import org.kie.api.KieBase;
import org.kie.api.cdi.KBase;
import org.kie.api.cdi.KSession;
import org.kie.api.runtime.KieSession;
import org.springframework.beans.factory.annotation.Autowired;
import org.springframework.stereotype.Service;

@Service("ruleService")
public class RuleService{
    @KSession("ksession")//注：这里的值与配置文件中的值是一样的
    KieSession ksession;

    @KBase("kbase")//注：这里的值与配置文件中的值是一样的
    KieBase kBase;

    @Autowired
    private KieBase KieBase;

    public String ruleKSession() {
        int count = ksession.fireAllRules();
        return "共执行了"+count+"规则";
    }

    public String ruleKBase() {
        KieSession kieSession = kBase.newKieSession();
        int count = kieSession.fireAllRules();
        return "共执行了"+count+"规则";
    }

    public String ruleAutowired() {
        KieSession kieSession = KieBase.newKieSession();
        int count = kieSession.fireAllRules();
        return "共执行了"+count+"规则";
    }
}
```

新项目添加到 tomcat 中，直接启动项目。通过控制台可以看出，tomcat 启动后初始化了规则，浏览器中输出 http://localhost:8080/rule/name，就会出现如图 8-3 所示的结果。

第 8 章 Spring 整合 Drools

图 8-3　tomcat 启动效果

这些调用都是通过 service 层进行操作的，研究 Drools 6.4 版本时，如果直接在 controller 层使用 Drools 相关的注解时是会报空指针的，但在 Drools 7.10 中是没有这个问题的。在浏览器输出 http://localhost:8080/rule/name2，就会出现如图 8-4 所示的结果。

图 8-4　controller 使用 Drools 效果

整合 Spring 或 Spring+Web 时会遇到一些很奇怪的问题。本例中所使用的工具是 IntelliJ IDEA，用 eclipse 做 Web 项目整合时可能会出现找不到文件路径的问题，其解决方案如下。

为了使项目默认部署到 tomcat 安装目录下的 webapps 中，选择 show view → servers 选项，找到需要修改的 tomcat 选项，并在其上右击。

① 停止 eclipse 内的 tomcat 服务器 (stop)。

② 删除该容器中部署的项目 (add and remove)。

③ 清除该容器相关数据 (clean)。

④ 打开 tomcat 的修改界面 (open)。

⑤ 找到 servers location，选择第二个 (User tomcat Installation)。

⑥ 修改 deploy path 为 webapps。

⑦ 保存关闭。

用 eclipse 会出现这样的问题，用 myeclipse 同样会有这样的问题，如图 8-5 所示。解决方案如下。

① clean 下，把 target、bin 目录删除。

② 找到 Context 并删除 "appBase="webapps"" 即可。

③ 暴力点删除 Context。

目的就是让项目只做一次初始化操作。

与 tomcat 进行整合时，规则很容易发生错误，规则在 junit 上测试没有问题，放在 tomcat 中就会出现问题，规则没有问题，而是编码出了问题，解决方案如下。

```
Duplicate method defaultConsequence(KnowledgeHelper) in type Rule_rule11576546135
The type Rule_rule11576546135 is already defined]]
        at org.springframework.beans.factory.support.AbstractAutowireCapableBeanFactory.initializeBean(AbstractAut
        at org.springframework.beans.factory.support.AbstractAutowireCapableBeanFactory.doCreateBean(AbstractAuto
        at org.springframework.beans.factory.support.AbstractAutowireCapableBeanFactory.createBean(AbstractAutowi
        at org.springframework.beans.factory.support.AbstractBeanFactory$1.getObject(AbstractBeanFactory.java:306
        at org.springframework.beans.factory.support.DefaultSingletonBeanRegistry.getSingleton(DefaultSingletonBea
        at org.springframework.beans.factory.support.AbstractBeanFactory.doGetBean(AbstractBeanFactory.java:302
        at org.springframework.beans.factory.support.AbstractBeanFactory.getBean(AbstractBeanFactory.java:197)
        at org.springframework.beans.factory.support.BeanDefinitionValueResolver.resolveReference(BeanDefinitionVa
        ... 39 more
Caused by: java.lang.RuntimeException: Error while creat
ing KieBase[Message [id=1, level=ERROR, path=person.drl, line=3, column=0
    text=Duplicate rule name: rule1], Message [id=2, level=ERROR, path=person.drl, line=3, column=0
    text=Rule Compilation error Duplicate method defaultConsequence(KnowledgeHelper) in type Rule_rule11576546135
Duplicate method defaultConsequence(KnowledgeHelper) in type Rule_rule11576546135
The type Rule_rule11576546135 is already defined]]
        at org.drools.compiler.kie.builder.impl.KieContainerImpl.getKieBase(KieContainerImpl.java:450)
        at org.kie.spring.KieObjectsResolver.resolveKBase(KieObjectsResolver.java:33)
        at org.kie.spring.factorybeans.KBaseFactoryBean.afterPropertiesSet(KBaseFactoryBean.java:169)
        at org.springframework.beans.factory.support.AbstractAutowireCapableBeanFactory.invokeInitMethods(Abstract
        at org.springframework.beans.factory.support.AbstractAutowireCapableBeanFactory.initializeBean(AbstractAut
        ... 46 more
```

图8-5　Myeclipse错误信息

Drools 的 when 中有中文时无法匹配到规则的问题：Drools 规则的 when 有中文，当用 tomcat 启动项目时，要配置编码，配置两处即可，一是 tomcat 的编码，这时 IDEA 控制台会出现乱码，二是解决控制台的乱码问题，打开 IDEA 的安装目录找到 idea.exe.vmoptions 文件 (64 位操作系统找 idea64.exe.vmoptions)，打开此文件加入 "-Dfile.encodeing=UTF-8" 重启 IDEA 即可，如图 8-6~ 图 8-8 所示。

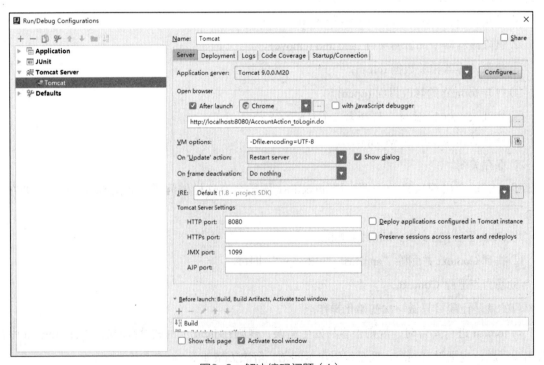

图8-6　解决编码问题（1）

第 8 章 Spring 整合 Drools

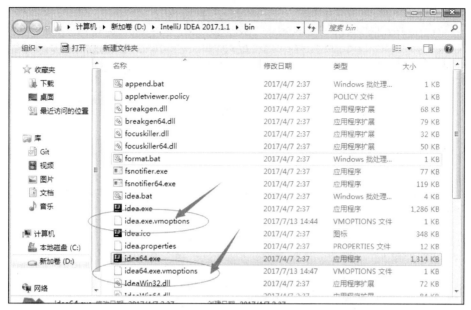

图8-7 解决编码问题（2）

图8-8 解决编码问题（3）

8.3 Drools整合Spring Boot

8.3.1 Spring Boot测试整合官方例子

（1）创建一个 Maven 项目。

下面是项目所应用的 pom.xml，其内容为：

```
<?xml version="1.0" encoding="UTF-8"?>
```

```xml
<project xmlns="http://maven.apache.org/POM/4.0.0"
         xmlns:xsi="http://www.w3.org/2001/XMLSchema-instance"
         xsi:schemaLocation="http://maven.apache.org/POM/4.0.0 http://maven.apache.org/xsd/maven-4.0.0.xsd">
    <modelVersion>4.0.0</modelVersion>

    <groupId>com.springBoot.test</groupId>
    <artifactId>droolssb</artifactId>
    <version>1.0-SNAPSHOT</version>

    <properties>
        <project.build.sourceEncoding>UTF-8</project.build.sourceEncoding>
        <!-- drools 规则引擎 版本 -->
        <drools.version>7.10.0.Final</drools.version>
    </properties>

    <!-- Maven POM文件继承 spring-boot-starter-parent -->
    <parent>
        <groupId>org.springframework.boot</groupId>
        <artifactId>spring-boot-starter-parent</artifactId>
        <version>1.5.7.RELEASE</version>
    </parent>
    <!-- 依赖项定义 -->
    <dependencies>
        <!-- start drools -->
        <dependency>
            <groupId>org.drools</groupId>
            <artifactId>drools-core</artifactId>
            <version>${drools.version}</version>
        </dependency>
        <dependency>
            <groupId>org.drools</groupId>
            <artifactId>drools-compiler</artifactId>
            <version>${drools.version}</version>
        </dependency>
        <dependency>
            <groupId>org.kie</groupId>
            <artifactId>kie-spring</artifactId>
            <version>${drools.version}</version>
        </dependency>
        <dependency>
            <groupId>org.kie</groupId>
            <artifactId>kie-internal</artifactId>
            <version>${drools.version}</version>
        </dependency>
        <dependency>
            <groupId>org.drools</groupId>
            <artifactId>drools-templates</artifactId>
```

```xml
            <version>${drools.version}</version>
        </dependency>
        <!-- end drools -->
        <!-- TEST begin -->
        <dependency>
            <groupId>junit</groupId>
            <artifactId>junit</artifactId>
            <version>4.11</version>
        </dependency>
        <!-- TEST end -->
        <!-- Spring boot start -->
        <dependency>
            <groupId>org.springframework.boot</groupId>
            <artifactId>spring-boot-starter-web</artifactId>
            <exclusions>
                <!-- 想要配置log4j2,就要先去除logging包 -->
                <exclusion>
                    <groupId>org.springframework.boot</groupId>
                    <artifactId>spring-boot-starter-logging</artifactId>
                </exclusion>
            </exclusions>
        </dependency>

        <dependency>
            <groupId>org.springframework.boot</groupId>
            <artifactId>spring-boot-starter-test</artifactId>
            <scope>test</scope>
        </dependency>

        <dependency>
            <groupId>org.springframework.boot</groupId>
            <artifactId>spring-boot-starter-data-redis</artifactId>
        </dependency>

        <dependency>
            <groupId>org.springframework.boot</groupId>
            <artifactId>spring-boot-starter-log4j2</artifactId>
        </dependency>
    </dependencies>
    <build>
        <plugins>
            <plugin>
                <groupId>org.springframework.boot</groupId>
                <artifactId>spring-boot-maven-plugin</artifactId>
                <configuration>
                    <fork>true</fork>
                </configuration>
            </plugin>
```

```
        </plugins>
    </build>
</project>
```

（2）分别创建各项包并创建相关类。

创建 TestController 控制类，其内容为：

```
package com.controller;

import com.service.TestService;
import org.springframework.beans.factory.annotation.Autowired;
import org.springframework.stereotype.Controller;
import org.springframework.web.bind.annotation.RequestMapping;
import org.springframework.web.bind.annotation.ResponseBody;

@RequestMapping("/test")
@Controller
public class TestController {
    @Autowired
    private TestService testService;

    @ResponseBody
    @RequestMapping("/test001")
    public void test() {
        testService.testService01();
    }
}
```

创建 Person 实例类，其内容为：

```
package com.model.work;

public class Person {
    private String name;

    public String getName() {
        return name;
    }

    public void setName(String name) {
        this.name = name;
    }
}
```

创建 TestService 接口类，其内容为：

```
package com.service;

import com.model.work.Person;
import org.kie.api.runtime.KieSession;
import org.springframework.stereotype.Service;
```

```java
import javax.annotation.Resource;

@Service
public class TestService {
    @Resource
    private KieSession kieSession;
    public void  testService01(){
        Person p=new Person();
        p.setName("张三");
        kieSession.insert(p);
        int ruleFiredCount = kieSession.fireAllRules();
    }
}
```

创建 DroolsConfiguration 工具类，其内容为：

```java
package com.util;

import java.io.IOException;

import org.kie.api.KieBase;
import org.kie.api.KieServices;
import org.kie.api.builder.KieBuilder;
import org.kie.api.builder.KieFileSystem;
import org.kie.api.builder.KieModule;
import org.kie.api.builder.KieRepository;
import org.kie.api.builder.ReleaseId;
import org.kie.api.runtime.KieContainer;
import org.kie.api.runtime.KieSession;
import org.kie.internal.io.ResourceFactory;
import org.kie.spring.KModuleBeanFactoryPostProcessor;
import org.springframework.boot.autoconfigure.condition.ConditionalOnMissingBean;
import org.springframework.context.annotation.Bean;
import org.springframework.context.annotation.Configuration;
import org.springframework.core.io.Resource;
import org.springframework.core.io.support.PathMatchingResourcePatternResolver;
import org.springframework.core.io.support.ResourcePatternResolver;

@Configuration
public class DroolsConfiguration {

    private static final String RULES_PATH = "rules/";

    @Bean
    @ConditionalOnMissingBean(KieFileSystem.class)
    public KieFileSystem kieFileSystem() throws IOException {
        KieFileSystem kieFileSystem = getKieServices().
```

```java
newKieFileSystem();
        for (Resource file : getRuleFiles()) {
    kieFileSystem.write(ResourceFactory.newClassPathResource(RULES_PATH + file.getFilename(), "UTF-8"));
        }
        return kieFileSystem;
    }

    private Resource[] getRuleFiles() throws IOException {
        ResourcePatternResolver resourcePatternResolver = new PathMatchingResourcePatternResolver();
        return resourcePatternResolver.getResources("classpath*:" + RULES_PATH + "**/*.*");
    }

    @Bean
    @ConditionalOnMissingBean(KieContainer.class)
    public KieContainer kieContainer() throws IOException {
        final KieRepository kieRepository = getKieServices().getRepository();
        kieRepository.addKieModule(new KieModule() {
            public ReleaseId getReleaseId() {
                return kieRepository.getDefaultReleaseId();
            }
        });
        KieBuilder kieBuilder = getKieServices().newKieBuilder(kieFileSystem());
        kieBuilder.buildAll();
        return getKieServices().newKieContainer(kieRepository.getDefaultReleaseId());
    }

    private KieServices getKieServices() {
        return KieServices.Factory.get();
    }

    @Bean
    @ConditionalOnMissingBean(KieBase.class)
    public KieBase KieBase() throws IOException {
        return kieContainer().getKieBase();
    }

    @Bean
    @ConditionalOnMissingBean(KieSession.class)
    public KieSession kieSession() throws IOException {
        return kieContainer().newKieSession();
    }

    @Bean
    @ConditionalOnMissingBean(KModuleBeanFactoryPostProcessor.class)
```

```
        public KModuleBeanFactoryPostProcessor kiePostProcessor() {
            return new KModuleBeanFactoryPostProcessor();
        }
}
```

创建 Application 启动类,其内容为:

```
package com;

import org.slf4j.Logger;
import org.slf4j.LoggerFactory;
import org.springframework.boot.SpringApplication;
import org.springframework.boot.autoconfigure.SpringBootApplication;

@SpringBootApplication
public class Application{
    protected static Logger logger=LoggerFactory.getLogger(Application.class);
        public static void main(String[] args) {
            SpringApplication.run(Application.class, args);
            logger.info("SpringBoot Start Success");
        }
}
```

(3)在 src/main/resources 资源文件创建目录 rules。

创建规则文件 myenum.drl,其内容为:

```
package rules

rule "规则测试SpringBoot原始整合方式"
    when
    then
        System.out.println("规则测试SpringBoot原始整合方式");
end
```

创建 myenum2.drl,其内容为:

```
package rules

import com.model.work.Person;

rule "规则测试insertPerson值"
    when
        $p:Person(name=="张三")
    then
        System.out.println("规则测试insertPerson值");
end
```

创建 application.properties 文件,其内容为:

```
server.port=8080
```

创建完成所有项目后进行编译，启动 Application 类在浏览器中访问地址 http://localhost:8080/test/test001，如果控制台输出了如下内容，那么证明请求是正确的。

规则测试SpringBoot原始整合方式
规则测试insertPerson值

8.3.2 Spring Boot测试整合

（1）创建一个 maven 项目，并引用相关 jar 包。

```
<properties>
    <project.build.sourceEncoding>UTF-8</project.build.sourceEncoding>
    <!-- drools 规则引擎 版本 -->
    <drools.version>7.10.0.Final</drools.version>
    <log4j2.version>2.5</log4j2.version>
</properties>
<!-- Maven POM文件继承 spring-boot-starter-parent -->
<parent>
    <groupId>org.springframework.boot</groupId>
    <artifactId>spring-boot-starter-parent</artifactId>
    <version>1.5.7.RELEASE</version>
</parent>
<!-- 依赖项定义 -->
<dependencies>
    <!-- start drools -->
    <dependency>
        <groupId>org.drools</groupId>
        <artifactId>drools-core</artifactId>
        <version>${drools.version}</version>
    </dependency>
    <dependency>
        <groupId>org.drools</groupId>
        <artifactId>drools-compiler</artifactId>
        <version>${drools.version}</version>
    </dependency>
    <dependency>
        <groupId>org.kie</groupId>
        <artifactId>kie-spring</artifactId>
        <version>${drools.version}</version>
    </dependency>
    <dependency>
        <groupId>org.kie</groupId>
        <artifactId>kie-internal</artifactId>
        <version>${drools.version}</version>
    </dependency>
    <dependency>
```

```xml
        <groupId>org.drools</groupId>
        <artifactId>drools-templates</artifactId>
        <version>${drools.version}</version>
    </dependency>
    <!-- end drools -->
    <!-- Spring boot start -->
    <dependency>
        <groupId>org.springframework.boot</groupId>
        <artifactId>spring-boot-starter-web</artifactId>
    </dependency>
    <dependency>
        <groupId>org.springframework.boot</groupId>
        <artifactId>spring-boot-starter-test</artifactId>
        <scope>test</scope>
    </dependency>
    <dependency>
        <groupId>org.springframework.boot</groupId>
        <artifactId>spring-boot-starter-data-redis</artifactId>
    </dependency>
    <!-- mysql连接 -->
    <dependency>
        <groupId>mysql</groupId>
        <artifactId>mysql-connector-java</artifactId>
        <version>5.1.30</version>
    </dependency>
    <!-- mybatis -->
    <dependency>
        <groupId>org.mybatis.spring.boot</groupId>
        <artifactId>mybatis-spring-boot-starter</artifactId>
        <version>1.1.1</version>
    </dependency>
    <dependency>
        <groupId>com.alibaba</groupId>
        <artifactId>fastjson</artifactId>
        <version>RELEASE</version>
    </dependency>
    <!--Gson-->
    <dependency>
        <groupId>com.google.code.gson</groupId>
        <artifactId>gson</artifactId>
    </dependency>
</dependencies>
<build>
    <plugins>
        <plugin>
            <groupId>org.springframework.boot</groupId>
            <artifactId>spring-boot-maven-plugin</artifactId>
            <configuration>
                <fork>true</fork>
```

```
            </configuration>
        </plugin>
    </plugins>
</build>
```

在上面的引用中，为什么会有 Redis 和 Mybatis 的引用，在这个例子中要对实际的应用场景做一个 demo 的说明。

（2）创建 Mysql 数据库，如图 8-9 所示。

图8-9　创建数据库（1）

（3）创建表，并添加数据，如图 8-10 和图 8-11 所示。

图8-10　创建数据库（2）

图8-11　添加数据

（4）创建相关包及类 PersonDao.java 和 mapper 配置文件，其内容为：

```
package com.dao;
```

```
import java.util.List;

public interface PersonDao {
    List  listAll();
}
```
```xml
<?xml version="1.0" encoding="UTF-8" ?>
<!DOCTYPE mapper PUBLIC "-//mybatis.org//DTD Mapper 3.0//EN"
"http://mybatis.org/dtd/mybatis-3-mapper.dtd" >
<mapper namespace="com.dao.PersonDao">
    <sql id="Base_Column_List">
        rule_name
    </sql>

    <select id="listAll" resultType="java.lang.String">
        select
            *
        from person
    </select>
</mapper>
```

创建 Spring Boot 配置文件，其内容为：

```
server.port=8888
#Mysql
spring.jpa.show-sql=true
spring.datasource.url=jdbc:mysql://locahost:3306/drools
spring.datasource.username=root
spring.datasource.password=Best_2017
spring.datasource.driver-class-name=com.mysql.jdbc.Driver
spring.datasource.max-active=10
spring.datasource.max-idle=5
spring.datasource.min-idle=0
```

创建日志配置文件，其内容为：

```
#定义LOG输出级别
log4j.rootLogger=DEBUG,Console,File
#定义日志输出目的地为控制台
log4j.appender.Console=org.apache.log4j.ConsoleAppender
log4j.appender.Console.Target=System.out
#可以灵活地指定日志输出格式，下面一行是指定具体的格式
log4j.appender.Console.layout = org.apache.log4j.PatternLayout
#log4j.appender.Console.layout.ConversionPattern=[%c] - %m%n
log4j.appender.Console.layout.ConversionPattern=%d %p [%c] - %m%n
#文件大小到达指定尺寸时产生一个新的文件
log4j.appender.File = org.apache.log4j.RollingFileAppender
#指定输出目录
log4j.appender.File.File = logs/promote.log
#定义文件最大大小
log4j.appender.File.MaxFileSize = 10MB
# 输出所有日志，如果换成DEBUG表示输出DEBUG以上级别日志
```

```
log4j.appender.File.Threshold = ALL
log4j.appender.File.layout = org.apache.log4j.PatternLayout
log4j.appender.File.layout.ConversionPattern =[%p] [%d{yyyy-MM-dd
HH\:mm\:ss}][%c]%m%n
###显示mybatis的SQL语句部分,类似于hibernate在控制台打印sql语句部分
log4j.logger.java.sql.ResultSet=INFO
log4j.logger.org.apache=INFO
log4j.logger.java.sql.Connection=DEBUG
log4j.logger.java.sql.Statement=DEBUG
log4j.logger.java.sql.PreparedStatement=DEBUG
```

创建 POJO, 其内容为:

```
package com.model;

public class Person {
    String name;
    String age;
    public String getName() {
        return name;
    }
    public void setName(String name) {
        this.name = name;
    }
    public String getAge() {
        return age;
    }
    public void setAge(String age) {
        this.age = age;
    }
}
```

创建 Service, 其内容为:

```
package com.service;

import com.dao.PersonDao;
import com.model.Person;
import org.springframework.stereotype.Service;

import javax.annotation.Resource;
import java.util.List;

@Service
public class PersonService {

    @Resource
    private PersonDao personDao;
    public List<Person> listPerson(){
        List<Person> list=personDao.listAll();
        return list;
```

        }
}
```

创建 controller，其内容为：

```java
package com.controller;

import com.model.Person;
import com.service.PersonService;
import org.springframework.stereotype.Controller;
import org.springframework.web.bind.annotation.RequestBody;
import org.springframework.web.bind.annotation.RequestMapping;
import org.springframework.web.bind.annotation.RequestMethod;
import org.springframework.web.bind.annotation.ResponseBody;

import javax.annotation.Resource;
import javax.servlet.http.HttpServletRequest;
import java.util.HashMap;
import java.util.List;
import java.util.Map;

@Controller
@RequestMapping("/person")
public class PersonController {
    @Resource
    private PersonService promoteService;
    @RequestMapping(value = "/list", method = RequestMethod.POST, produces = "application/json")
    @ResponseBody
    public Map<String, Object> toshopping(HttpServletRequest request, @RequestBody String requestBody) {
        Map<String, Object> map = new HashMap<String, Object>();
        List<Person> personList= promoteService.listPerson();
        map.put("personList", personList);
        return map;
    }
}
```

（5）创建 Spring Boot 启动类，其内容为：

```java
package com;

import org.apache.ibatis.session.SqlSessionFactory;
import org.apache.tomcat.jdbc.pool.DataSource;
import org.mybatis.spring.SqlSessionFactoryBean;
import org.mybatis.spring.annotation.MapperScan;
import org.slf4j.Logger;
import org.slf4j.LoggerFactory;
import org.springframework.boot.SpringApplication;
import org.springframework.boot.autoconfigure.SpringBootApplication;
import org.springframework.boot.context.properties.ConfigurationProperties;
```
```

```java
import org.springframework.context.annotation.Bean;
import org.springframework.core.io.support.PathMatchingResourcePatternR
esolver;
import org.springframework.jdbc.datasource.DataSourceTransactionManager;
import org.springframework.transaction.PlatformTransactionManager;


@SpringBootApplication
@MapperScan("com.dao")
public class Application {

    protected static Logger
logger=LoggerFactory.getLogger(Application.class);

    @Bean
    @ConfigurationProperties(prefix="spring.datasource")
    public DataSource dataSource() {
       return new DataSource();
    }
    @Bean
    public SqlSessionFactory sqlSessionFactoryBean() throws Exception {

        SqlSessionFactoryBean sqlSessionFactoryBean = new
SqlSessionFactoryBean();
        sqlSessionFactoryBean.setDataSource(dataSource());

        PathMatchingResourcePatternResolver resolver = new
PathMatchingResourcePatternResolver();

sqlSessionFactoryBean.setMapperLocations(resolver.getResources("classpa
th*:mapper/*.xml"));
        return sqlSessionFactoryBean.getObject();
    }
    @Bean
    public PlatformTransactionManager transactionManager() {
       return new DataSourceTransactionManager(dataSource());
    }
    public static void main(String[] args) {
         SpringApplication.run(Application.class, args);
         logger.info("SpringBoot Start Success");
    }
}
```

运行启动类，查看控制台日志输出，如图 8-12 所示。

图8-12　Spring Boot启动

通过 postman 进行访问，如图 8-13 所示。

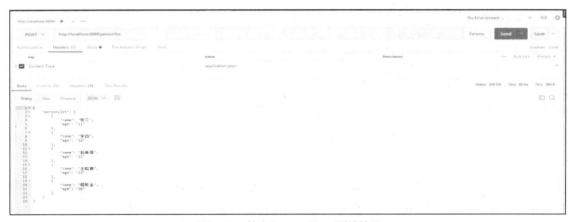

图8-13　整合Spring Boot测试结果

配置规则工具类，其内容为：

```
package com.rule;

import org.kie.api.KieBase;
import org.kie.api.io.ResourceType;
import org.kie.internal.utils.KieHelper;

public class NewKieBase {
    //将业务规则写到规则库中
    public static KieBase ruleKieBase(String rule) {//rule值就是动态传入的规则内容
        KieHelper helper = new KieHelper();
        KieBase KieBase = null;
        try {
            helper.addContent(rule,ResourceType.DRL);
//为了省事，直接将rule写成activityRule()
            KieBase = helper.build();
        } catch (Exception e) {
```

```
            e.printStackTrace();
        }
        return KieBase;
    }
}
```

编写规则接口方法，其内容为：

```
package com.rule;

public class RuleExecute {
    public static String activityRule() {
        StringBuffer ruleDrl = new StringBuffer();
        ruleDrl.append("package rules \n ");
        ruleDrl.append("import   com.model.Person; \n");
        ruleDrl.append(" rule     \'person_1\' \n");
        ruleDrl.append(" no-loop true \n");
        ruleDrl.append(" salience  10 \n");
        ruleDrl.append(" when \n");
        ruleDrl.append("   $p:Person(name==\'张三\',age==\'22\' )\n ");
        ruleDrl.append(" then \n");
        ruleDrl.append("   modify($p){ setName(\'张小三\' )} \n");
        ruleDrl.append("end \n");
        return ruleDrl.toString();
    }
}
```

Drools 实现类，主要用到创建 KieBase 的方式。创建 KieBase，而并非直接创建 Session，不管是有状态的还是无状态的，都是为了更好地进行创建和完成编辑规则，其内容为：

```
package com.service;

import org.kie.api.KieBase;
import org.springframework.stereotype.Service;

import static com.rule.NewKieBase.ruleKieBase;

@Service
public class DroolsService {
    /**
     * 创建KieSession
     * @return
     */
    public KieBase newKieBase() {
        KieBase KieBase = ruleKieBase();
        return KieBase;
    }
}
```

重新编辑 PersonService，其内容为：

```java
package com.service;

import com.dao.PersonDao;
import com.model.Person;
import org.kie.api.KieBase;
import org.kie.api.runtime.KieSession;
import org.springframework.stereotype.Service;

import javax.annotation.Resource;
import java.util.List;

@Service
public class PersonService {
    @Resource
    private PersonDao personDao;
    @Resource
    private DroolsService droolsService;

    public List<Person> listPerson(){
        KieBase KieBase=droolsService.newKieBase();
        List<Person> list=personDao.listAll();
        for(Person person:list){
            KieSession kieSession= KieBase.newKieSession();
            kieSession.insert(person);
            int i=kieSession.fireAllRules();
            kieSession.dispose();
        }
        return list;
    }
}
```

重新启动服务，再次通过 postman 调用服务，结果如图 8-14 所示。

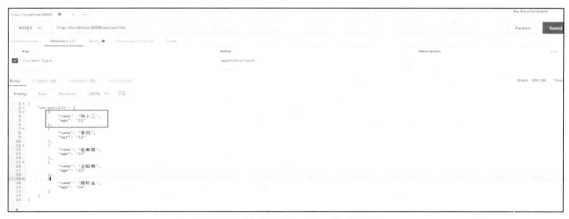

图8-14　整合Spring Boot修改效果

通过图 8-14 可以看出，之前的张三已经改为张小三，证明成功利用规则进行了修改、业务判断。

上面只是测试的例子,与 Spring Boot 整合的方式很多,但多数是接口提供应用,如果使用有状态的 Session 要注意迭代的问题。对于规则库来说,规则构建一次就好,是可以通过一些方式进行封装保存的。

8.3.3 Spring Boot实战整合

Spring Boot 作为现在比较流行的框架,它与 Drools 的整合使用也是必须掌握的,前文讲了与 Spring 整合,与其说是整合还不如说只是将这两个不相关的工具放到一个项目中,Spring Boot 只是处理本职工作,而 Drools 同样也是如此。接下来介绍 Spring Boot 中的 Drools 是如何使用的,本例中使用的是 Spring Boot 2.0 版本。为了更容易理解 Drools 在项目中的应用,下面通过一个实战的例子进行讲述。

项目背景: 针对电商平台优惠券的动态添加。

项目环境: Windows 10、JDK 8、Spring Boot 2.0.1、Drools 7.10.0、模板引擎、MySQL 5、MyBatis。

项目开发工具: IntelliJ IDEA 2017.2。

创建一个 Drools Boot 的项目,并添加如下文件。配置 pom.xml 文本,其内容为:

```xml
<?xml version="1.0" encoding="UTF-8"?>
<project xmlns="http://maven.apache.org/POM/4.0.0"
         xmlns:xsi="http://www.w3.org/2001/XMLSchema-instance"
         xsi:schemaLocation="http://maven.apache.org/POM/4.0.0 http://maven.apache.org/xsd/maven-4.0.0.xsd">
    <modelVersion>4.0.0</modelVersion>

    <groupId>com.droolsBoot</groupId>
    <artifactId>droolsBoot</artifactId>
    <version>1.0-SNAPSHOT</version>

    <properties>
        <project.build.sourceEncoding>UTF-8</project.build.sourceEncoding>
        <project.reporting.outputEncoding>UTF-8</project.reporting.outputEncoding>
        <java.version>1.8</java.version>
        <drools.version>7.10.0.Final</drools.version>
    </properties>

    <!-- Maven POM文件继承 spring-boot-starter-parent -->
    <parent>
        <groupId>org.springframework.boot</groupId>
        <artifactId>spring-boot-starter-parent</artifactId>
        <version>2.0.1.RELEASE</version>
    </parent>
```

```xml
<dependencies>
    <dependency>
        <groupId>org.drools</groupId>
        <artifactId>drools-compiler</artifactId>
        <version>${drools.version}</version>
    </dependency>
    <dependency>
        <groupId>junit</groupId>
        <artifactId>junit</artifactId>
        <version>4.11</version>
    </dependency>

    <!-- Spring boot start -->
    <dependency>
        <groupId>org.springframework.boot</groupId>
        <artifactId>spring-boot-starter-web</artifactId>
        <exclusions>
            <exclusion>
                <groupId>org.springframework.boot</groupId>
                <artifactId>spring-boot-starter-logging</artifactId>
            </exclusion>
        </exclusions>
    </dependency>

    <dependency>
        <groupId>org.springframework.boot</groupId>
        <artifactId>spring-boot-starter-log4j2</artifactId>
    </dependency>

    <dependency>
        <groupId>org.springframework.boot</groupId>
        <artifactId>spring-boot-starter-thymeleaf</artifactId>
    </dependency>

    <!-- mysql连接 -->
    <dependency>
        <groupId>mysql</groupId>
        <artifactId>mysql-connector-java</artifactId>
        <version>5.1.30</version>
    </dependency>
    <!-- mybatis -->
    <dependency>
        <groupId>org.mybatis.spring.boot</groupId>
        <artifactId>mybatis-spring-boot-starter</artifactId>
        <version>1.3.0</version>
    </dependency>
    <dependency>
        <groupId>com.alibaba</groupId>
```

```xml
            <artifactId>fastjson</artifactId>
            <version>RELEASE</version>
        </dependency>
        <dependency>
            <groupId>org.antlr</groupId>
            <artifactId>ST4</artifactId>
            <version>4.0.8</version>
        </dependency>
    </dependencies>
    <build>
        <plugins>
            <plugin>
                <groupId>org.springframework.boot</groupId>
                <artifactId>spring-boot-maven-plugin</artifactId>
                <configuration>
                    <fork>true</fork>
                </configuration>
            </plugin>
        </plugins>
    </build>
</project>
```

创建application.properties文件，目录为resources，其内容为：

```
#激活哪一个环境的配置文件
spring.profiles.active=dev
```

创建application-dev.properties文件，目录为resources，其内容为：

```
server.port=9999
#Mysql
spring.jpa.show-sql=true
spring.datasource.url=jdbc:mysql://localhost:3306/drools
spring.datasource.username=root
spring.datasource.password=root
spring.datasource.driver-class-name=com.mysql.jdbc.Driver
#mybatis配置
#Mapper.xml所在的位置
mybatis.mapper-locations=classpath*:mapper/*Mapper.xml
#entity扫描的包名
mybatis.type-aliases-package=com.droolsBoot.model
spring.thymeleaf.prefix=classpath:/templates/
```

创建两个静态页面index.html，目录为resources/templates，其内容为：

```html
<html lang="en">
<head>
    <title>Getting Started: Serving Web Content</title>
    <meta http-equiv="Content-Type" content="text/html; charset=UTF-8" />
</head>
<body>
```

```html
<form action="/promotion/ediePromote" method="POST" enctype="multipart/
form-data">
    优惠券金额：<input type="text" name="money"/> <br>
    优惠券名称：<input type="text" name="rulename"/> <br>
    <input type="submit" />
</form>
</body>
</html>
```

创建 shopping.html，目录为 resources/templates，其内容为：

```html
<html lang="en">
<head>
    <title>Getting Started: Serving Web Content</title>
    <meta http-equiv="Content-Type" content="text/html; charset=UTF-8"/>
</head>
<body>
<form action="/promotion/toShopping" method="POST"
enctype="multipart/form-data">
    购物车总金额：<input type="text" name="money"/>
    <input type="submit"/>
</form>
</body>
</html>
```

创建 promote_rule 表，导出的 sql 文件，其内容为：

```
/*
Navicat MySQL Data Transfer

Source Server         : drools
Source Server Version : 50129
Source Host           : localhost:3306
Source Database       : drools

Target Server Type    : MYSQL
Target Server Version : 50129
File Encoding         : 65001

Date: 2018-05-18 17:20:13
*/

SET FOREIGN_KEY_CHECKS=0;

-- ----------------------------
-- Table structure for promote_rule
-- ----------------------------
DROP TABLE IF EXISTS 'promote_rule';
CREATE TABLE 'promote_rule' (
  'id' int(11) NOT NULL AUTO_INCREMENT,
  'promote_code' varchar(15) NOT NULL COMMENT '优惠券编码',
```

```
  'promote_rule' text COMMENT '优惠规则',
  'promote_name' varchar(255) DEFAULT NULL COMMENT '优惠券名称',
  PRIMARY KEY ('id','promote_code')
) ENGINE=InnoDB AUTO_INCREMENT=18 DEFAULT CHARSET=utf8;
```

创建 java 包 com.droolsBoot，并分别创建 controller、dao、model、service、util 包。

创建 ShopController.java 文件，目录为 controller，其内容为：

```
package com.droolsBoot.controller;

import com.droolsBoot.service.PromoteEdieService;
import org.springframework.beans.factory.annotation.Autowired;
import org.springframework.stereotype.Controller;
import org.springframework.web.bind.annotation.RequestMapping;
import org.springframework.web.bind.annotation.RequestMethod;
import org.springframework.web.bind.annotation.ResponseBody;

import java.util.Map;


@Controller
@RequestMapping("/promotion")
public class ShopController {

    @Autowired
    private PromoteEdieService promoteEdieService;


    @RequestMapping("/greeting")
    public String greeting() {
        return "index";
    }


    @RequestMapping("/shop")
    public String shop() {
        return "shopping";
    }


    /**
     * 编辑促销活动
     *
     * @return 结果
     */
    @RequestMapping(value = "/ediePromote", method = RequestMethod.POST,
produces = "application/json")
    @ResponseBody
    public void addPromote(String money, String rulename) {
        promoteEdieService.ediePromomteMap(money, rulename);
```

```
    }
    /**
     * 购物车
     *
     * @return 返回结果
     */
    @RequestMapping(value = "/toShopping", method = RequestMethod.POST,
produces = "application/json")
    @ResponseBody
    public Map<String, Object> toShopping(String money) {
        Map<String, Object> data = promoteEdieService.toShopping(money);
        return data;
    }
}
```

创建 PromoteExecuteDao.java 文件，目录为 dao，其内容为：

```
package com.droolsBoot.dao;

import com.droolsBoot.model.PromoteExecute;
import org.springframework.stereotype.Repository;

@Repository
public interface PromoteExecuteDao {
    /**
     * 插入促销基础信息
     */
    int insertPromoteExecute(PromoteExecute promote);
}
```

创建 PromoteExecute.java，目录为 model，其内容为：

```
package com.droolsBoot.model;

import org.kie.api.KieBase;
import org.kie.api.runtime.StatelessKieSession;

import java.util.List;

import static com.droolsBoot.service.NewKieBase.ruleKieBase;

public class PromoteExecute {
    //促销编号
    private String promoteCode;
    //业务Kbase
    private KieBase workKbase;
    //业务session
    private StatelessKieSession workSession;
    //规则内容
```

```java
private String WorkContent;
//促销规则名称:
private List<String> ruleName;

private String promoteName;

public String getPromoteName() {
    return promoteName;
}

public void setPromoteName(String promoteName) {
    this.promoteName = promoteName;
}

public String getWorkContent() {
    return WorkContent;
}

public void setWorkContent(String workContent) {
    WorkContent = workContent;
}

public List<String> getRuleName() {
    return ruleName;
}

public void setRuleName(List<String> ruleName) {
    this.ruleName = ruleName;
}


public String getPromoteCode() {
    return promoteCode;
}

public void setPromoteCode(String promoteCode) {
    this.promoteCode = promoteCode;
}

public KieBase getWorkKbase() {
    if (this.workKbase == null) {
        this.setWorkKbase();
    }
    return workKbase;
}

public void setWorkKbase() {
    this.workKbase = ruleKieBase(this.getWorkContent());
}
```

```java
    public StatelessKieSession getWorkSession() {
        if (this.workSession == null) {
            this.setWorkSession();
        }
        return workSession;
    }

    public void setWorkSession() {
        if (null != this.getWorkKbase()) {
            this.workSession = this.getWorkKbase().newStatelessKieSession();
        }
    }
}
```

该类是 Drools 的工具类，用来存放规则库、规则会话、规则内容等，使其更方便地管理和使用。

创建 RuleResult.java 文件，目录为 model，其内容为：

```java
package com.droolsBoot.model;

import java.util.ArrayList;
import java.util.List;

/**
 * describe：返回值实体
 *
 * @author laizhihui
 * @date 2018/04/19
 */
public class RuleResult {
    //参加活动商品优惠后的价格
    private double finallyMoney;
    //参加活动的名称
    private double moneySum;

    public double getMoneySum() {
        return moneySum;
    }
    public void setMoneySum(double moneySum) {
        this.moneySum = moneySum;
    }
    private List<String> promoteName = new ArrayList<>();


    public double getFinallyMoney() {
        return finallyMoney;
    }
    public void setFinallyMoney(double finallyMoney) {
```

```java
        this.finallyMoney = finallyMoney;
    }

    public List<String> getPromoteName() {
        return promoteName;
    }

    public void setPromoteName(String promoteName) {
        this.promoteName.add(promoteName);
    }
}
```

创建 DrlExecute.java 文件，目录为 service，其内容为：

```java
package com.droolsBoot.service;


import com.droolsBoot.model.PromoteExecute;
import com.droolsBoot.model.RuleResult;
import org.kie.internal.command.CommandFactory;
import org.slf4j.Logger;
import org.slf4j.LoggerFactory;

import java.text.DecimalFormat;
import java.util.*;

/**
 * describe: 规则工具类
 *
 * @author laizhihui
 */
public class DrlExecute {
    private static DecimalFormat df = new DecimalFormat("######0.00");
    protected static Logger logger = LoggerFactory.getLogger(DrlExecute.class);
    /**
     * 判断购物车中所有参加的活动商品
     *
     * @return 结果
     */
    static RuleResult rulePromote(PromoteExecute promoteExecute, Double moneySum) {
        // 判断业务规则是否存在
        RuleResult ruleresult = new RuleResult();
        //统计所有参加活动商品的件数和金额
        ruleresult.setMoneySum(moneySum);//返回优惠前的价格
        logger.info("优惠前的价格" + moneySum);
        //统计完成后再将参数insert促销规则中
        List cmdCondition = new ArrayList<>();
        cmdCondition.add(CommandFactory.newInsert(ruleresult));
```

```
promoteExecute.getWorkSession().execute(CommandFactory.newBatchExecution(cmdCondition));
        logger.info("优惠后的价格" + ruleresult.getFinallyMoney());
        return ruleresult;
    }
}
```

该类是用来操作规则的工具类，将规则操作单独写一个工具类，方便扩展和管理。

创建 NewKieBase.java 文件，目录为 service，其内容为：

```
package com.droolsBoot.service;

import org.kie.api.KieBase;
import org.kie.api.io.ResourceType;
import org.kie.internal.utils.KieHelper;

public class NewKieBase {
    //将业务规则写到规则库中
    public static KieBase ruleKieBase(String rule) {
        KieHelper helper = new KieHelper();
        try {
            helper.addContent(rule, ResourceType.DRL);
            return helper.build();
        } catch (Exception e) {
            e.printStackTrace();
            throw new RuntimeException("规则初始化失败");
        }
    }
}
```

该类是用来初始化规则库的，是 KIE 的一个工具类。这里对 KieHelper 进行一个引用和简单封装。

创建 PromoteEdieService.java 文件，目录为 service，其内容为：

```
package com.droolsBoot.service;


import com.alibaba.fastjson.JSONObject;
import com.droolsBoot.dao.PromoteExecuteDao;
import com.droolsBoot.model.PromoteExecute;
import com.droolsBoot.model.RuleResult;
import com.droolsBoot.util.UUIDUtil;
import org.springframework.beans.factory.annotation.Autowired;
import org.springframework.stereotype.Service;
import org.springframework.transaction.annotation.Transactional;

import java.util.ArrayList;
import java.util.HashMap;
import java.util.List;
import java.util.Map;
```

```java
@Service
public class PromoteEdieService {

    @Autowired
    private PromoteExecuteDao promoteExecuteDao;
    @Autowired
    private PromoteNeaten promoteNeaten;

    private Map<String, PromoteExecute> promoteExecuteMap;

    /**
     * 生成优惠券
     */
    @Transactional
    public void ediePromomteMap(String money, String rulename) {
        if (this.promoteExecuteMap == null) {
            promoteExecuteMap = new HashMap<>();
        }
        PromoteExecute promoteExecute = new PromoteExecute();
        double v = Double.parseDouble(money);
        String rule = UUIDUtil.rule(ruleWorkMap(rulename, v));
        String promoteCode = UUIDUtil.typeJoinTime();
        promoteExecute.setPromoteCode(promoteCode);
        promoteExecute.setWorkContent(rule);
        promoteExecute.setPromoteName(rulename);
        //插入优惠券
        int i = promoteExecuteDao.insertPromoteExecute(promoteExecute);
        if (i > 0) {
            PromoteExecute execute = promoteNeaten.editRule(rule);
            this.promoteExecuteMap.put(promoteCode, execute);
        }
    }

    /**
     * 购物车计算
     *
     * @return
     */
    public Map<String, Object> toShopping(String moneySum) {
        //购物车请求信息
        Map<String, Object> map = new HashMap<>();
        double v = Double.parseDouble(moneySum);
        List<Object> pn = new ArrayList<>();
        if (this.promoteExecuteMap != null) {
            //证明有优惠券
            for (Map.Entry<String, PromoteExecute> codes : this.promoteExecuteMap.entrySet()) {
                RuleResult ruleResult =
```

```java
DrlExecute.rulePromote(codes.getValue(), v);
            v = ruleResult.getFinallyMoney();
            pn.add(ruleResult);
        }
    }
    map.put("moneySumYuanJia",moneySum);
    map.put("youhuiquanjiegou", pn);
    return map;
}

/**
 * 组合业务规则Json方法
 *
 * @return 结果
 */
private String ruleWorkMap(String name, Double money) {
    Map<String, Object> map = new HashMap<>();
    //组合Rule部分
    Map<String, Object> rule = new HashMap<>();
    rule.put("name", name);
    map.put("rule", rule);
    //组合 规则When部分
    Map<String, Object> when = new HashMap<>();
    map.put("condition", when);
    //组合 规则Then部分
    Map<String, Object> then = new HashMap<>();
    then.put("money", money);
    map.put("action", then);
    //组合规则When And Then 部分
    return JSONObject.toJSONString(map);
    }
}
```

该类是业务层的服务器，主要负责生成优惠券、使用优惠券、计算结果等。

创建 PromoteNeaten.java 文件，目录为 service，其内容为：

```java
package com.droolsBoot.service;

import com.droolsBoot.model.PromoteExecute;
import org.springframework.stereotype.Service;

/**
 * describe: 规则库生成
 *
 * @author laizhihui
 * @date 2018/03/01
 */
@Service
```

```java
public class PromoteNeaten {

    /**
     * 初始化指定的规则
     *
     * @param rule 促销实体数据
     * @return 返回值
     */
    public PromoteExecute editRule(String rule) throws RuntimeException {
        PromoteExecute promoteExecute = new PromoteExecute();
        promoteExecute.setWorkContent(rule);//促销业务规则
        //规则库 初始化
        promoteExecute.getWorkSession();
        return promoteExecute;
    }
}
```

创建 LossMoneyTemplate.java 文件，目录为 util，其内容为：

```java
package com.droolsBoot.util;
/**
 * describe:
 *
 * @author laizhihui
 * @date 2018/04/25
 */
public class LossMoneyTemplate {
    public static final String workMoneyST = "wordImport(rules) ::=<<\n" +
            "package com.promote\n" +
            "\n" +
            "import\tcom.droolsBoot.model.RuleResult;\n" +
            "<rules; separator=\"\\n\\n\">\n" +
            ">>\n" +
            "\n" +
            "ruleValue(condition,action,rule) ::=<<\n" +
            "rule \"<rule.name>\"\n" +
            "\tno-loop true\n" +
            "\t\twhen\n" +
            "\t\t    $r:RuleResult(true)\n" +
            " \t\tthen\n" +
            "            modify($r){\n" +
            "                setPromoteName(drools.getRule().getName())<if(action)>,\n" +
            "                setFinallyMoney($r.getMoneySum() - <action.money><endif>)\n" +
            "            }\n" +
            "end\n" +
```

```
            ">>\n";
}
```

该类是一个规则模板工具类，通过传参数即可生成规则内容，将规则内容存放在数据库或生成规则文件，为规则库的构建设定基础。

创建 StringJointUtil.java 文件，目录为 util，其内容为：

```java
package com.droolsBoot.util;

import java.text.SimpleDateFormat;
import java.util.Date;
public class StringJointUtil {
    private final static SimpleDateFormat sdfOne = new SimpleDateFormat("yyyy-MM-dd HH:mm:ss");
    private final static SimpleDateFormat sdfTwo = new SimpleDateFormat("yyyy-MM");
    private final static SimpleDateFormat sdfThree = new SimpleDateFormat("yyyyMMdd");

    /**
     * 日期转换
     * @param date
     * @return
     */
    public static String dateToString(Date date){
        String dateString = sdfOne.format(date);
        return dateString;
    }

    /**
     * 日期转换
     * @param date
     * @return
     */
    public static String dateToStringTwo(Date date){
        String dateString = sdfTwo.format(date);
        return dateString;
    }

    /**
     * 日期转换
     * @param date
     * @return
     */
    public static String dateToStringThree(Date date){
        String dateString = sdfThree.format(date);
        return dateString;
    }
```

```
}
```

创建 UUIDUtil.java 文件，目录为 util，其内容为（注意加粗部分——模板传入）：

```java
package com.droolsBoot.util;

import com.alibaba.fastjson.JSONObject;
import org.apache.commons.codec.digest.DigestUtils;
import org.stringtemplate.v4.ST;
import org.stringtemplate.v4.STGroup;
import org.stringtemplate.v4.STGroupString;

import java.util.Date;
import java.util.UUID;

import static com.droolsBoot.util.LossMoneyTemplate.workMoneyST;


/**
 * describe:
 *
 * @author laizhihui
 * @date 2018/03/07
 */
public class UUIDUtil {

    /**
     * 生成原始UUID
     *
     * @return
     */
    private static String UUIDString() {
        String uuid = UUID.randomUUID().toString();
        return uuid;
    }

    /**
     * 生成格式化UUID
     *
     * @return
     */
    public static String UUIDFormatString(String replace) {
        String uuid = UUID.randomUUID().toString().replaceAll(replace,
"");
        return uuid;
    }

    /**
     * 值加密
```

```java
     *
     * @return
     */
    public static String MD5AndUUID() {
        //时间戳
        String timeJab = String.valueOf(System.currentTimeMillis());
        //UUID+时间戳
        String concat = UUIDString().concat(timeJab);
        return DigestUtils.md5Hex(concat);
    }

    /**
     * 生成不重复促销编码
     *
     * @return
     */
    public static String typeJoinTime() {
        String dateNowStr = StringJointUtil.dateToStringThree(new Date());
        Integer math = (int) ((Math.random() * 9 + 1) * 1000000);
        String code = dateNowStr.concat(math.toString());
        return code;

    }

    public static String rule(String json) {
        String rule = ruleWordExchangsST(json);
        return rule;
    }

    /**
     * 规则业务生成
     */
    public static String ruleWordExchangsST(String json) {
        STGroup group = new STGroupString(workMoneyST);
        ST stFile = group.getInstanceOf("wordImport");
        ST stRule = group.getInstanceOf("ruleValue");
        JSONObject jsonObject = JSONObject.parseObject(json);
        JSONObject condition = jsonObject.getJSONObject("condition");
        JSONObject action = jsonObject.getJSONObject("action");
        JSONObject rule = jsonObject.getJSONObject("rule");
        stRule.add("condition", condition);
        stRule.add("action", action);
        stRule.add("rule", rule);
        stFile.add("rules", stRule);
        String result = stFile.render();
        return result;
    }
}
```

创建 Spring Boot 启动类 Application.java，目录为 com/droolsBoot，内容为：

```java
package com.droolsBoot;

import org.mybatis.spring.annotation.MapperScan;
import org.slf4j.Logger;
import org.slf4j.LoggerFactory;
import org.springframework.boot.SpringApplication;
import org.springframework.boot.autoconfigure.SpringBootApplication;
import org.springframework.transaction.annotation.EnableTransactionManagement;


@SpringBootApplication
@EnableTransactionManagement
@MapperScan("com.droolsBoot.dao")
public class Application {
    protected static Logger logger=LoggerFactory.getLogger(Application.class);

    public static void main(String[] args) {
        SpringApplication.run(Application.class, args);
        logger.info("SpringBoot Start Success");
    }
}
```

创建 Mybatis 配置文件 PromoteExecuteMapper.xml，目录为 resources/mapper，其内容为：

```xml
<?xml version="1.0" encoding="UTF-8" ?>
<!DOCTYPE mapper PUBLIC "-//mybatis.org//DTD Mapper 3.0//EN"
 "http://mybatis.org/dtd/mybatis-3-mapper.dtd" >
<mapper namespace="com.droolsBoot.dao.PromoteExecuteDao">
    <resultMap id="BaseResultMap" type="com.droolsBoot.model.PromoteExecute">
        <result column="promote_code" property="promoteCode" jdbcType="VARCHAR"/>
        <result column="promote_rule" property="WorkContent" jdbcType="VARCHAR"/>
        <result column="promote_name" property="promoteName" jdbcType="VARCHAR"/>
    </resultMap>

    <insert id="insertPromoteExecute" parameterType="com.droolsBoot.model.PromoteExecute">
        insert into promote_rule(
        promote_code,
        promote_rule,
        promote_name
        )
        values
```

```
            (
            #{promoteCode},
            #{WorkContent},
            #{promoteName}
            )
    </insert>
</mapper>
```

在浏览器中输入 http://localhost:9999/promotion/shop，出现如图 8-15 所示的页面。

图8-15　购物车总金额页面

首先测试当前购物车中是没有任何优惠券的。在购物车总金额文本框中输入 500，单击"提交"按钮，结果如图 8-16 所示。

图8-16　未有优惠券的页面效果

从图 8-16 中的结果可以看出，此时购物车需要支持的金额是 500 元，并没有优惠。在浏览器中输入 "ttp://localhost:9999/promotion/greeting"　出现如图 8-17 所示的页面，并在优惠券金额中输入 "30"，优惠券名称中输入 "优惠券 30 元"。

图8-17　设置优惠券页面

其次进行测试，在购物车页面总金额文本框中输入 500，单击"提交"按钮，结果如图 8-18 所示。

图8-18　使用优惠券后的效果

查询数据库 promote_rule 表，其中多了一条数据，也就是刚才创建的优惠券，如图 8-19 所示。

图8-19 数据库中的业务场景

测试过程中,并没有重启服务器,一切都是动态操作的业务,这也是使用规则引擎的魅力。代码很简单,项目中有 3 个新概念,分别是模板引擎、KieHelper、自定义 Kie 聚合类 PromoteExecute.java。模板引擎是一个新的技术,本教程中只有测试使用,如果读者有兴趣可以进行源码的学习或自行上网学习。KieHelper[①] 是 Kie 自己封装的工具类,官方没有讲述,第 15 章源码分析中会对 KieHelper 做详细说明。聚合类是核心类之一,该类主要属性包括了 KieBase、StatelessKieSession(无状态会话)、规则内容,其目的是更方便地操作规则库,将这些值放在内存静态变量中,使用时只需对该类进行一个遍历引用就可以实现规则的调用,大大减少了创建规则相关对象的内存开销。

通过本节及整合 Spring 中不难看出,与其说是整合还不如说只是将这两个不相关的功能放到一个项目中,Spring Boot 和 Spring 都只是处理本职工作,而 Drools 同样如此。

通过 3 个小节分别讲述了 3 种不同的方式,官方例子相比 KieHelper 来说使用起来比较烦琐,但它也提供了更方便的管理,分离了 Drools 各项功能 API,给编程人员带来了很大的便利,KieHelper 是笔者比较倾向使用的。可以通过第一个测试整合说明一下自己进行的封装,毕竟符合自己的开发风格才是最好的。

在进行 Spring、SpringMvc 和 MyBatis 整合时,SSM 是 SSM,Drools 是 Drools,只不过提供了接口。Spring Boot 整合时,Spring Boot 是 Spring Boot,Drools 是 Drools,两者只是关系上的调用。

使用 Drools 时,不要添加 Spring Boot 的热启动,这会导致一个很严重的问题,即规则库生成失败,经常出现的空指针的问题。其原因是规则库在构建时会在系统数据库中生成一个 .class 的规则类文件,这时 Spring Boot 会认为项目发生了变化,热启动会被激活。将已经加载或初始化好的规则库再次清空或再次初始化,这一点在部署中一定要注意。

---

① KieHelper 是一个工具类,是 Kie 自己封装的工具类,简化了规则的构建,可快速生成规则容器、规则库。

第9章
KieSession状态

KieSession 有两种状态，俗称有状态和无状态。整合 Spring Boot 实战中和介绍 Global 时提到过无状态的使用。之前所有的测试用例几乎都是通过有状态的 KieSession 进行阐述。

KieSession 默认是有状态的，之所以单独讲解 KieSession 的状态，是因为它在实际项目中是非常重要的。通过 KieContainer 获取两种不同状态的 KieSession，其内容为：

```
...
KieBase kBase1 = kContainer.getKieBase("kbXXX");
KieSession kieSession1 = kContainer.newKieSession("knXXX");
StatelessKieSession kieSession2 = kContainer.newStatelessKieSession("knXXX ")
...
```

KieSession1 是有状态的，KieSession2 是无状态的，所有 KieContainer 在声明 KieSession 时需要指定具体的 Kie 会话 name。如果 KieSession 请求的类型 KieContainer 与 kmodule.xml 文件中声明的类型不符，则会报错。

修改 kmodule.xml 配置文件，并添加如下配置（注意加粗部分）：

```xml
<kbase name="isKiesession" packages="rules.rulesHello">
    <ksession name="isKiesession" type="stateless"/>
</kbase>
```

创建 RuleKieSession.java 文件，目录为 com/rulesConstraint，其内容为：

```java
package com.ruleKiesession;

import com.pojo.Person;
import org.junit.Test;
import org.kie.api.KieServices;
import org.kie.api.runtime.KieContainer;
import org.kie.api.runtime.KieSession;

public class RuleKieSession {
    @Test
    public void testkieSessionType (){
        KieServices kss = KieServices.Factory.get();
        KieContainer kc = kss.getKieClasspathContainer();
        KieSession ks = kc.newKieSession("isKieSession");
        Person person = new Person();
        person.setName("张三");
        person.setAge(30);
        ks.insert(person);
        int count = ks.fireAllRules();
        System.out.println("总执行了" + count + "条规则");
        ks.dispose();
    }
}
```

执行 testkieSessionType() 方法，结果会报类型不匹配的异常，如图 9-1 所示。注意，由于

KieBase 的 KieSession 被标记为默认值，可以在 KieContainer 没有传递任何名称的情况下获得它们。

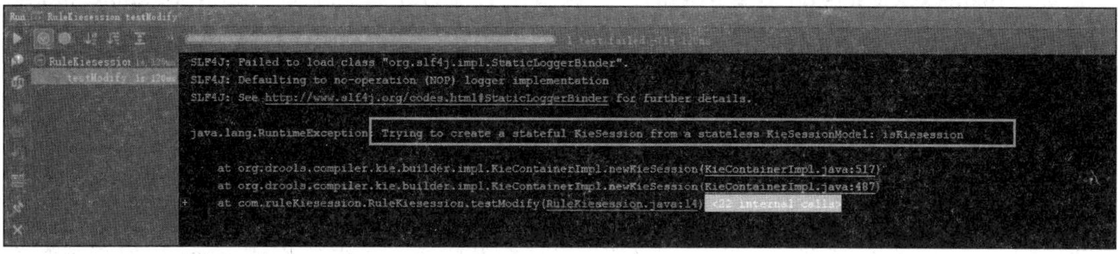

图9-1 类型不匹配

9.1 有状态的KieSession

KieSession 会在多次与规则引擎进行交互中，维护会话的状态。定义 KieSession，在 kmodule.xml 文件中定义 type 为 stateful，其代码为：

```
<kbase name="isKiesession" packages="rules.rulesHello">
  <ksession name="isKiesession" type="stateful"/>
</kbase>
```

stateful 是 type 属性的默认值，所以可以忽略不写，也就是如果不特意指定当前 KieSession 的类型是其他状态的，默认只能创建有状态的 KieSession，否则会报类型异常。获取 KieSession 实例"KieSession statefulSession = KieContainer.newKieSession("stateful_session");"可以在 KieSession 中执行一些操作。如果需要清理 KieSession 维护的状态，调用 dispose() 方法；如果在调用规则时不调用 dispose() 方法，则 KieSession.insert(Object) 会产生迭代方式插入的（笛卡儿积）。这个问题会在 Kiserver[①] 中出现。

9.2 无状态的StatelessKieSession

无状态的 KieSession 与有状态的 KieSession 无论是 API 的操作还是状态的保留都大不相同，StatelessKieSession 封装了 KieSession，是一个独立的功能，其主要目的是提供了更多的服务方式，使用 StatelessKieSession 时不需要再调用 dispose() 方法。而且无状态会话不支持迭代插入，有状态调用规则从 Java 代码调用 fireAllRules() 方法开始；无状态调用规则从 Java 调用 execute() 方法开始。该方法是一次执行的，它将在内部实例化 KieSession，并调用 fireAllRules() 方法，然后在 finally 中

---

① Kiserver 是 Kie 提供的引用 kjar 的一个服务器，是一个 Web 接口。

调用 dispose() 方法。所以在 execute() 之后不能再调用会话的其他操作。

无状态 StatelessKieSession 需要声明时，必须在配置文件中进行配置，修改 kmodule.xml 配置文件并添加如下配置，其内容为（注意加粗部分）：

```xml
<kbase name="isKbase" packages="rules.isKiesession">
    <ksession name="kiesession" type="stateful" />
    <ksession name="stateless" type="stateless" />
</kbase>
```

代码中，"type="stateful","是可以省略的，默认为有状态。无状态 KieSession 需要单独配置，即"type="stateless""，<KieSession> 标签中还可以添加其他属性的配置，具体内容将在后面章节中讲解。

声明了 stateless 的 KieSession 类型是无状态后，就可以在代码中进行创建了。定义方式与有状态的差不多，只是类型不一样，其内容为（注意加粗部分）：

```java
@Test
public void testStatelessKIESession() {
    KieServices kss = KieServices.Factory.get();
    KieContainer kc = kss.getKieClasspathContainer();
    StatelessKieSession kieSession = kc.newStatelessKieSession("stateless");
    kieSession.execute(new Object());
}
```

execute 是将用户数据通过命名的方式进行传输的，其参数共有两种类型：对象和集合。

创建规则文件 kiesession.drl，目录为 rules/isKieSession，其内容为：

```
package rules.isKiesession
import com.pojo.Person;
import com.pojo.School;

    rule "无状态规则测试Person"
        when
            $p:Person();
        then
            System.out.println(drools.getRule().getName()+$p.getName());
    end
```

编辑 RuleKiesession.java 文件，添加 testStatelessKIESession 方法，其内容为：

```java
@Test
public void testStatelessKIESession() {
    KieServices kss = KieServices.Factory.get();
    KieContainer kc = kss.getKieClasspathContainer();
    StatelessKieSession kieSession = kc.newStatelessKieSession("stateless");
    Person person = new Person();
    person.setName("张三");
```

```
        person.setAge(30);
        kieSession.execute(person);
}
```

执行 testStatelessKIESession() 方法，结果如图 9-2 所示。

图9-2　无状态KieSession测试效果

如果执行一次 insert 有多个 Fact 事实，其是否与 KieSession 一样，那么执行两次 execute(Object) 呢？

创建规则文件 kiesession.drl，并添加规则，其内容为：

```
rule "无状态规则测试School"
    when
        $p:School();
    then
        System.out.println(drools.getRule().getName()+$p.getClassName());
    end
```

编辑 RuleKieSession.java 文件，添加 testStatelessKIESession2() 方法，其内容为：

```
@Test
public void testStatelessKIESession2() {
    KieServices kss = KieServices.Factory.get();
    KieContainer kc = kss.getKieClasspathContainer();
    StatelessKieSession kieSession = kc.newStatelessKieSession("stateless");
    Person person = new Person();
    person.setName("张三");
    School school=new School();
    school.setClassName("一班");
    kieSession.execute(person);
    kieSession.execute(school);
}
```

执行 testStatelessKIESession2() 方法，结果如图 9-3 所示。

图9-3 测试无状态插入两个Fact测试

结果似乎是输出了，但稍稍变化一下规则后，就会证明当前的想法是错的。

编辑规则文件 kiesession.drl，并添加规则，其内容为：

```
rule "无状态规则测试School Person"
    when
        $s:School();
        $p:Person();
    then
System.out.println(drools.getRule().getName()+$s.getClassName()+"-------"+$p.getName());
end
```

再次执行 testStatelessKIESession2() 方法，结果与图 9-3 一样。上述例子很好地证明这一点，如果是有状态的 KieSession 同时 insert 两个 Fact 事实对象，加上规则体的 LHS 部分只有简单的约束，是应该执行这 3 个规则的。而无状态却不可以，原因则需要追述到源码中。

源码目录为 StatelessKnowledgeSessionImpl.java，其内容为：

```
public void execute(Object object) {
    StatefulKnowledgeSession ksession = newWorkingMemory();
    try {
        ksession.insert( object );
        ksession.fireAllRules();
    } finally {
        dispose(ksession);
    }
}
```

分析 execute 的执行过程。

（1）每次执行一次 execute 方法，都会创建一个新的 StatefulKnowledgeSession。

（2）KieSession 执行 fireAllRules() 方法。

（3）每次执行完成后都会调用 "dispose(ksession);" 方法。

总结：无状态的规则不会出现类似迭代（产生笛卡儿积）的问题，它会适时创建和清空当前的 KieSession。

以上都是概念性的知识，那么如何通过无状态的 insert 多个对象呢？前面介绍无状态的 KieSession 时提到了，它有两种方式：Object 和集合。与其说是集合还不如说是通过 Drools 的 API

批量执行命令的方式进行数据处理。编辑 RuleKieSession.java 文件，添加 testStatelessKIESession2() 方法，其内容为：

```
@Test
public void testStatelessKIESession3() {
    KieServices kss = KieServices.Factory.get();
    KieContainer kc = kss.getKieClasspathContainer();
    StatelessKieSession kieSession = kc.newStatelessKieSession("stateless");
    Person person = new Person();
    person.setName("张三");
    School school=new School();
    school.setClassName("一班");
    List cmds = new ArrayList();
    cmds.add( CommandFactory.newInsert(person) );
    cmds.add( CommandFactory.newInsert(school) );
    kieSession.execute( CommandFactory.newBatchExecution( cmds ));
}
```

执行 testStatelessKIESession3() 方法，结果如图 9-4 所示。

图9-4　无状态批量插入Fact效果

Drools 规则引擎技术指南

第四篇

高级篇

第10章

Drools高级用法

10.1 决策表

决策表是规则文件的一种变形，是以 xls/xlsx 为扩展名的文件，是指通过 Excel 完成对规则的匹配。它是一种"精确而紧凑"的条件逻辑方式，非常适合业务场景规则。决策表并非新的技术概念（在软件术语中），其应用领域非常广泛，已经有很多企业在使用，通过实践证明决策表在某些应用中是可行的。

通俗地讲，决策表就是向电子表格中输入特定的值，并加载到 Drools 规则库的一种数据驱动的规则方法。如果存在可表示为规则模板和数据的规则，其决策表数据如图 10-1 所示。

	A	B	C	D	E	F
1	**RuleSet**	rulesTwo.isXls				
2	**RuleTable 测试规则**					
3	CONDITION	ACTION				
4						
5	eval(true);	System.out.println("$1");				
6						
7		测试规则_7				
8		测试规则_8				
9		测试规则_9				
10						
11						

图10-1 决策表数据

决策表调用与版本有密切关系，在 Drools 5.x 版本中，决策表的用法是不一样的，它需要指定文件类型与具体文件，但在 Drools 6.0 版本以后，因为加入了 KIE 的概念，优化了相当多的代码。所以调用决策表的方式与调用 drl 规则文件是一样的。

创建决策表文件 tableXls.xls，目录为 rulesTwo/isXls，其内容如图 10-1 所示。

修改 kmodule.xml 配置文件，并添加如下配置：

```xml
<kbase name="isXls"  packages="rulesTwo.isXls">
    <ksession name="isXls" />
</kbase>
```

创建 RulesTable.java 文件，目录为 comTwo.DecisionTable，其内容为：

```java
package comTwo.DecisionTable;

import com.pojo.Person;
import org.junit.Test;
import org.kie.api.KieServices;
import org.kie.api.runtime.KieContainer;
import org.kie.api.runtime.KieSession;

public class RulesTable {
    @Test
    public void testisXls() {
```

```
        KieServices kss = KieServices.Factory.get();
        KieContainer kc = kss.getKieClasspathContainer();
        KieSession ks = kc.newKieSession("isXls");
        int count = ks.fireAllRules();
        System.out.println("总执行了" + count + "条规则");
        ks.dispose();
    }
}
```

执行 testisXls () 方法，结果如图 10-2 所示。

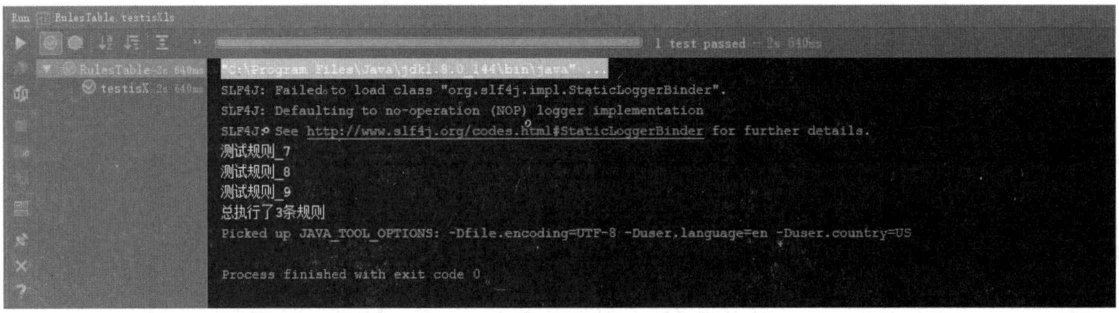

图10-2 决策表测试效果

从图 10-2 中可以看出，分别有关键字 RuleSet、RuleTable、CONDITION、ACTION，其含义如下。

（1）RuleSet 和 drl 文件中的 package 功能是一样的，为必填项。

（2）RuleTable 和 drl 文件中规则名称的功能是一样的，为必填项。

（3）CONDITION 和规则体中的 LHS 部分功能是一样的，为必填项。

（4）ACTION 和规则体中的 RHS 部分功能是一样的，为必填项。

CONDITION 下面两行则表示 LHS 部分（指图 10-1 中的第 4、5 行），第 3 行则为注释行（指图 10-1 中的第 6 行），不计为规则部分，从第四行开始 (指图 10-2 中第 7 行)，每一行表示一条规则。图 10-1 中自第 7 行开始共有 3 行，即 3 个规则。所以执行了 testisXls() 命令后，控制台会输出共执行了 3 条规则。

通过上述分析，相信读者对 Drools 的决策表已经有了一个初步的认识。不难发现，决策表拥有自己的一套语法关键字，Drools 也为程序设计提供了一套能验证决策表生成 Drl 文件内容的工具，通过生成的 Drl 内容来检查决策表是否正确。

编辑 Pom.xml 文件，其内容为：

```xml
<dependency>
    <groupId>org.drools</groupId>
    <artifactId>drools-decisiontables</artifactId>
    <version>RELEASE</version>
</dependency>
```

编辑 RulesTable.java 文件,并添加 verificationDT() 方法,其内容为:

```java
@Test
public void  verificationDT() throws FileNotFoundException {
    File file = new File(
"D:\\project\\drools\\src\\main\\resources\\rulesTwo\\isXls\\tableXls.xls");
    InputStream is = new FileInputStream(file);
    SpreadsheetCompiler converter = new SpreadsheetCompiler();
    String drl = converter.compile(is, InputType.XLS);
    System.out.println(drl);
}
```

执行 verificationDT() 方法,结果如图 10-3 所示。

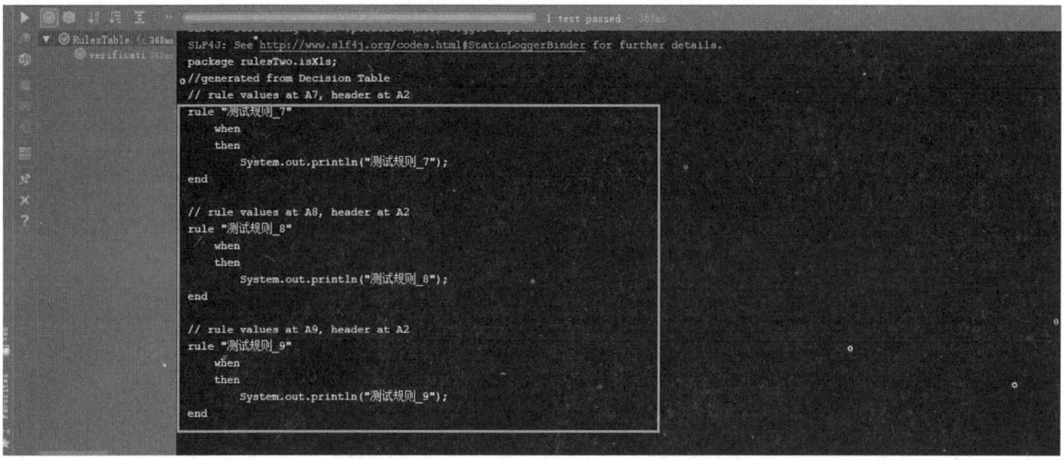

图10-3 决策表信息转换成Drl语法

该功能可以验证决策表是否正确,将其内容转换成规则内容,以方便程序设计员更容易理解决策表生成的规则。

决策表中有一个很重要的功能——占位符,占位符功能分为两种形式:$param 和 $1,$2…。决策表是 Excel,而 Excel 是由单元格组成的,每一个单元格是 Excel 的最小单位,而 $param 占位符的作用是获取每一个单元格的内容,如图 10-4 所示。

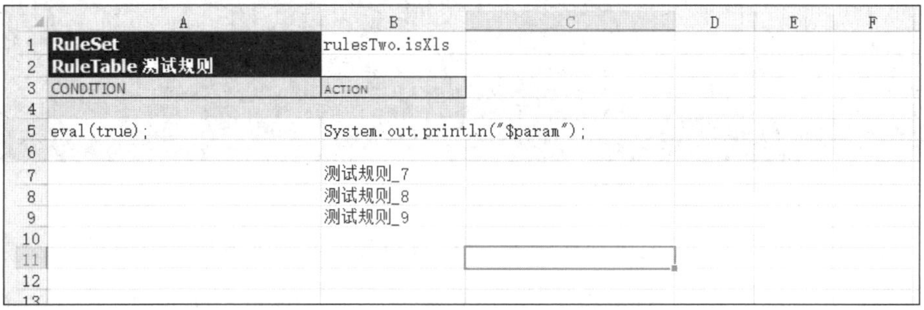

图10-4 决策表占位符的另一种写法

测试结果与图 10-2 中一样，这是因为 $param 占位符只获取当前单元格的信息，修改 Excel 内容，在 B7~B9 单元格中添加"，测试占位符"，并将 $param 换成"$1,$2"，再次执行 testisXls() 方法，输出结果如图 10-5 所示。

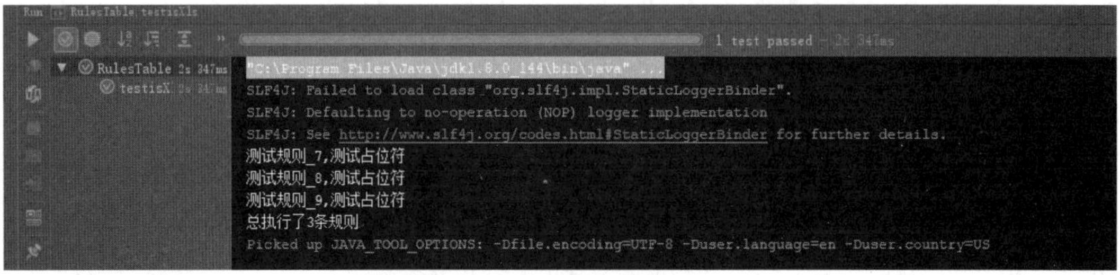

图10-5　$符号占位符使用效果

既然使用了逗号，那么可以做这样的猜想：是否通过逗号可以分别输出单元格中的内容？这时 $1 占位符就起了作用，修改 B5 单元格的内容，如图 10-6 所示。

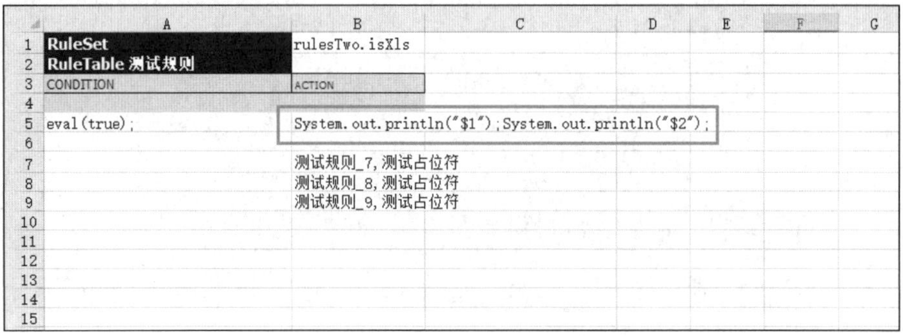

图10-6　决策表占位符使用效果

执行 testisXls() 方法，结果如图 10-7 所示。

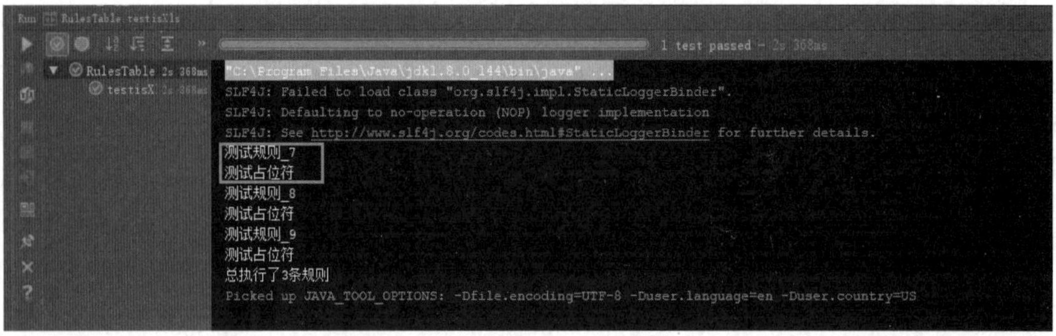

图10-7　测试决策表占位符效果

$1 占位符与 $param 占位符的区别在于 $1 可以再切分单元格内容值，但前提是必须以英文逗号作为分隔符，$1 可以是多个，若单元格中的内容是 a、b、c，则 $1 表示 a、$2 表示 b、$3 表示 c，顺序是没有强制要求的。

决策表关键字与规则文件关键字很相似,大体可以分为两个部分:决策表部分和规则表部分,如表 10-1 和表 10-2 所示。

表10-1 决策表关键字

关键字	是否必填	值	说 明
RuleSet	是	string	在这个单元的右边单元中包含RuleSet的名称,与drl文件中的package是一样的
Sequential	否	true/false	值为true或false,若是true,则确保规则按照从表格的上面到下面的顺序执行(规则触发是从上朝下,若是false则是乱序)
Import	否	string	导入所需要的引用类或方法,如要导入规则库中类的列表(逗号隔开)
Functions	否	string	功能与标准的drl函数相同,包含有返回值与无返回值两种。如果要定义多个函数,就在Functions后面以逗号作为分隔符定义多个函数
Variables	否	string	功能与标准的drl全局变量相同。如果要定义多个变量,就在Variables后面以逗号作为分隔符 定义多个函数
Queries	否	string	功能与标准的DRL查询相同。如果要定义多个函数,就在Queries后面以逗号作为分隔符 定义多个查询
RuleTable	是	string	表示规则名,在RuleTable后直接写规则名的前缀,不用另写一列,规则以行号为规则名

表10-2 规则表部分

关键字	是否必填	值	说 明
CONDITION	是	string	指明该列将被用于判断CONDITION(代表条件)相当于drl中的when,每个规则表至少有一个
ACTION	是	string	指明该列将被用于推断,简单理解为结果,相当于drl中的then。ACTION与CONDITION是平行的
PRIORITY	否	int	指明该列的值将被设置为该规则行的'salience'值。覆盖'Sequential'标志。但注意,若在ruleSet下设置了sequential的值为true,则PRIORITY将不起作用;若sequential设置为false或不设置,则PRIORITY生效
NAME	否	string	指明该列的值将被设置为从哪行产生的规则名称
NO-LOOP	否	true/false	指明这个规则不允许循环。为了使这个选项能正常运行,在该单元格中必须是让该选项生效的一个值(true 或 false)。如果该单元格保持为空,那么这个选项将不会为该行设置,与drl文件中的no-loop属性功能是一样的

关键字	是否必填	值	说 明
ACTIVATION-GROUP	否	string	在这个列中单元格的值，指出该规则行属于特定的活动分组。一个活动组意味着在命名组中的规则只有一条会被引发（首条规则引发，中止其他规则活动），与drl中的含义是一样的
AGENDA-GROUP	否	string	在这个列中单元格的值，指出该规则行属于特定的议程组，可以理解成获取焦点（这是一种在规则组之间控制流的方法），与drl中的含义是一样的
RULEFLOW-GROUP	否	string	在这个列中单元格的值，指出该规则行属于特定的规则流组

决策表中有详细的说明，规则以 RuleSet 为开始，也就是说通过 RuleSet 来判断决策表是从哪里开始的，而且在 RuleSet 中可以添加注释，如图 10-8 所示。

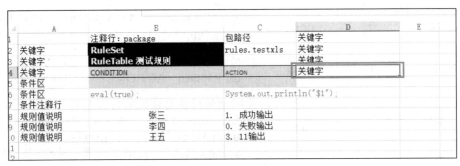

图10-8　决策表中添加注释

但是这里要提醒读者，不能在所在行及其相邻例下方 CONDITION/ACTION 写注释，否则就会报错，如图 10-9 所示，报错信息如图 10-10 和图 10-11 所示。

图10-9　决策表添加注释的错误效果

图10-10 决策表注释结果报错信息（1）

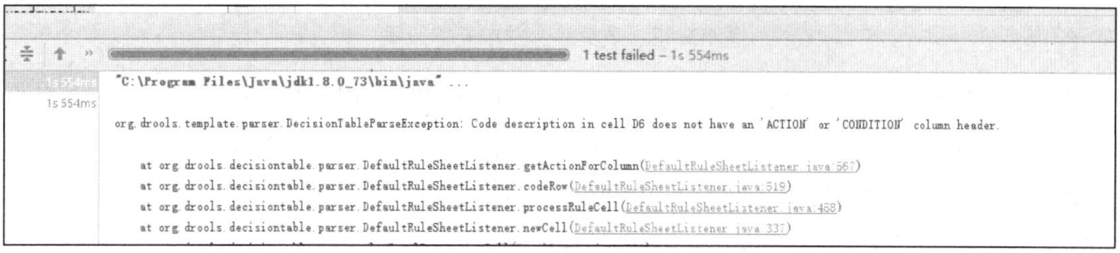

图10-11 决策表注释结果报错信息（2）

Condition 与 Action 可以交叉使用，如图 10-12 所示。

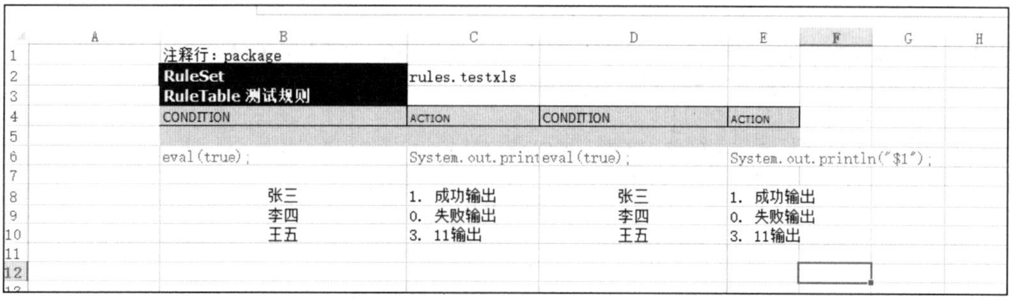

图10-12 决策表交叉使用的 Condition 与 Action

调用决策表的另一种方式是通过 Drools 工具类 KieHelper。

编辑 RulesTable.java 文件，并添加 testKieHelper() 方法，其内容为（注意加粗部分）：

```
@Test
public void testKieHelper() {
    Resource dis =
ResourceFactory.newClassPathResource("rulesTwo/isXls/tableXls.xls",
RulesTable.class);
    KieHelper helper = new KieHelper();
    helper.addResource(dis, ResourceType.DTABLE);
    KieSession ksession = helper.build().newKieSession();
    int i = ksession.fireAllRules();
    System.out.println("        " + i + "次");
    ksession.dispose();
}
```

下面介绍决策表的特殊用法，创建一个决策表文件，如图 10-13 所示。

	A	B	C	D
1	RuleSet	rulesTwo.isXls2		
2	Import	com.pojo.Person;		
3	RuleTable 测试规则			
4	CONDITION	CONDITION	CONDITION	ACTION
5	$p:Person			
6	name=="$1"	age>=$1	age<$1	System.out.println("$1");
7	判断名字	判断年龄大于	判断年龄小于	
8	张三	20	50	测试规则张三输出条件
9	李四		60	测试规则李四输出条件
10	王五	30		测试规则王五输出条件

图10-13　决策表合并单元格效果

从图 10-13 中可以看出，A5 单元格是一个合并单元格，它的作用是当前决策表的 A~C 列的条件均受 Person 的约束。第 7 行是注释行，B8 与 C9 是合并单元格，自第 8 行开始，每一行即是一条规则。先将决策表转成的 drl 内容输出一下看看效果。

编辑 RulesTable.java 文件，并添加 verificationDT2() 方法，其内容为：

```java
@Test
public void  verificationDT2() throws FileNotFoundException {
    File file = new File(
"E:\\drools\\src\\main\\resources\\rulesTwo\\isXls\\tableXlsS.xls");
    InputStream is = new FileInputStream(file);
    SpreadsheetCompiler converter = new SpreadsheetCompiler();
    String drl = converter.compile(is, InputType.XLS);
    System.out.println(drl);
}
```

执行 verificationDT2() 方法，其结果为：

```
package rulesTwo.isXls2;
//generated from Decision Table
import com.pojo.Person;;
// rule values at A8, header at A3
rule "测试规则_8"
    when
        $p:Person(name=="张三", age>=20, age<50)
    then
        System.out.println("测试规则张三输出条件");
end

// rule values at A9, header at A3
rule "测试规则_9"
    when
        $p:Person(name=="李四", age>=20, age<60)
    then
        System.out.println("测试规则李四输出条件");
end
```

```
// rule values at A10, header at A3
rule "测试规则_10"
    when
            $p:Person(name=="王五", age>=30, age<60)
    then
            System.out.println("测试规则王五输出条件");
end
```

输出结果就已经证明了合并单元格的作用，它代表着条件重复使用，或者说决策表 Excel 文件看起来更明了。虽然它合并了，但所选中规则行中的规则还是一条，不管是约束 when 还是结果 then 都是一样的。

10.2 DSL领域语言

Drools 领域语言又称为自然语言，它是业务人员通过 dslr 文件编写的规则文件，通过用文字描述来实现业务规则。DSL 是 Drools 提供的一种可通过非正规的规则语法编写的规则文件，使用 DSL 领域语言时，需要创建 *.dsl 和 *.dslr 两个规则文件，核心是 *.dsl 文件。DSL 句子可以作为条件元素和结果动作的"模板"，在规则中重复使用。

DSL 的运行与其他规则文件是一样的，它只是一个编译构建的过程，需要特殊的解析器与转换器。

DSL 机制是允许定制 conditional expressions（条件表达式 LHS）和 consequence actions（结果值，RHS），也可以替换全局变量。下面对官方的例子做一个分析。

[when]Something is {colour}=Something(colour=="{colour}")

分析一下这段代码的关键字，"[when]"指表达式的作用域，即它属于一条规则的 LHS。在 [作用域] 后面的部分是使用在该规则中的表达式（通常是一个自然语言表达式）。""=""的右边部分是映射到规则语言（当然它的格式取决于 LHS）。

上述的说明可能不太容易理解，换个角度，关键字 [when] 是 dsl 的特定语法，而 when 后面的值是一个对应关系，也有点类似 Spring Boot 中的配置文件，如等号的左边为 Key、右边为 Value。这样理解起来就容易理解了，虽然这样理解可能不会有更深刻的认知，但是结合实际使用就更容易理解了。

创建规则文件 personDSL.dsl，目录为 rulesTwo/isDSL，其内容为（注意加粗部分）：

```
[when][]小于或等于=<=
[when][]是===
[when][]年龄=age
[when][]名字=name
```

```
[when][]- {field:\w*} ={field}
[when]学生办找一个人=$p:Person()
[then]学校决定将你安排到"{className}"=$p.setClassName("{className}");
```

创建规则文件 personDslr.dslr，目录为 rulesTwo/isDSL，其内容为（注意加粗部分）：

```
package rulesTwo.isDSL;
import com.pojo.Person;
expander personDSL.dsl
rule "dslTest"
    when
        学生办找一个人
            - 年龄 小于或等于 30
            - 名字 是 "张三"
    then
        学校决定将你安排到"三班"
end
```

修改 kmodule.xml 配置文件，并添加如下配置：

```xml
<kbase name="isDSL" packages="rulesTwo.isDSL">
    <ksession name="isDSL"/>
</kbase>
```

创建 RulesDSL.java 文件，其目录为 comTwo.DSL，其内容为：

```java
package comTwo.DSL;

import com.pojo.Person;
import org.junit.Test;
import org.kie.api.KieServices;
import org.kie.api.runtime.KieContainer;
import org.kie.api.runtime.KieSession;

public class RulesDSL {
    @Test
    public void testIsDSL() {
        KieServices kss = KieServices.Factory.get();
        KieContainer kc = kss.getKieClasspathContainer();
        KieSession ks = kc.newKieSession("isDSL");
        Person person=new Person();
        person.setName("张三");
        person.setAge(20);
        ks.insert(person);
        int count = ks.fireAllRules();
        System.out.println("总执行了" + count + "条规则");
        System.out.println("总执行了" + person.getClassName() + "条规则");
        ks.dispose();
    }
}
```

执行 testIsDSL() 方法，结果如图 10-14 所示。

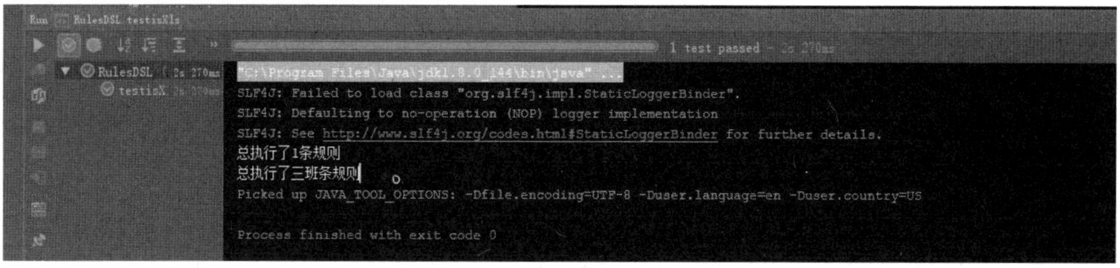

图10-14 领域语言的测试效果

通过上述的例子，相信读者对 DSL 领域语言有了一个清晰的认识，DSL 文件被理解为模板是因为它可以应用在多个 *.dslr 文件中。与之前讲的 drl 是不一样的，DSL 是通过业务人员能看明白的文字进行判断，并在 LHS 和 RHS 部分中体现出来。规则文件 drl 中每一个规则体的 then 部分都最好以 ";" 结尾，在 DSL 文件中也不例外，因为 [then] 正是规则文件中 RHS 部分的映射。

DSL 表达式（条件部分，等号左边）作为业务人员使用的模板在 dslr 文件中，它所对应的值（等号右边）是可以通过正则表达式进行匹配的。但要注意，这里使用正则表达式时要通过 "\" 进行转意。在 DSL 中可以使用正则表达式，为编辑过程提供了极大的方便。如何更好地应用正则表达式在自然语言中，是接下来讨论的问题。为保证代码的完整性，创建 regular 文件夹，并复制上例中的两个领域语言文件，编辑 personDslr.dslr 文件，目录为 rulesTwo/isDSL/regular，其内容为（注意加粗部分）：

```
package rulesTwo.isDSL.regular;
import com.pojo.Person;
expander personDSL.dsl

rule "dslTest"
    when
        学生办找一个人
            - 年龄 小于或等于 30
            - 名字 是 "张三"
    then
        学校决定将你安排到"四班"
        打印出 33ww33
end
```

编辑 personDSL.dsl 文件，目录为 rulesTwo/isDSL/regular，其内容为（注意加粗部分）：

```
[when][]小于或等于=<=
[when][]是===
[when]学生办找一个人=$p:Person()
[when][]名字=name
[when][]年龄=age
[when][]学名=name
[when][]- {operator} {value:\d*}={operator} {value}
[then]学校决定将你安排到"{className}"=$p.setClassName("{className}");
[then]打印出 {names:[_a-zA-Z0-9^"]+}=System.out.println("{names}");
```

找一个在线正则检验工具进行一次验证，如果不能确定自己写的正则表达式正确，可以进行验证，如图 10-15 所示。

图10-15　验证正则表达式的正确性

修改 kmodule.xml 配置文件，并添加如下配置：

```xml
<kbase name="regular" packages="rulesTwo.isDSL.regular">
    <ksession name="regular"/>
</kbase>
```

编辑 RulesDSL.java 文件，并添加 testIsDSLregular() 方法，其内容为：

```java
@Test
public void testIsDSLregular() {
    KieServices kss = KieServices.Factory.get();
    KieContainer kc = kss.getKieClasspathContainer();
    KieSession ks = kc.newKieSession("regular");
    Person person=new Person();
    person.setName("张三");
    person.setAge(20);
    ks.insert(person);
    int count = ks.fireAllRules();
    System.out.println("总执行了" + count + "条规则");
    System.out.println("总执行了" + person.getClassName() + "条规则");
    ks.dispose();
}
```

执行 testIsDSLregular() 方法，结果如图 10-16 所示。

图10-16　使用正则表达式的DSL效果

再次编辑 personDSL.dsl 文件，目录为 rulesTwo/isDSL/regular，其内容为（注意加粗部分）：

```
[when][]小于或等于=<=
[when][]是===
[when]学生办找一个人=$p:Person()
[when][]名字=name
[when][]年龄=age
[when][]学名=name
[when][]- {operator} {value:\d*}={operator} {value}
[then]学校决定将你安排到"{className}"=$p.setClassName("{className}");
[then]打印出 {names:[_a-zA-Z0-9^"]}=System.out.println("{names}");
```

执行 testIsDSLregular() 方法，结果如图 10-17 所示。

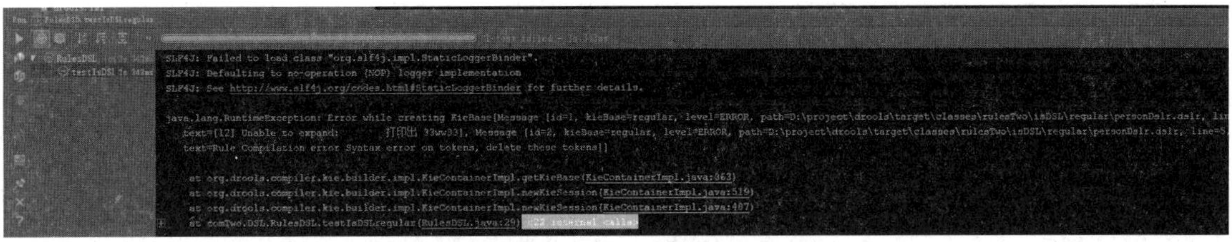

图10-17　错误使用正则表达式的DSL

通过图 10-17 中的报错信息，虽然正则表达式是正确的，但该正则是用来匹配单个的字母或数字的，而测试用例中是将"33ww33"整体作为一个参数，这时的正则表达式就应该把该参数作为一个值进行正则匹配。

为了更方便地进行值的正则匹配，通常的做法是写一个比较全面的可匹配任何值的正则表达式，当然这并不是绝对的，具体设计还需要根据具体业务，如指定参数是数字或字母，或限制参数长度。在实际项目中使用时，需要先定义 dsl 文件，该文件与模板功能相似，并为业务人员提供只有 dsl 文件中所展现的值进行匹配，否则规则会报错。举一个相对通用的 dsl 文件内容，目的是通过正则表达式匹配出更多的参数。

```
[when][]- {field:[\u4e00-\u9fa5_a-zA-Z0-9^"]+} {operator} {value:[\u4e00-\u9fa5_a-zA-Z0-9^"]+}={field}
[then][]1 {field:[\u4e00-\u9fa5_a-zA-Z0-9^"]+} {operator} {value:[\u4e00-\u9fa5_a-zA-Z0-9^"]+} =System.out.println("{field} +++++"); System.out.println("{operator} ++++"); System.out.println({value});
```

使用 GUI 编辑 DSL，或者用文本编辑，[作用域]项目在每个映射行的开始。它指示该语句或文字是否应用到 LHS、RHS，或者是一个关键字。有效的值有 [condition][consequence] 和 [keyword]（[when] 和 [then] 分别与 [condition] 和 [consequence] 相同）。当使用 [keyword] 时，它可以映射语言的任何关键字，例如，"rule"或"end"可以设置成其他关键字来使用。通常，它只用于希望使用非英语规则语言时（最好将它影射成单个文字）。

下面通过一个例子来说明。创建 scope 文件夹，将第一个测试用例 copy 到该目录下，编辑

personDslr.dslr 文件，目录为 rulesTwo/isDSL/scope，其内容为（注意加粗部分）：

```
package rulesTwo.isDSL.scope;
import com.pojo.Person;
expander personDSL.dsl
规则开始 "dslTest"
    when
        学生办找一个人
            - 年龄 小于或等于 30
            - 名字 是 "张三"
    then
        学校决定将你安排到"11班"
规则结束
```

编辑 personDSL.dsl 文件，其目录为 rulesTwo/isDSL/scope，内容为（注意加粗部分）：

```
[keyword]规则开始=rule
[keyword]规则结束=end
[when][]小于或等于=<=
[when][]是===
[when][]年龄=age
[when][]名字=name
[when][]- {field:\w*} ={field}
[when]学生办找一个人=$p:Person()
[then]学校决定将你安排到"{className}"=$p.setClassName("{className}");
```

修改 kmodule.xml 配置文件，并添加如下配置：

```
<kbase name="scope" packages="rulesTwo.isDSL.scope">
    <ksession name="scope"/>
</kbase>
```

编辑 RulesDSL.java 文件，并添加 testIsDSLscope() 方法，其内容为：

```
@Test
public void testIsDSLscope() {
    KieServices kss = KieServices.Factory.get();
    KieContainer kc = kss.getKieClasspathContainer();
    KieSession ks = kc.newKieSession("scope");
    Person person=new Person();
    person.setName("张三");
    person.setAge(20);
    ks.insert(person);
    int count = ks.fireAllRules();
    System.out.println("总执行了" + count + "条规则");
    System.out.println("总执行了" + person.getClassName() + "条规则");
    ks.dispose();
}
```

执行 testIsDSLscope() 方法，结果如图 10-18 所示。

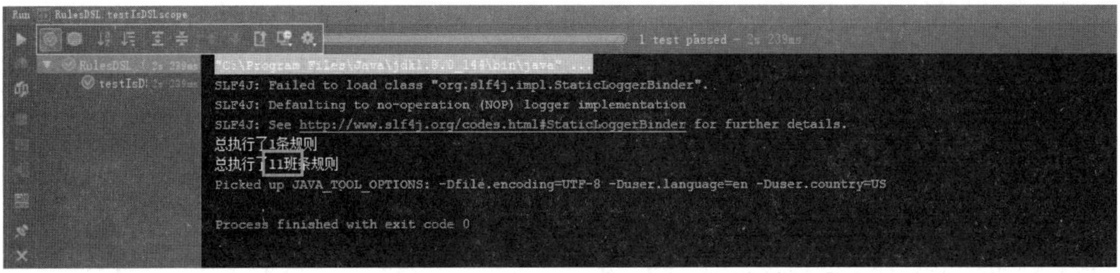

图 10-18　作用域测试效果

有一些 dsl 具有特殊性，主要是根据关键字作用功能进行区别，如下面的 dsl 代码：

```
[keyword]规则开始=rule
[keyword]规则结束=end
[when][]小于或等于=<=
[when][]是===
[when][]年龄=age
[when][]名字=name
[when][]- {field:\w*} ={field}
[when]学生办找一个人=$p:Person()
[then]名字=$p.getName();
[then]学校决定将你安排到"{className}"=$p.setClassName("{className}");
```

上述代码的两个名称是不冲突的，相反，如果在规则体 RHS 部分设置其他在 dsl 中没有映射的参数，就会报无法解析该类型，如图 10-19 所示。

图 10-19　作用域的错误使用

作用域是为了限定等号左边中引用变量是否符合该正则表达式。不管是在 [when] 还是 [then] 中，其名称是不冲突的，在 dsl 文件中可以看到名称已经被替换成 name 和 $p.getName()。在使用一些关键符号时，这里用官方例子"if (foo) {doSomething();}"说明，如果直接写，DSL 会认为是等号左边的定义，if 语法就会出错，这时需要使用"\"进行转意。如以下代码：

```
[then]do something= if (foo) \{ doSomething(); \}
```

DSL 领域语言理解起来很简单，就是在 DSL 文件中做映射，从而实现让业务人员可以认识和操作的自然语言。dslr 文件与 drl 很相似，只不过内容可由业务人员自己定义，在 dslr 文件中有一个比较重要的文件，即 expander personDSL.dsl。

强调"-"的使用，"-"可以看作是 Person 类型中属性添加条件约束，"-"年龄小于或等于 30 其实就是 drl 文件中所写的 "Person(age<=30)"。

DSL 文件中还可以使用逻辑判断，内容为：

```
[when][]小于或等于=<
[when][]小于=<
[when][]大于或等于==>
[when][]大于=>
[when][]等于===
[when][]并=&&
[when][]或=||
[when][]或者=or
[when][]并且=and
[when][]- {field:[\u4e00-\u9fa5_a-zA-Z0-9^"]+} {operator} {value:[\
u4e00-\u9fa5_a-zA-Z0-9^"]+}={field} {operator} {value}
```

分析 DSL 的执行步骤如下。

（1）解析器将获得指定的表达式，并提取出现在输入中的 {colour}（令牌）相匹配的值。

（2）利用映射等号相应的 {colour}。

（3）插入字符串替换任何被整个表达式匹配线的 DSL 规则文件。

简单来说，就是将等号左边的 {colour} 作为模板参数，传值给等号右边 {colour} 的一个值的解析过程，等号左边的整体自然语言表达式是由等号右边规则语言表达式提供解析和匹配的。

10.3 规则模板

规则引擎中，Drools 提供了一个规则模板的概念，规则模板是指规则条件比较值是可变的，且可生成多个规则进行规则调用。规则模板可以分为两种：官方提到的以 drt 扩展名 +xls(源数据) 的方式和 API 模板赋值方式。

第一种：官方提供的例子，通过规则模板 drt 文件 + 模板数据源 xls 文件，创建一个 ExampleCheese.xls 文件，目录为 rulesTwo/isDrt，其内容如图 10-20 所示。

	A	B	C	D	E	F
1						
2		年龄	className	log		
3		50	一班	1		
4		40	二班	2		
5		30	三班	3		
6		20	四班	4		
7						
8						

图10-20 规则模板数据源

图 10-20 中的颜色与边框是非必需的，位置也是随意写的，但必须是相邻行、列。创建 Cheese.drt 文件，目录为 rulesTwo/isDrt，其内容为：

```
template header
age
className

package rulesTwo.isDrt;
import com.pojo.Person;

template "cheesefans"

rule "Cheese fans_@{row.rowNumber}"
no-loop true
    when
        $p: Person(age == @{age})
    then
        $p.setClassName("@{className}");
        update($p);
end
end template
```

修改 kmodule.xml 配置文件，并添加如下配置（注意加粗部分）：

```
<kbase name="isDrt" packages="rulesTwo.isDrt">
    <ruleTemplate dtable="rulesTwo/isDrt/ExampleCheese.xls"
                  template="rulesTwo/isDrt/Cheese.drt"
                  row="3" col="2"/>
    <ksession name="isDrt"/>
</kbase>
```

创建 RulesDRT.java 文件，目录为 comTwo.DRT，其内容为：

```
package comTwo.DRT;

import com.pojo.Person;
import org.junit.Test;
import org.kie.api.KieServices;
import org.kie.api.runtime.KieContainer;
import org.kie.api.runtime.KieSession;
```

```java
public class RulesDRT {
    @Test
    public void testIsDRT() {
        KieServices kss = KieServices.Factory.get();
        KieContainer kc = kss.getKieClasspathContainer();
        KieSession ks = kc.newKieSession("isDrt");
        Person person=new Person();
        person.setName("张三");
        person.setAge(50);
        ks.insert(person);
        int count = ks.fireAllRules();
        System.out.println("总执行了" + count + "条规则");
        System.out.println("总执行了" + person.getClassName() + "条规则");
        ks.dispose();
    }
}
```

执行testIsDRT()方法，结果如图10-21所示。

图10-21　规则模板的使用效果

通过图10-21中的结果，成功将xls中的数据输出到控制台。这里的Cheese.drt文件是规则模板文件，为了方便理解。

```
template header
age
className
template "cheesefans"
@{row.rowNumber}
@{age}
"@{className}"
end template
```

其他代码与之前所讲的drl文件是相同的，规则模板文件以template header开头，以end template结尾。age表示Excel文件中的列，这里指如图10-20中的列。当前测试用例中，命名第一列为"age"。className功能与age相似，不能与age同行。"template "cheesefans""为模板名称，是唯一的，关键字template表示一个规则模板的开始，一个规则模板中可以有多个模板。"@{row.rowNumber}"是模板中的一个函数，用来让规则名唯一。"@{age}""@{className}""是对应age、className的，是最终获取数据源中参数值的，这里@{className}加了引号，是非常重要的，规则在进行推断或给出结论时所操作的Fact事实对象的属性类型必须是对应的，"@{age}"是数字，

"@{className}"是字符串，所以在使用的时候需要注意。而 xls 中的列数据要与规则模板的类型包括 Java 的类型保持一致，否则就会报错。但要注意的一点，如果用的是 String 类型，在模板中引用必须要加引号（""），由于"@{age}"是数字类型，所以可以省略不加。

ExampleCheese.xls 文件是规则模板中的数据源文件，将指定行、列开始的数据取出赋值给规则模板中的模板数据。图 10-20 中数据从 B3 开始，而 B2、C2 只是注释行。特意将 age 对应的行写成年龄，将 className 对应的行写成 className，是为了更容易理解，这只是注释部分，而且是必填信息。规则模板数据源文件其实很好理解，重要的是对应的值类型应转换正确，一定要防止类型转换错误。

Kmodule.xml 文件中规则模板的配置比较特殊，其内容为：

```
<ruleTemplate dtable="rulesTwo/isDrt/ExampleCheese.xls"
              template="rulesTwo/isDrt/Cheese.drt"
              row="3" col="2"/>
```

从 row/col 开始的行、列起，每一行都是一条规则，更准确地说，就是将 xls 中的数据填写到模板文件中。但编写 xls 有几个细节要注意，下面就对配置文件进行分析。

（1）xls 中的具体行数据是根据配置文件 row 值来确定的，若设置为 row="1"，则表示从第一行开始取数据，若设置成 row="3"，则从第三行开始取数据，这里设置的是第三行，所以是从第三行开始的，当调用规则模板时，如果 Person 的 age 为 50，就会输出一班。

（2）xls 中的具体列数据是根据配置文件 col 值来确定的，若设置为 col="1"，则表示从第一列开始取数据，若设置成 col="2"，则是从第二列开始取数据。这里设置 col="2"，所以是从第二列开始的，当调用规则模板时，只要匹配成功，就会输出后面的班级名称了。

（3）为什么只取到了 C3 的数据，而 D3 的 log 行为却没有取到呢，是在 drt 中只设置了两个占位符的原因，可以将 row/col 看作坐标的起点，从起点开始横向读取数据，当占位符用完时，再转到起点的下一行开始获取数据。

（4）row/col 表示行/列，如果引用 xls，该值就不能省略。

具体的 Java 调用，其实与执行规则是一样的，只是在配置文件中是不同的，ruleTemplate 规则模板，dtable 引用 xls 文件表示二维表。template 为具体的模板文件，上述代码中设置的是 row="3",col="2"，就表示从 B3 单元格取数据，第一行为一条规则，直到没有行为止，因为这里有 4 行数据，所以就形成了如下代码所示的 4 个规则。

```
package rulesTwo.isDrt;
import com.pojo.Person;

rule "Cheese fans_1"
no-loop true
    when
        $p: Person(age == 50)
    then
```

```
        $p.setClassName("一班");
        update($p);
end

rule "Cheese fans_2"
no-loop true
    when
        $p: Person(age == 40)
    then
        $p.setClassName("二班");
        update($p);
end

rule "Cheese fans_3"
no-loop true
    when
        $p: Person(age == 40)
    then
        $p.setClassName("三班");
        update($p);
end

rule "Cheese fans_4"
no-loop true
    when
        $p: Person(age == 20)
    then
        $p.setClassName("四班");
        update($p);
end
```

  这个规则模板也算是决策表的一种，官方的说法与决策表（不一定需要电子表格）相关的是"规则模板"（drools-templates 模块中）。它们使用任何表格式的数据源作为一个规则数据源，并将数据依次填入模板中生成多条规则。这让用户使用起来更方便，而且实例在现有的数据库中管理，代价是预先开发产生规则的模板。利用规则模板，不仅可以使数据与规则分离，而且可以用这种决策表（规则模板 +Excel）的方式做如下几点。

  （1）存储数据在数据库中（或任何其他格式）。
  （2）根据数据的值有条件地产生规则。
  （3）规则的任何部分使用数据（如条件运算符、类名、属性名）。
  （4）在相同的数据上运行不同的模板。

  以上就是第一种方式的规则模板内容，规则模板的官方说明也基本是通过 Excel 的方式进行说明的，可能官方认为这种方式更容易理解。相比第一种方式而言，程序员会更喜欢第二种。第二种方式是在阅读源码时找到的，规则模板文件还是之前的文件，编辑 RulesDRT.java 文件，并添加 testIsDRTAPI() 方法，其内容为（注意加粗部分）：

```java
@Test
public void testIsDRTAPI() {
    ObjectDataCompiler converter = new ObjectDataCompiler();
    //赋值给模板属性
    Person person = new Person();
    person.setAge(50);
    person.setClassName("一班");
    Collection<Person> cfl = new ArrayList<>();
    cfl.add(person);//每add一次，就代码一条规则
    InputStream dis = null;
    try {
        dis = ResourceFactory.newClassPathResource("rulesTwo/isDrt/Cheese.drt",
RulesDRT.class).getInputStream();
    } catch (IOException e) {
        e.printStackTrace();
    }
    String drl = converter.compile(cfl, dis);
    System.out.println(drl);
    KieHelper helper = new KieHelper();
    helper.addContent(drl, ResourceType.DRL);
    KieSession ksession = helper.build().newKieSession();
    Person ps = new Person();
    ps.setAge(50);
    person.setName("张三");
    ksession.insert(ps);
    int i = ksession.fireAllRules();
    System.out.println(ps.getClassName() + "    " + i + "次");
    ksession.dispose();
}
```

执行 testIsDRTAPI() 方法，结果如图 10-22 所示。

图10-22　规则模板的第二种使用

通过这种 API 的方式，可以对代码进行一个封装。

ObjectDataCompiler：调用规则模板的一个工具类。

Collection<Person> cfl：指给规则模板赋值的参数，一个元素就代表一个规则。

ResourceFactory：加载规则模板文件。

converter.compile(cfl, dis)：把规则文件与参数合并，生成 drl 文件。

规则模板可以传一些基本类型，至于集合或对象，则是传 toString() 方法，所以在模板中取不到相应的集合内容和地址。规则模板属性中，如果用的是 JavaBean，就要注意 JavaBean 的属性引用，调用的是 get 方法。例如，用 bName 和 bSex 作成员变量，在 Java 代码中生成的 get 方法是 getbSet() 和 getbName()，在属性中引用时会引用不到，所以得出的一个结论：JavaBean 属性做模板属性时，要注意生成的 get 方法，get 后的第一个字母必须大写。此外，在模板中，其实属性引用的是无参的有返回值的方法，也可以在模板中直接写 get 方法等。

规则模板是很强大的，本节又一次提到了占位符的概念，那是不是可以将其理解成或改造成动态规则呢？规则模板是可以这样使用的，类似在 API 中给模板的占位符赋值。在规则模板中也可以将条件赋值，这样就能实现动态规则。因此，使用规则模板是实现动态规则的方式之一。

10.4 规则流

Rule Flow 为 Drools 平台提供规则 + 流程的功能。一个业务流程或工作流都是通过一个流程图进行工作的，该流程图描述了一系列需要执行步骤的顺序。这使规则流程可以应对更多的复杂组合。Drools Flow 允许用户使用这些流程来定制、执行和监控（一部分）其业务逻辑。Drools Flow 流程的核心是以 JBPM 为标准的流程语言，可以在任何 Java 应用（作为一个简单的 Java 组件）或一个服务器环境的模式中运行。

所谓规则流，其实就是规则 + 流程，简单理解为规则是规则、流程是流程，两者可独立存在，通过代码就能很容易看出来。

规则流可以通过 Workbench 或本机 IDE 插件进行构建，当然也能使用 activiti，但默认的 activiti 所用到的是 Drools 5.x 版本。官方所推荐的也是 JBPM 标准。

讲述规则流就要先学习安装它的插件，这里选择的是 Eclipse，安装 Drools 的插件有两种方式。第一种是使用 link 文件（直接复制也可以，但有风险），在浏览器输入 "https://www.drools.org/" 并找到 Download 导航栏，找到 Drools and jBPM tools 插件并下载，如图 10-23 和图 10-24 所示。

第 10 章 Drools 高级用法

Name	Description	Download
Drools Engine	Drools Expert is the rule engine and Drools Fusion does complex event processing (CEP). Distribution zip contains binaries, examples, sources and javadocs.	Distribution ZIP
Drools and jBPM integration	Drools and jBPM integration with third party project like Spring. Distribution zip contains binaries, examples and sources.	Distribution ZIP
Drools Workbench	Drools Workbench is the web application and repository to govern Drools and jBPM assets. See documentation for details about installation.	WildFly 11 WAR EAP 7 WAR Tomcat 8 WAR
Drools and jBPM tools	Eclipse plugins and support for Drools, jBPM and Guvnor functionality. Distribution zip contains binaries and sources.	Distribution ZIP
KIE Execution Server	Standalone execution server that can be used to remotely execute rules using REST, JMS or Java interface. Distribution zip contains WAR files for all supported containers.	Distribution ZIP

图10-23　Eclipse插件下载

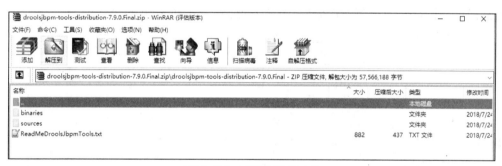

图10-24　下载压缩包内容

在图 10-24 中，文件内容分别是插件包、源码包、说明文档，然后在 Eclipse 的安装目录中创建 drools 文件夹，这个文件夹可自定义，本地的 Eclipse 安装目录是 D:\eclipse\eclipse，找到压缩包"binaries/org.drools.updatesite"目录，将目录下的 features 和 plugins 文件夹放在 drools 文件夹中，返回 Eclipse 的安装目录，新建 links 文件夹，并创建 drools.link 文件，其内容为：

```
path=D:/eclipse/eclipse/drools
```

然后重启 Eclipse，可以在插件管理中看到新建的文件夹，如图 10-25 所示。

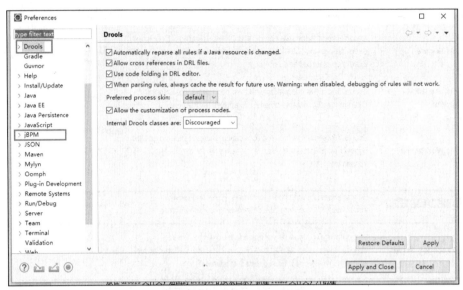

图10-25　重启后的Eclipse

任意打开一个规则文件，发现图标与内容都发现了变化，如图10-26所示。

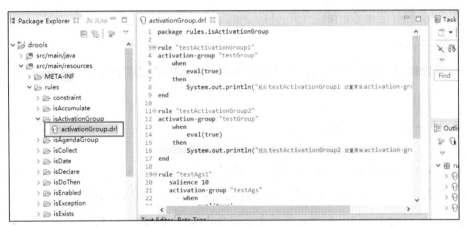

图10-26　插件安装后的效果

第二种方式是通过 Eclipse 的 Install new software 进行安装，具体步骤如下。

（1）通过 Eclipse 在线下载插件，如图 10-27 所示。

（2）在 Name 与 Location 中分别输入 "Drools" "http://download.jboss.org/drools/release/7.10.0.Final/org.drools.updatesite/"，一直单击 "下一步" 按钮直到重启 Eclipse。安装完成后，即可创建规则流程文件。

创建 ruleFlow.bpmn 文件，目录为 rulesTwo/isFlow，如图 10-28 所示。

第 10 章 Drools 高级用法

图10-27　通过Eclipse在线下载插件

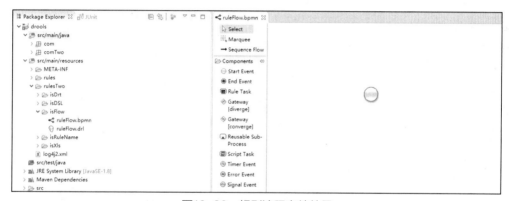

图10-28　规则流程文件效果

分别介绍在规则流中要使用的基本元素，如图 10-29 所示。

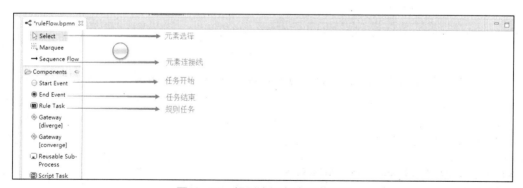

图10-29　规则流程文件元素说明

编辑 .bpmn 需要使用 properties 窗口，如图 10-30 所示。

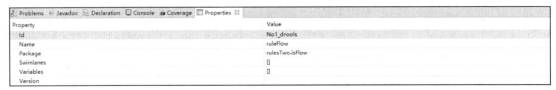

图10-30 properties设置

（1）不选中任何组件，全局配置，分别设置规则流程文件"IDNo1_drools,Package:rulesTwo.isFlow"，ID在当前规则库中的逻辑路径是唯一的，这点与规则文件rule name是一样的，package指规则流程文件的逻辑路径，与规则文件中的package功能一样，这里建议将该值与规则文件保持一致，效果如图10-31所示。

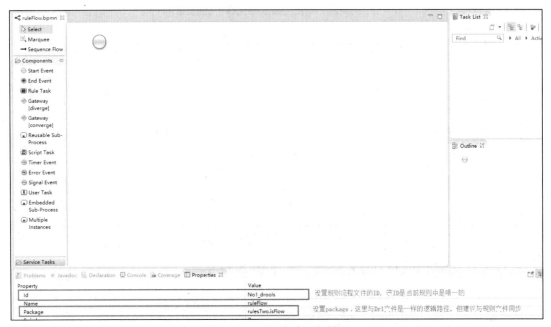

图10-31 设置全局属性

（2）创建规则文件ruleFlow.drl，目录为rulesTwo/isFlow，其内容为（注意加粗部分）：

```
package rulesTwo.isFlow

rule "规则流的第一个例子"
dialect "mvel"
ruleflow-group "No1_drools_RuleFlow"
    when

    then
        System.out.println("输出第一个规则流");
end

rule "没有使用规则流属性的规则"
```

```
    when
    then
        System.out.println("没有使用规则流属性的规则");
end
```

（3）配置任务元素组件，通过元素组件画出如图10-32所示的效果，并选中规则任务元素组件，设置"RuleFlowGroup：No1_drools_RuleFlow"选项，该值要与规则文件中的 ruleflow-group 保持一致。

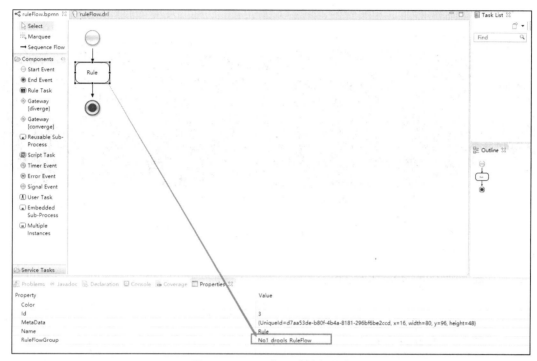

图10-32　设置规则流分组信息

修改 kmodule.xml 配置文件，并添加如下配置：

```
<kbase name="isFlow" packages="rulesTwo.isFlow">
    <ksession name="isFlow"/>
</kbase>
```

创建 RulesFlow.java 文件，目录为 comTwo.Flow，其内容为（注意加粗部分）：

```
package comTwo.Flow;

import org.junit.Test;
import org.kie.api.KieServices;
import org.kie.api.runtime.KieContainer;
import org.kie.api.runtime.KieSession;

public class RulesFlow {
    @Test
    public void testIsFlow() {
```

```
        KieServices kss = KieServices.Factory.get();
        KieContainer kc = kss.getKieClasspathContainer();
        KieSession ks = kc.newKieSession("isFlow");
        ks.startProcess("No1_drools");//流程启动
        ks.dispose();
    }
}
```

执行 testIsFlow() 方法，其结果如图 10-33 所示。

图10-33　执行规则流程异常

出现这个结果的原因是少引用了一个 jar 包，导致无法对流程文件进行编译，无法加载到规则库中，其解决方案如下。

编辑 pom.xml 文件，并添加如下内容：

```
<dependency>
    <groupId>org.jbpm</groupId>
    <artifactId>jbpm-test</artifactId>
    <version>${drools.version}</version>
</dependency>
```

再次执行 testIsFlow() 方法，结果如图 10-34 所示。

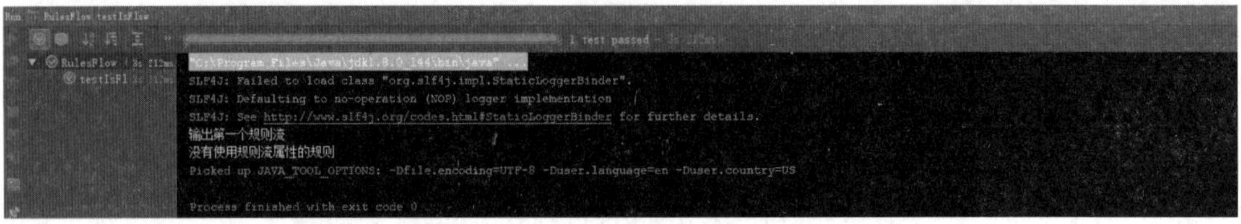

图10-34　规则流执行成功的效果

版本差异性对比，测试 Drools 6.4 版本时，如果不在执行调用规则代码中添加如下代码，那么规则是不会执行的。但在 Drools 7.10 版本中是不需要添加如下代码的。

```
int count = ks.fireAllRules();//规则调用
System.out.println("总执行了"+count+"条规则");
```

Drools 还提供了一个 API 调用规则流的方式，编辑 RulesFlow.java 文件，并添加 testIsFlowAPI() 方法，其内容为（注意加粗部分）：

```
@Test
public  void testIsFlowAPI() throws IOException {
    Resource bpmn = 
ResourceFactory.newClassPathResource("rulesTwo/isFlow/ruleFlow.bpmn",
RulesFlow.class);
    Resource drl = 
ResourceFactory.newClassPathResource("rulesTwo/isFlow/ruleFlow.drl",Rules-
Flow.class);
    KieHelper helper = new KieHelper();
    helper.addResource(bpmn, ResourceType.BPMN2);
    helper.addResource(drl,ResourceType.DRL);
    KieSession ksession = helper.build().newKieSession();
    ksession.startProcess("No1_drools");
    ksession.dispose();
}
```

执行 testIsFlowAPI() 方法，结果与图 10-34 是一样的。

如果要让规则流程文件变得更美观，可以选择如图 10-35 所示的效果。

图10-35　美化规则流程文件

10.4.1　规则流中的事件

使用规则流时，需要对规则体中属性定义有一个深入的认知，通过测试用例分别对规则流属性的设置进行详细说明。

（1）在规则文件中设置两个规则体，ruleflow-group 参数相同。创建规则文件 flowNameSame.drl，目录为 rulesTwo/isFlow/flowNameSame，其内容为（注意加粗部分）：

```
package rulesTwo.isFlow.flowNameSame

rule "测试规则流参数相同规则1"
ruleflow-group "FlowNameSame"
    when
    then
        System.out.println(drools.getRule().getName());
```

```
end

rule "测试规则流参数相同规则2"
ruleflow-group "FlowNameSame"
    when
    then
        System.out.println(drools.getRule().getName());
end

rule "区别规则流参数相同的规则"
ruleflow-group "FlowNameNotSame"
    when
    then
        System.out.println(drools.getRule().getName());
end
```

创建规则文件 flowNameSame.bpmn，目录为 rulesTwo/isFlow/flowNameSame，其效果如图 10-36 所示。

图10-36　设置规则流程全局属性

设置元素组件并添加 RuleFlowGroup：FlowNameSame，效果如图 10-37 所示。

修改 kmodule.xml 配置文件，并添加如下配置：

```xml
<kbase name="flowNameSame" packages="rulesTwo.isFlow.flowNameSame">
    <ksession name="flowNameSame"/>
</kbase>
```

编辑 RulesFlow.java 文件，添加 flowNameSame() 方法，其内容为（注意加粗部分）：

```java
@Test
public void flowNameSame() {
    KieServices kss = KieServices.Factory.get();
    KieContainer kc = kss.getKieClasspathContainer();
    KieSession ks = kc.newKieSession("flowNameSame");
    ks.startProcess("flowNameSame");//流程启动
    ks.dispose();
}
```

执行flowNameSame()方法，结果如图10-38所示。

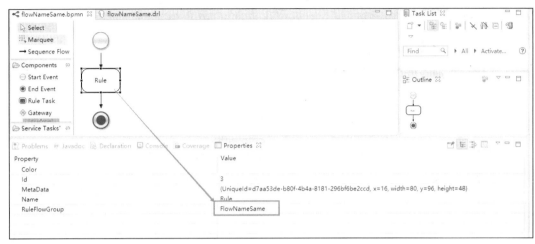

图10-37　设置规则流程属性

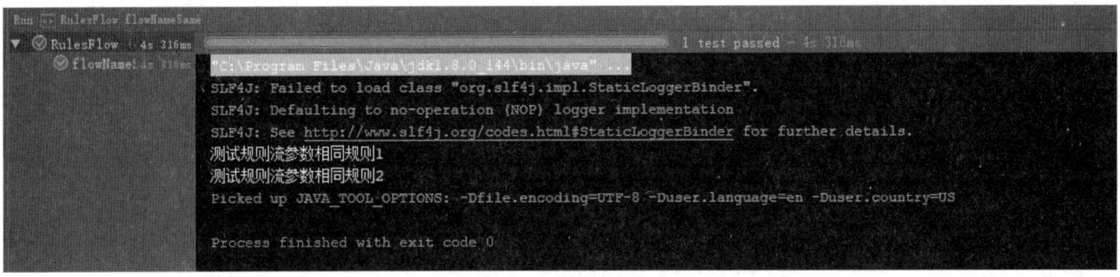

图10-38　规则流测试结果

总结：当流程文件元素组件只有一个，多个规则体属性 ruleflow-group 相同，且流程文件元素组件的 RuleFlowGroup 属性与规则体的属性相同时，则会执行满足条件的规则。

（2）规则文件中设置两个规则体，ruleflow-group 参数相同。如设置多个规则元素组件，应使用第一种情况的规则文件，因为 ruleflow-group 参数相同。

```
package rulesTwo.isFlow.flowNameSame

rule "测试规则流参数相同规则1"
ruleflow-group "FlowNameSame"
    when
    then
        System.out.println(drools.getRule().getName());
end

rule "测试规则流参数相同规则2"
ruleflow-group "FlowNameSame"
    when
```

```
        then
                System.out.println(drools.getRule().getName());
end

rule "区别规则流参数相同的规则"
ruleflow-group "FlowNameNotSame"
        when
        then
                System.out.println(drools.getRule().getName());
end
```

创建规则文件 flowNameSames.bpmn，目录为 rulesTwo/isFlow/flowNameSame，效果如图 10-39 所示，与 flowNameSame.bpmn 比较只多了一个规则元素组件，其他是一致的。

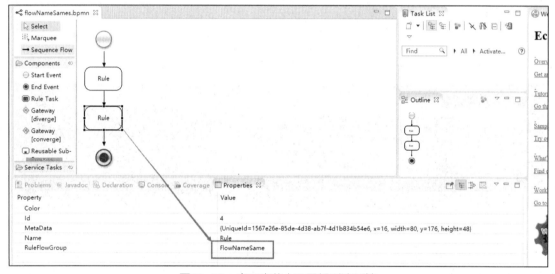

图10-39　多元素节点配置规则流属性

编辑 RulesFlow.java 文件，添加 flowNameSame() 方法，其内容为（注意加粗部分）：

```
@Test
public void flowNameSames() throws IOException {
    Resource bpmn =
ResourceFactory.newClassPathResource("rulesTwo/isFlow/flowNameSame/
flowNameSames.bpmn", RulesFlow.class);
    Resource drl =
ResourceFactory.newClassPathResource("rulesTwo/isFlow/flowNameSame/
flowNameSame.drl",RulesFlow.class);
    KieHelper helper = new KieHelper();
    helper.addResource(bpmn, ResourceType.BPMN2);
    helper.addResource(drl,ResourceType.DRL);
    KieSession ksession = helper.build().newKieSession();
    ksession.startProcess("flowNameSame");
    ksession.dispose();
}
```

执行 flowNameSames() 方法，结果与图 10-39 中是一样的。

总结：当流程文件有多个规则元素组件，该规则元素组件的 RuleFlowGroup 相同，多个规则体属性 ruleflow-group 相同，且流程文件元素组件的 RuleFlowGroup 属性与规则体的属性相同时，则会执行满足条件的规则，并且只会执行一遍，不会因规则元素组件的增多而执行多次规则。但凡是规则流参数相同的规则体都会受 salience 的影响，这点需要格外注意。

（3）在规则文件中设置两个规则体，ruleflow-group 参数不相同。设置多个规则元素组件，规则元素组件 RuleFlowGroup 不相同。

创建规则文件 flowNameNotSame.drl，目录为 rulesTwo/isFlow/flowNameNotSame，其内容为（注意加粗部分）：

```
package rulesTwo.isFlow.flowNameNotSame

rule "测试规则流参数不相同规则1"
ruleflow-group "FlowNameNotSame1"
    when
    then
        System.out.println(drools.getRule().getName());
end

rule "测试规则流参数不相同规则2"
ruleflow-group "FlowNameNotSame2"
    when
    then
        System.out.println(drools.getRule().getName());
end
```

创建规则文件 flowNameNotSame.bpmn，目录为 rulesTwo/isFlow/flowNameNotSame，其效果如图 10-40 所示。

图10-40　设置规则流全局属性

分别为规则元素组件设置 RuleFlowGroup 值，内容包括 RuleFlowGroup：FlowNameNotSame1、RuleFlowGroup：FlowNameNotSame2。编辑 RulesFlow.java 文件，添加 flowNameNotSame() 方法，其内容为：

```
@Test
public  void flowNameNotSame() throws IOException {
    Resource bpmn = ResourceFactory.newClassPathResource("rulesTwo/isFlow/flowNameNotSame/flowNameNotSame.bpmn", RulesFlow.class);
    Resource drl = ResourceFactory.newClassPathResource("rulesTwo/isFlow/flowNameNotSame/flowNameNotSame.drl",RulesFlow.class);
    KieHelper helper = new KieHelper();
    helper.addResource(bpmn, ResourceType.BPMN2);
    helper.addResource(drl,ResourceType.DRL);
    KieSession ksession = helper.build().newKieSession();
    ksession.startProcess("flowNameNotSame");
    ksession.dispose();
}
```

执行flowNameNotSame()方法，结果如图10-41所示。

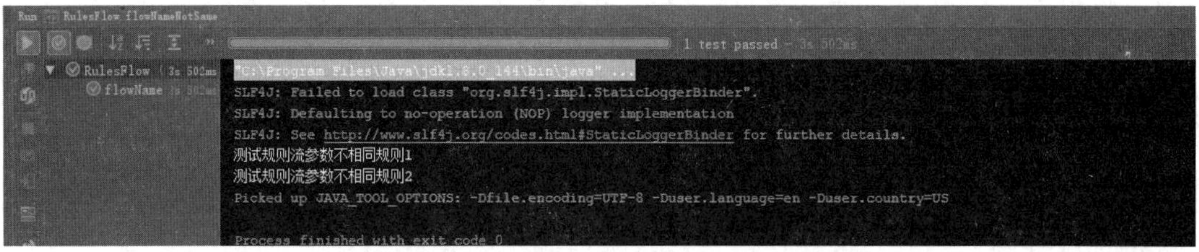

图10-41　规则流测试效果（1）

如果将流程文件中规则元素组件的 RuleFlowGroup 值颠倒一下，再次执行 flowNameNotSame() 方法，结果如图 10-42 所示。

图10-42　规则流测试效果（2）

总结：当流程文件有多个规则元素组件，并且该规则元素组件的 RuleFlowGroup 是不相同的，多个规则体属性 ruleflow-group 与规则元素组件一一对应时，则会执行满足条件的规则。就算在规则体中设置了 salience 属性，规则元素组件也不会受其影响而改变规则的执行顺序。

（4）流程文件中规则元素组件的 RuleFlowGroup 值与规则文件中规则体属性 ruleflow-group 没有匹配时，则规则不会被执行。

（5）当规则文件与流程文件设置的第一种情况相同时，为规则体设置相同的 activation-group 属性，具体分为两种。为保证代码的完整性，新建两个规则文件。

① 将两个规则设置为 activation-group 属性和规则流属性，第三个规则不设置属性。创建 flowNameSameACT1.drl 文件目录与 flowNameSame.drl 文件相同，其内容为：

```
package rulesTwo.isFlow.flowNameSame

rule "测试规则流参数相同规则1"
ruleflow-group "FlowNameSame"
activation-group "actSame"
salience 50
    when
    then
        System.out.println(drools.getRule().getName());
end

rule "测试规则流参数相同规则2"
ruleflow-group "FlowNameSame"
activation-group "actSame"
salience 90
    when
    then
        System.out.println(drools.getRule().getName());
end

rule "未设置规则流参数的规则"
    when
    then
        System.out.println(drools.getRule().getName());
end
```

编辑 RulesFlow.java 文件，添加 flowNameActSame() 方法，其内容为：

```
@Test
public  void flowNameActSame() throws IOException {
    Resource bpmn = ResourceFactory.newClassPathResource("rulesTwo/isFlow/flowNameSame/flowNameSame.bpmn", RulesFlow.class);
    Resource drl = ResourceFactory.newClassPathResource("rulesTwo/isFlow/flowNameSame/flowNameSameACT1.drl",RulesFlow.class);
    KieHelper helper = new KieHelper();
    helper.addResource(bpmn, ResourceType.BPMN2);
    helper.addResource(drl,ResourceType.DRL);
    KieSession ksession = helper.build().newKieSession();
    ksession.startProcess("flowNameSame");
    ksession.dispose();
}
```

执行 flowNameActSame() 方法，结果如图 10-43 所示。

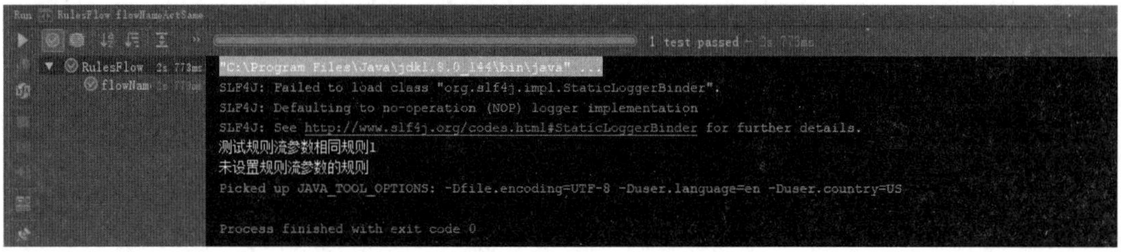

图10-43 规则流测试效果

② 两个规则设置了activation-group和"actsame",为没有设置规则流属性的设置优先级salience 10,创建flowNameSameACT2.drl文件,目录与flowNameSame.drl相同,其内容为:

```
package rulesTwo.isFlow.flowNameSame

rule "测试规则流参数相同规则1"
ruleflow-group "FlowNameSame"
activation-group "actSame"
    when
    then
        System.out.println(drools.getRule().getName());
end

rule "测试规则流参数相同规则2"
ruleflow-group "FlowNameSame"
activation-group "actSame"
    when
    then
        System.out.println(drools.getRule().getName());
end

rule "未设置规则流参数的规则"
salience 10
activation-group "actSame"
    when
    then
        System.out.println(drools.getRule().getName());
end
```

编辑RulesFlow.java文件,添加flowNameActSalienceSame()方法,其内容为:

```
@Test
public void flowNameActSalienceSame() throws IOException {
    Resource bpmn = ResourceFactory.newClassPathResource("rulesTwo/isFlow/flowNameSame/flowNameSame.bpmn", RulesFlow.class);
    Resource drl = ResourceFactory.newClassPathResource("rulesTwo/isFlow/flowNameSame/flowNameSameACT2.drl",RulesFlow.class);
    KieHelper helper = new KieHelper();
    helper.addResource(bpmn, ResourceType.BPMN2);
    helper.addResource(drl,ResourceType.DRL);
```

```
        KieSession ksession = helper.build().newKieSession();
        ksession.startProcess("flowNameSame");
        ksession.dispose();
}
```

执行 flowNameActSalienceSame() 方法,结果如图 10-44 所示。

图10-44　规则流测试效果

总结:规则流属性会受 activation-group 属性的影响,以(1)为例分析,结果会输出两个规则体的名称,但加上 activation-group 属性后,就会受分组属性的影响。因此,规则流也会受 salience 属性的影响。

(6)为规则体相同的设置 activation-group 属性。但规则流属性不相同时,结合(3)的流程文件,创建 flowNameSameACT2.drl 文件,目录与 flowNameSame.drl 相同,其内容为:

```
package rulesTwo.isFlow.flowNameNotSame

rule "测试规则流参数不相同规则1"
ruleflow-group "FlowNameNotSame1"
activation-group "actSame"
    when
    then
        System.out.println(drools.getRule().getName());
end

rule "测试规则流参数不相同规则2"
ruleflow-group "FlowNameNotSame2"
activation-group "actSame"
    when
    then
        System.out.println(drools.getRule().getName());
end

rule "未设置规则流参数的规则"
    when
    then
        System.out.println(drools.getRule().getName());
end
```

编辑 RulesFlow.java 文件,添加 flowNameNotSameAct() 方法,其内容为:

```
@Test
public void flowNameNotSameAct() throws IOException {
    Resource bpmn = 
ResourceFactory.newClassPathResource("rulesTwo/isFlow/flowNameNotSame/
flowNameNotSame.bpmn", RulesFlow.class);
    Resource drl = 
ResourceFactory.newClassPathResource("rulesTwo/isFlow/flowNameNotSame/
flowNameSameACT1.drl",RulesFlow.class);
    KieHelper helper = new KieHelper();
    helper.addResource(bpmn, ResourceType.BPMN2);
    helper.addResource(drl,ResourceType.DRL);
    KieSession ksession = helper.build().newKieSession();
    ksession.startProcess("flowNameNotSame");
    ksession.dispose();
}
```

执行 flowNameNotSameAct() 方法，结果如图 10-45 所示。

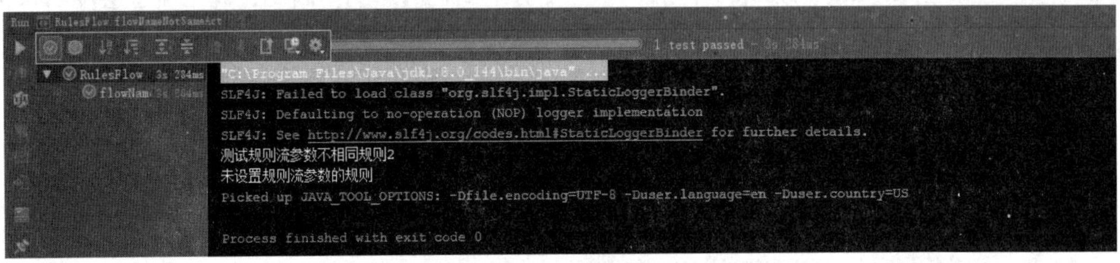

图10-45　规则流测试效果

总结：规则流属性会受 activation-group 属性的影响，以（3）为例分析，结果会输出两个规则体的名称，但加上 activation-group 属性后，就会受分组属性的影响。因为在流程文件规则元素组件上先为其执行了 FlowNameNotSame2，所以结果才会如图 10-45 所示。

综上所述，使用规则流属性时必须添加流程文件，规则体中规则流属性要与流程文件规则元素组件的 RuleFlowGroup 一一对应。规则流的执行会受其他属性的影响，并且还会受流程文件元素组件顺序的影响。

10.4.2　规则流中的Java代码

规则流中调用 Java 代码其实就是调用 Java 方法，这里与规则 function 函数很相似，但这并非是规则流属性的功能，而是流程元素中的知识点，如果对流程文件或工作流特别熟悉，可以跳过本节的内容。

创建 FlowToJava.java 文件，目录为 comTwo.flow，内容为流程中要调用 Java 代码需要加静态关键字，或者将对象创建出来，通过对象的变量名来使用对象中的方法。

```
package comTwo.flow;
```

```java
public class FlowToJava {
    public static void flowToJava01() {
        System.out.println("输出第一个通过流程调用Java方法");
    }
}
```

创建规则文件 flowToJava.drl，目录为 rulesTwo/isFlow/flowJava，其内容为：

```
package rulesTwo.isFlow.flowJava

rule "规则流调用Java方法第一个例子"
ruleflow-group "flowToJava"
    when
    then
        System.out.println(drools.getRule().getName());
end
```

创建流程文件 flowToJava.bpmn，如图 10-46 所示。

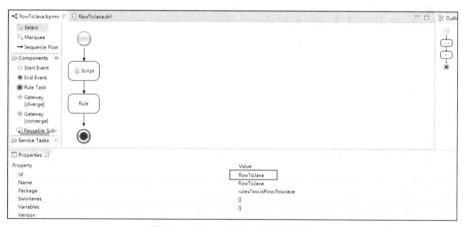

图10-46　设置规则流全局属性

设置规则元素组件的 RuleFlowGroup 值为 flowToJava，如图 10-47 所示。

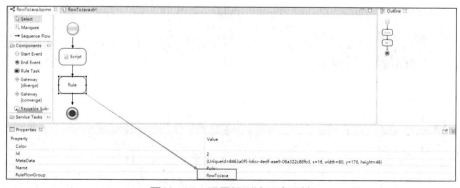

图10-47　设置规则流元素属性

设置 Script 元素组件，如图 10-48 和图 10-49 所示。

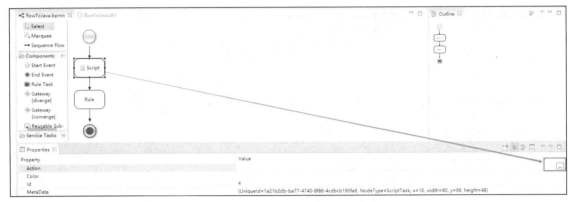

图10-48　设置Script元素过程（1）

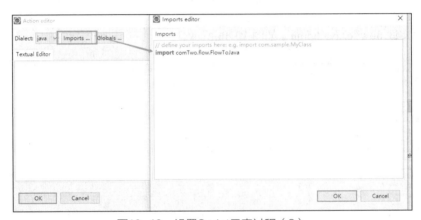

图10-49　设置Script元素过程（2）

设置调用Java的方法时要注意，需要通过类的全名"."方法输入，以"；"结束。如果不加"；"结束，则会报如图10-50所示的问题；如果不添加类的命名"."方法，则会报如图10-51所示的问题。

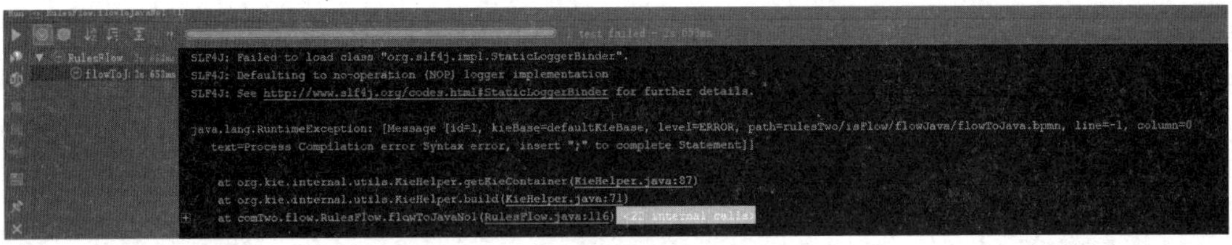

图10-50　不设置结束符号的异常

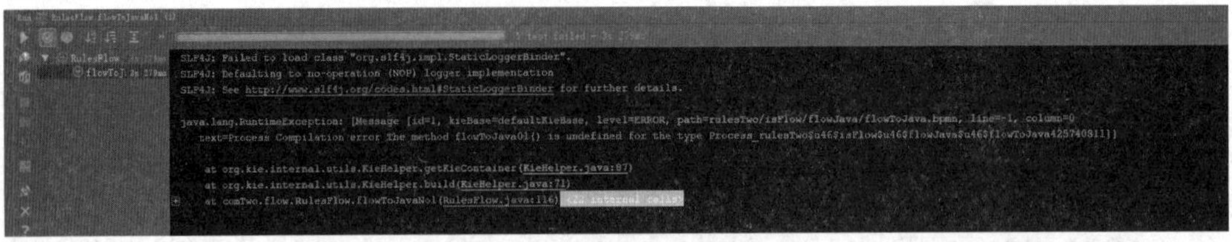

图10-51　不设置命令符号的异常

编辑 RulesFlow.java 文件，添加 flowToJavaNo1() 方法，其内容为：

```
@Test
public void flowToJavaNo1() throws IOException {
    Resource bpmn = ResourceFactory.newClassPathResource("rulesTwo/isFlow/flowJava/flowToJava.bpmn", RulesFlow.class);
    Resource drl = ResourceFactory.newClassPathResource("rulesTwo/isFlow/flowJava/flowToJava.drl", RulesFlow.class);
    KieHelper helper = new KieHelper();
    helper.addResource(bpmn, ResourceType.BPMN2);
    helper.addResource(drl, ResourceType.DRL);
    KieSession ksession = helper.build().newKieSession();
    ksession.startProcess("flowToJava");
    ksession.dispose();
}
```

执行 flowToJavaNo1() 方法，结果如图 10-52 所示。

图10-52　规则流调用Java代码的效果

还有一种非静态方法的调用，就是通常用到的单例模式。

创建 FlowToJavaTwo.java 文件，目录为 comTwo.flow，其内容为：

```
package comTwo.flow;

public class FlowToJavaTwo {
    private static final FlowToJavaTwo FLOW=new FlowToJavaTwo();
    public static FlowToJavaTwo getFlow(){
        return FLOW;
    }
    public   void flowToJava01() {
        System.out.println("第二种通过单例模式调用Java方法");
    }
}
```

编辑流程文件 flowToJava.bpmn，并添加新元素，如图 10-53 所示。

重复上述步骤，与上述不同的地方是，当前元素引用是通过"FlowToJavaTwo.getFlow().flowToJava01();"来完成的，如图 10-54 所示。

执行 flowToJavaNo1() 方法，结果如图 10-55 所示。

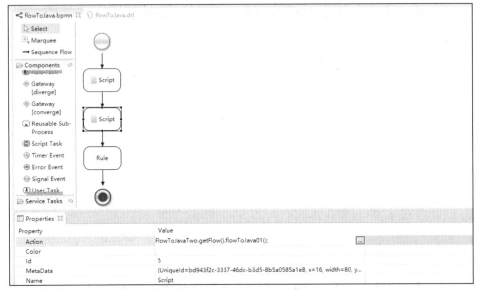

图10-53 添加新元素

图10-54 使用单例模式的配置

图10-55 配置单例模式的效果

既然有流程调用Java代码，那一定会涉及传参的调用方式，接下来对带有参数的调用Java代码进行测试。编辑FlowToJava.java文件，添加flowToJava02()方法，其内容为：

```
public static void flowToJava02(String value) {
    System.out.println("输出第一个通过流程调用Java带参数的方法 value=" + value);
}
```

新建 flowToJavaVal.bpmn 文件，目录为 rulesTwo/isFlow/flowJava，并设置 bpmn 文件的 ID 为 flowToJavaVal，如图 10-56 所示。

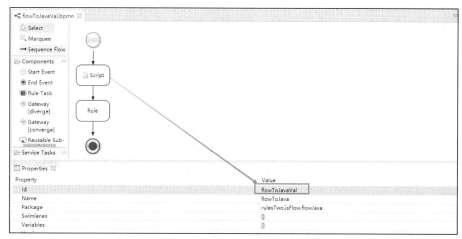

图10-56　配置流程元素

设置 Script 元素组件调用 Java 代码，如图 10-57 所示（注意在 () 中的 value）。

图10-57　设置流程元素访问的方法

选择流程文件空白区，为流程文件设置变量，该变量是全局性的，设置变量名如图 10-58 所示。

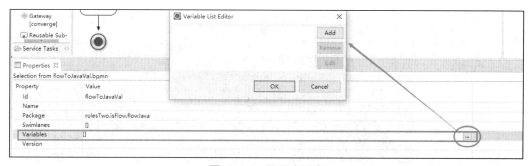

图10-58　设置变量名

设置全局变量的 Name，如图 10-59 所示，该 Name 的参数要与 Script 的方法参数名一致。

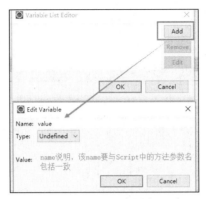

图10-59　设置全局变量

编辑 RulesFlow.java 文件，添加 flowToJavaNo1() 方法，其内容为（注意加粗部分）：

```
@Test
public void flowToJavaNo2() throws IOException {
    Resource bpmn = ResourceFactory.newClassPathResource("rulesTwo/isFlow/flowJava/flowToJavaVal.bpmn", RulesFlow.class);
    Resource drl = ResourceFactory.newClassPathResource("rulesTwo/isFlow/flowJava/flowToJava.drl", RulesFlow.class);
    KieHelper helper = new KieHelper();
    helper.addResource(bpmn, ResourceType.BPMN2);
    helper.addResource(drl, ResourceType.DRL);
    KieSession ksession = helper.build().newKieSession();
    Map map = new HashMap<>();
    map.put("value", "成功输出value");
    ksession.startProcess("flowToJavaVal", map);
    ksession.dispose();
}
```

执行 flowToJavaNo2() 方法，结果如图 10-60 所示。

图10-60　带参数的流程配置

规则流中调用 Java 代码其实是通过流程元素组件 Script 来完成的，通过 Properties 窗口对该元素组件进行编辑，通过关键字 import 类进行引用。需要注意一点，这里需要引用类的全路径，与 Java 引用类是一样的。而且引用一次，是全局性的，类似类文件中引用一样，可以在该类的任何方法中使用，在流程文件中也是一样的；使用流程传递参数，该变量也是全局性的，传参数时注意在 Script 元素组件中的方法参数要与 bpmn 设置 Variables 的 Name 一致，并且该参数还要与 Java 代码 map 的 Key 一致，类型也要保持一致，其代码为"KieSession.startProcess("flowToJava Val", map)"。

10.4.3 规则流中的网关

规则流中的网关也就是流程中的分支，了解它并深入学习是很有必要的，在实际应用中也是非常广泛的。网关可以分为两种：分离网关和合并网关。为了适应官方文档上的方式，先将 bpmn 的展现方式做一个变化，其步骤如图 10-61 和图 10-62 所示。

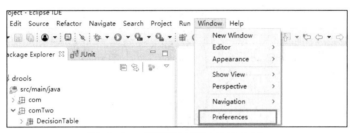

图10-61　设置插件显示效果（1）

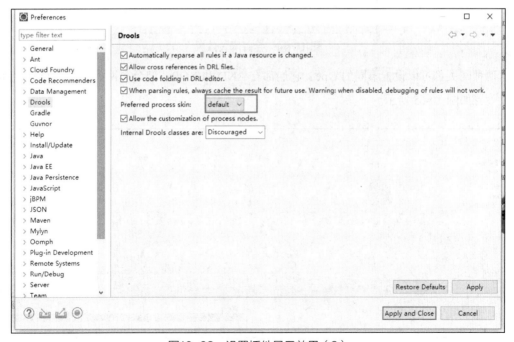

图10-62　设置插件显示效果（2）

分离网关效果如图 10-63 所示，该网关是用来做分支的。

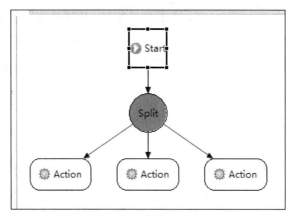

图10-63　分离网关效果

合并网关效果如图 10-64 所示，该网关是用来汇总的。

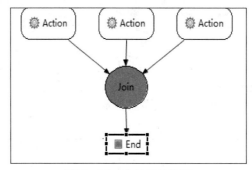

图10-64　合并网关效果

这两种网关都可以设置不同的 type，它们都有 AND 和 XOR，但分离网关多一种 OR。分离网关类型设置如图 10-65 所示。

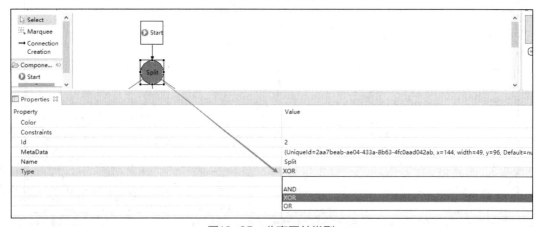

图10-65　分离网关类型

合并网关类型设置如图 10-66 所示。

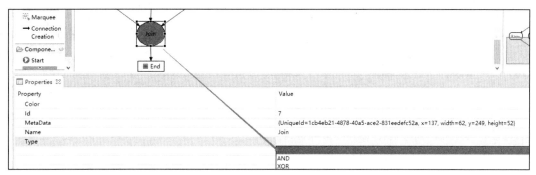

图10-66 合并网关类型

介绍完这两种网关的基本知识，下面对其类型分别进行测试。创建 separation.bpmn 文件，目录为 rulesTwo/isFlow/Gateway，如图 10-67 所示。

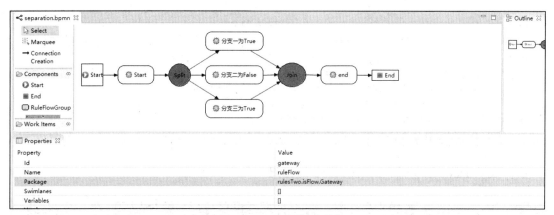

图10-67 设置完整测试流程

分离网关默认为 AND，但这里还是需要将 type 设置成 AND，如图 10-68 所示。

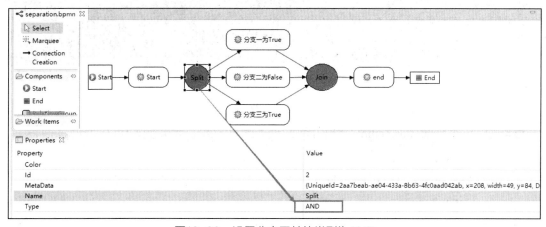

图10-68 设置分离网关的类型为AND

将合并网关的 type 设置成 AND，效果如图 10-69 所示。

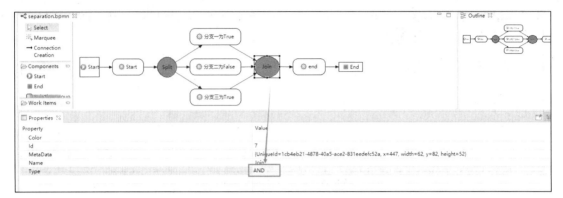

图10-69　设置合并网关的类型为AND

编辑 RulesFlow.java 文件，添加 flowGateway() 方法，其内容为（注意加粗部分）：

```
@Test
public void flowGateway() throws IOException {
    Resource bpmn = ResourceFactory.newClassPathResource("rulesTwo/isFlow/Gateway/separation.bpmn", RulesFlow.class);
    KieHelper helper = new KieHelper();
    helper.addResource(bpmn, ResourceType.BPMN2);
    KieSession ksession = helper.build().newKieSession();
    ksession.startProcess("gateway");
    ksession.dispose();
}
```

执行 flowGateway() 方法，结果如图 10-70 所示。

图10-70　规则流网关效果

当所有网关都为 AND 时，流程元素组件都将会被访问。

编辑规则文件 separation.bpmn，将合并网关 type 改为 XOR，效果如图 10-71 所示。

执行 flowGateway() 方法，结果如图 10-72 所示。

总结：当分离网关为 AND，合并网关为 XOR 时，只会执行其中一个流向，这点与规则属性 activation-group 很相似。

将分离网关的 type 设置成 XOR，效果如图 10-73 所示。

第 10 章 Drools 高级用法

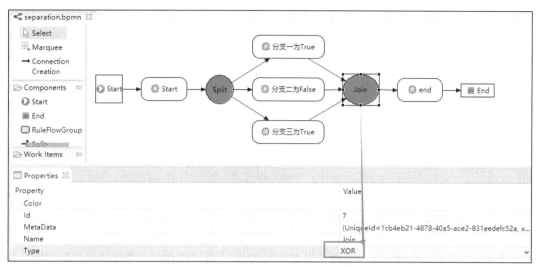

图10-71　设置合并网关类型为XOR

图10-72　规则流网关效果

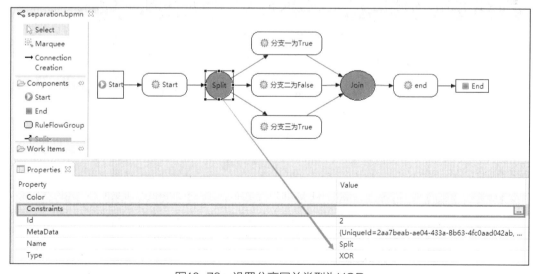

图10-73　设置分离网关类型为XOR

多了一个 Constraints 可选参数，单击该参数进行配置，如图 10-74 所示。

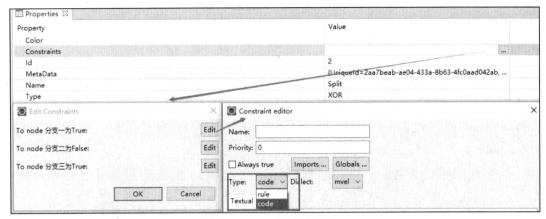

图10-74　设置具体参数

从图 10-74 中可以看到，将分离网关设置在 XOR 后，是可以在每一条连接线进行设置，并且设置该连接线又会分为以下几种情况。

（1）name：表示连接线的名称。

（2）priority：表示连接线的优先级，值越小优先级越高。

（3）Always true：表示永远为 true。

（4）Type：表示以什么方式进行判断结果，这里使用 rule 类型。

（5）Textual Editor：表示输入代码表达式的区域。

（6）Imports：表示可以引用其他类或方法。

（7）Globals：表示可以设置规则全局变量。

介绍完其作用后，分别对这 3 个分支进行设置，效果如图 10-75~ 图 10-79 所示。

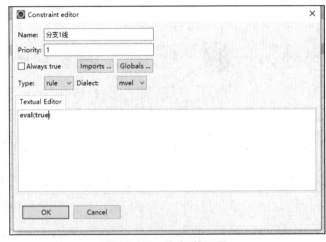

图10-75　分支1线设置

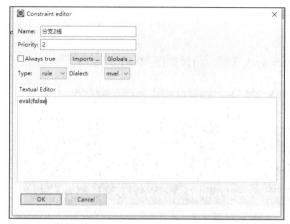
图10-76　分支2线设置

图10-77　分支3线设置

图10-78　3个分支编辑界面

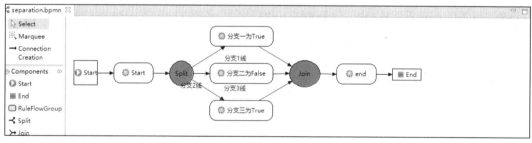
图10-79　3个分支连线效果

将合并网关的 type 设置成 AND，执行 flowGateway() 方法，结果如图 10-80 所示。

图10-80　分离网关XOR效果

总结：当分离网关为 XOR，合并网关为 AND 时，无论分离网关是全 True，还是有一个为 True 时，都不会执行到元素组件 End。

将合并网关的 type 设置成 XOR，执行 flowGateway() 方法，结果如图 10-81 所示。

图10-81　分离网关XOR效果

总结：当分离网关为 XOR，合并网关为 XOR 时，分支中有两个为 True，一个为 False 的情况下，只会有一个被执行，这是受优先级设置的影响。这证明只有分支为 True 的才会被执行并有且只有其中一个流向被执行。如果设置分支都为 False，则会报如图 10-82 所示的错误。

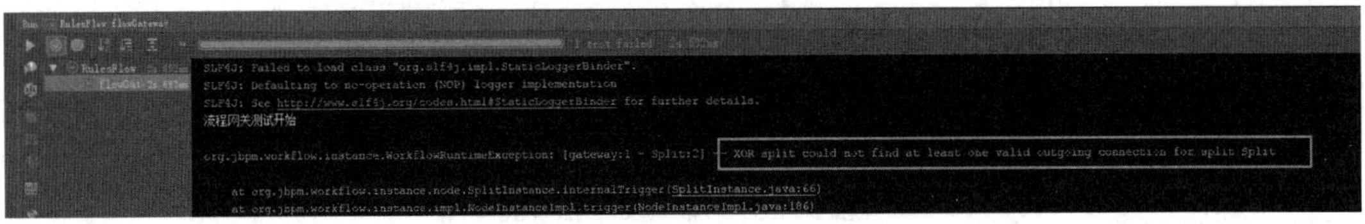

图10-82　设置错误的规则流网关

将分离网关的 type 设置成 OR，效果如图 10-83 所示，还原 3 个分支的初始化值。

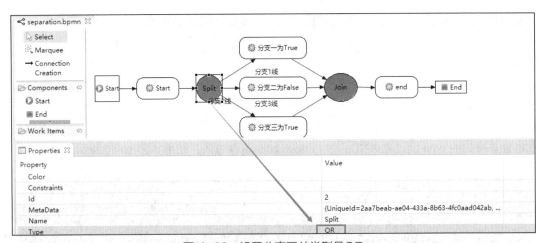

图10-83　设置分离网关类型是OR

将合并网关的 type 设置成 AND，执行 flowGateway() 方法，结果如图 10-84 所示。

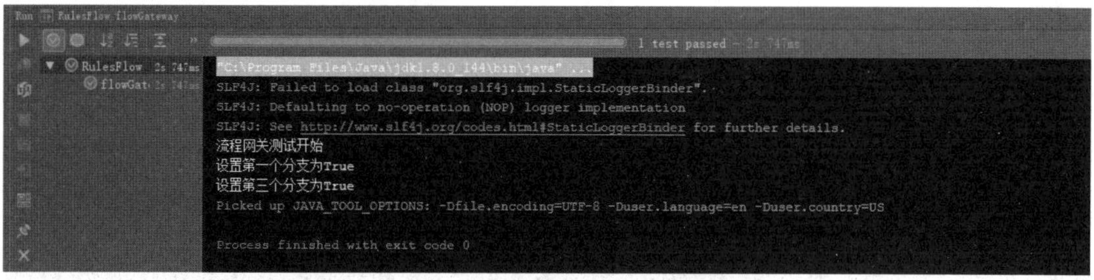

图10-84　设置分离网关OR效果

将所有的分支都设置为 eavl(True)，执行 flowGateway() 方法，结果如图 10-85 所示。

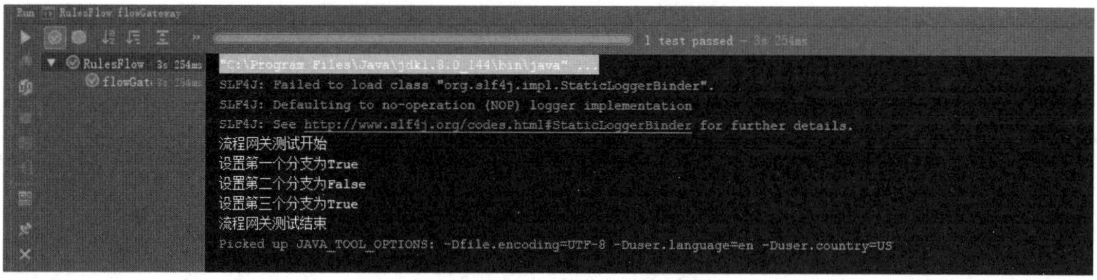

图10-85　分支设置eavl(True)

总结：当分离网关为 OR，合并网关为 AND 时，只有全部为 True 的情况下，所有流向才会被汇总，这也符合 AND 的逻辑与功能，如果分离网关所有分支为 False 时，则会报如图 10-86 所示的错误。

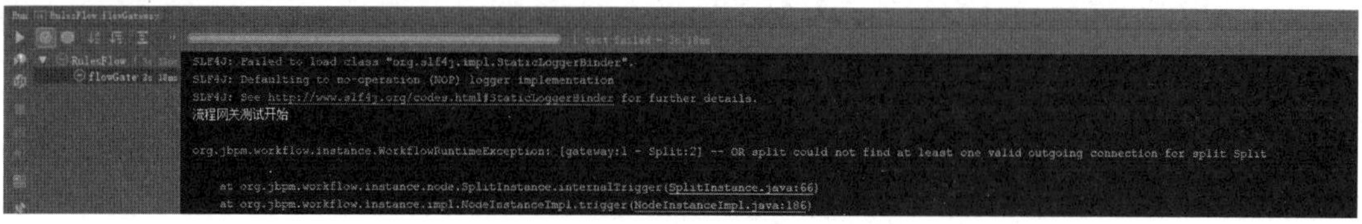

图10-86　错误的使用网关配置

将合并网关的 type 设置成 XOR，还原 3 个分支的初始化值，执行 flowGateway() 方法，结果如图 10-87 所示。

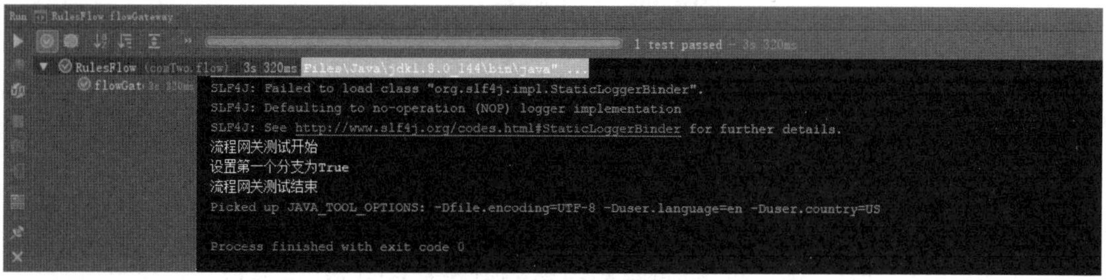

图10-87　规则流程测试效果

将其中两个分支都设置为 eavl(False)，执行 flowGateway() 方法，结果如图 10-88 所示。

图10-88　规则流程测试效果

总结：当分离网关为 OR，合并网关为 XOR 时，只会执行分支为 True 的流向，并且只会汇总其中一个，这是受优先级影响。

规则流最难理解的部分就是其网关了，结合规则体属性后，还会影响到输出结果，规则与流程之间都是相互作用的，使用规则流时一定要小心，如果在执行过程中可以通过属性控制，尽量不要使用规则流。

10.5　规则构建过程

对 Drools 语法和使用有了一定的理解后，就可以对规则执行过程进行研究，在本节中先有个简单的认识，对今后阅读源码会有帮助。Drools 的构建主要包括如图 10-89 所示的知识点。

图10-89　Drools构建

本节中对最开始的例子 Hello Wrold 进行了一个说明。从测试用例中复制一部分代码，作为主

要分析的构建过程,其内容为:

```
KieServices kss = KieServices.Factory.get();
KieContainer kc = kss.getKieClasspathContainer();
KieSession ks = kc.newKieSession("testhelloworld");
```

1. KieServices说明

KieServices 是一个接口,其主要结构如图 10-90 所示。

图10-90 KieServices结构

通过这些方法可以对 Kie 进行构建和使用,如可以获取 KieContainer、利用 KieContainer 来创建 KieBase 和 KieSession 等信息,可以获取 KieRepository 对象、利用 KieRepository 来管理

KieModule 等。

KieServices 是一个核心，通过它操作各种对象来完成规则构建、管理和执行等。它既是线程安全的，又是一个单例的，如图 10-91 所示。通过它可以访问到另一个 KieServices 提供的服务。一般用的代码为：

```
KieServices kieServices = KieServices.Factory.get();
```

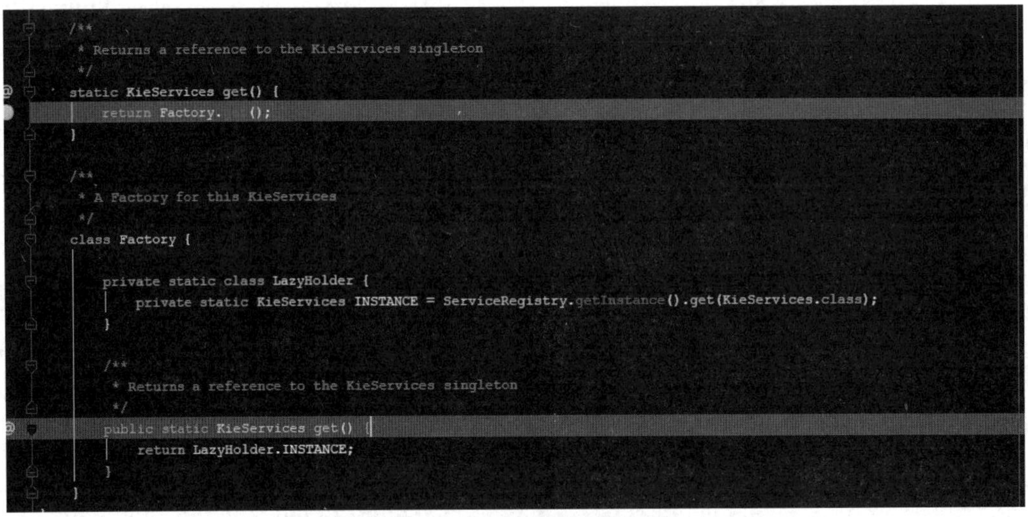

图10-91　KieServices中的单例

2.KieContainer 说明

KieContainer 是一个接口，是所有 KieBase 的容器，其主要结构如图 10-92 所示。

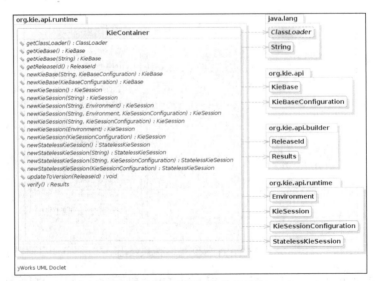

图10-92　KieContainer结构

通过提供的这些方法，可以很方便地操作 KieBase 和 KieSession。经常通过".newKieSession
("testhelloworld");"获取 KieSession。一直强调的事情中 KieSession 会话是基于 KieBase 规则库的，
所以通过查看源码，可以发现其创建是离不开 KieBase 的，如图 10-93 所示。

图10-93　KieSession的创建离不开KieBase

3. KieBase 说明

KieBase 是一个知识库（规则库），包含若干的规则、流程、函数等，在 Drools 中主要就是规则和方法，KieBase 本身并不包含运行时的相关数据，如果需要执行 KieBase 中的规则，就需要根据 KieBase 创建 KieSession。一般情况下都是直接根据 KieContainer 创建的。KieBase 是一个接口，其结构如图 10-94 所示。

KieBase 可以对规则进行基本操作，如删除规则、删除查询、删除函数等。

4. KieSession 说明

KieSession 就是一个与 Drools 引擎打交道的会话，基于 KieBase 创建，它会包含运行时数据，以及 Fact 事实对象，并对运行时数据事实进行规则运算。通过 KieContainer 创建 KieSession 是一种较为方便的做法，其实本质上是从 KieBase 中创建出来，且默认为是一个有状态的（前面讲 kmodule.xml 配置时已经有过说明）。KieSession 作为一个接口，它同时还继承了 StatefulRuleSession 接口、StatefulProcessSession 接口、CommandExecutor 接口、KieRuntime 接口。这里讲到了有状态与无状态两种 KieSession。

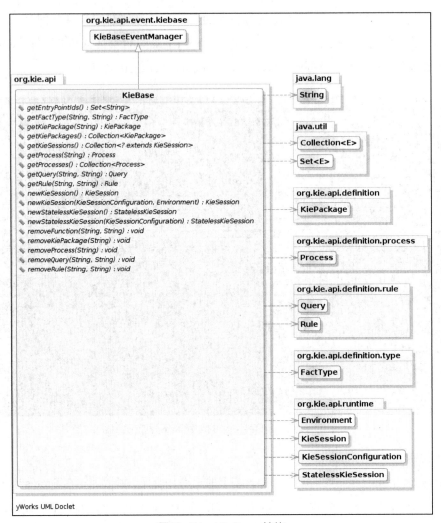

图10-94　KieBase结构

想要成功使用Drools规则引擎需要大量的工作，本节只简单讲述在测试用例中所用到的基础知识点，通过上述描述的信息，得出以下总结。

规则内容是规则内容，规则文件是规则文件，从决策表、规则模板中得知，规则文件只是存放规则内容的一个容器，就算只有规则内容，规则照样是可以执行的。

规则流中流程是流程、规则是规则，这也证明了KIE功能的强大。规则流属性只能在流程文件规则元素组件对应中使用，而流程文件却可以单独运行。规则流中的网关就很好地证明了这一点。

Spring是Spring，Drools是Drools。Drools与Spring整合时，发现基本上它们是各司其职、互不影响。唯一有所影响的是使用Web项目时，通过Spring将规则进行了初始化操作。

Spring Boot是Srping Boot，Drools是Drools，Drools与Spring Boot整合时，与整合Spring时相似，而且用到的是KieHelper，同样使用Drools。简单来说，Spring Boot的启动只是为了让Drools所用到的JVM值存活更长时间。

根据上述得出，Drools 规则引擎是项目中的一部分，可以是一个独立的模块。在实际应用中，Drools 项目可以对外提供接口，这也非常符合现在微服务的设计思想。

10.6 Drools事件监听

事件监听是一项很有用的功能，它可以提供规则引擎执行过程中的记录，能更容易地掌握 Drools 的执行结果。Drools 提供了 3 种类型的监听器，分别是 RuleRuntimeEventListener、AgendaEventListener 和 ProcessEventListener。根据官方的说明使用事件监听主要分为两大类型，分别为配置文件和 API code。配置文件可为分为 kmodule.xml 和 Spring.xml 两种。通过 API code 比较直观，而且相对比较通用，可扩展。

RuleRuntimeEventListener 为规则运行时的监听，它是一个接口类型，提供了 3 个接口方法，如图 10-95 所示。

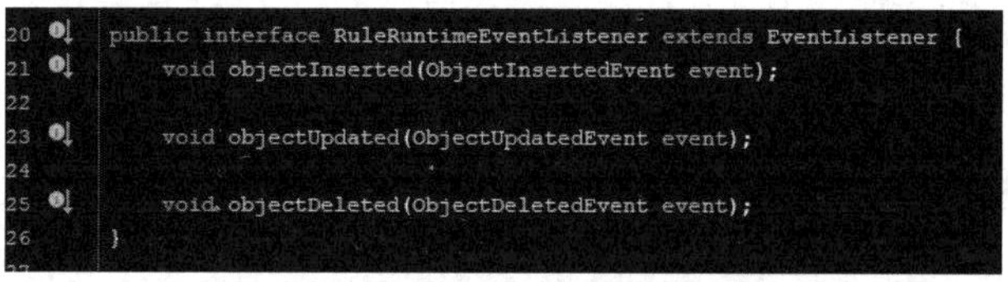

```
public interface RuleRuntimeEventListener extends EventListener {
    void objectInserted(ObjectInsertedEvent event);

    void objectUpdated(ObjectUpdatedEvent event);

    void objectDeleted(ObjectDeletedEvent event);
}
```

图10-95 监听器接口结构

从 3 个接口方法可以看出，监听的是 insert、update、delete 的操作，这 3 个操作在规则中也是经常用到的，尤其是 update(modify)。通过测试用例分别来测试这 3 个接口，创建 isRuleRunTime.drl 文件，目录为 resources/rulesTwo/Listener，其内容为：

```
package rulesTwo.Listener
import com.pojo.Person;

rule "测试Insert"
no-loop true
    when
    then
        Person person=new Person("张三",50);
        insert(person);
        System.out.println(drools.getRule().getName());
end

rule "测试Update"
no-loop true
    when
        $p:Person(name=="张三")
```

```
then
    modify($p){
    setName("李四")
    }
    System.out.println(drools.getRule().getName());
end

rule "测试Delete"
no-loop true
    when
        $p:Person(name=="李四")
    then
        delete($p);
        System.out.println(drools.getRule().getName());
end
```

为防止死循环需添加 no-loop 属性，创建 RuleTimeImpl.java 文件，目录为 comTwo.Listener，其内容为：

```
package comTwo.Listener;

import org.kie.api.event.rule.ObjectDeletedEvent;
import org.kie.api.event.rule.ObjectInsertedEvent;
import org.kie.api.event.rule.ObjectUpdatedEvent;
import org.kie.api.event.rule.RuleRuntimeEventListener;

public class RuleTimeImpl implements RuleRuntimeEventListener {
    @Override
    public void objectInserted(ObjectInsertedEvent event) {
        System.out.println("输出Insert监听方法");
    }
    @Override
    public void objectUpdated(ObjectUpdatedEvent event) {
        System.out.println("输出Update监听方法");
    }
    @Override
    public void objectDeleted(ObjectDeletedEvent event) {
        System.out.println("输出Delete监听方法");
    }
}
```

创建一个 RuleTimeImpl 类，实现 RuleRuntimeEventListener 接口，并重写 RuleRuntimeEventListener 中的 3 个接口方法。创建 TestRuleListener.java 文件，目录为 comTwo/Listener，其内容为（注意加粗部分）：

```
package comTwo.Listener;

import org.junit.Test;
import org.kie.api.io.Resource;
```

```java
import org.kie.api.io.ResourceType;
import org.kie.api.runtime.KieSession;
import org.kie.internal.io.ResourceFactory;
import org.kie.internal.utils.KieHelper;

public class TestRuleListener {

    @Test
    public void testRuleIUDListener(){
        Resource drl = ResourceFactory.newClassPathResource("rulesTwo/Listener/isRuleRunTime.drl", TestRuleListener.class);
        KieHelper helper = new KieHelper();
        helper.addResource(drl, ResourceType.DRL);
        KieSession ksession = helper.build().newKieSession();
        ksession.addEventListener(new RuleTimeImpl());
        int count = ksession.fireAllRules();
        System.out.println("总执行了" + count + "条规则");
        ksession.dispose();
    }
}
```

执行 testRuleIUDListener() 方法，结果如图 10-96 所示。

图10-96　使用监听器的效果

重写了 RuleRuntimeEventListener 接口文件中的 3 个接口后，发现它还提供了 3 个参数，分别是 ObjectInsertedEvent、ObjectUpdatedEvent 和 ObjectDeletedEvent。这 3 个接口继承了 RuleRuntimeEvent 空接口，其相同的地方：都有 "FactHandle getFactHandle();" "Object getObject();" "Rule getRule();"，其中 ObjectUpdatedEvent 多了 "Object getOldObject();" 接口。

接口中有 Rule 与 FactHandle，这两个是比较常用的，Java 代码中的 insert、update、delete 的返回值都是 FactHandle。所以可以先明确其操作了什么，而 Rule 是需要了解的。下面列举一些常用的接口。

event.getRule().getName()：指执行的规则名。

event.getRule().getPackageName()：指执行的规则名在 package 的路径。

event.getRule().getId()：指执行的规则名 rule 的参数在同包下是唯一的。

event.getRule().getNamespace()：指执行的规则名在 package 的路径。

event.getRule().getKnowledgeType()：指知识库类型（目前来看，感觉没有什么太大的作用）。

event.getRule().getMetaData()：指元数据相关的一些监听，返回值是一个 map。

AgendaEventListener：议程事项的监听。它是一个接口类型，提供了 10 个接口方法，详见以下代码：

```
void matchCreated(MatchCreatedEvent event);
void matchCancelled(MatchCancelledEvent event);
void beforeMatchFired(BeforeMatchFiredEvent event);
void afterMatchFired(AfterMatchFiredEvent event);
void agendaGroupPopped(AgendaGroupPoppedEvent event);
void agendaGroupPushed(AgendaGroupPushedEvent event);
void beforeRuleFlowGroupActivated(RuleFlowGroupActivatedEvent event);
void afterRuleFlowGroupActivated(RuleFlowGroupActivatedEvent event);
void beforeRuleFlowGroupDeactivated(RuleFlowGroupDeactivatedEvent event);
void afterRuleFlowGroupDeactivated(RuleFlowGroupDeactivatedEvent event);
```

理解这个监听比较简单，直接通过接口方法名就能知道是做什么的，如调用规则前、调用规则后、监听规则流的调用前与调用后等。使用该监听器时需要注意，它与规则运行时的监听相似，就是在指定一些规则的激活分组、规则流属性时，部分接口方法才会被执行，这点至关重要。以规则流为例，如果规则体中没有规则流属性，那么又怎么会去监听规则流的调用前与调用后呢。

ProcessEventListener：流程事项的监听。它是一个接口类型，提供了 12 个接口方法，详见以下代码：

```
void beforeProcessStarted(ProcessStartedEvent event);
void afterProcessStarted(ProcessStartedEvent event);
void beforeProcessCompleted(ProcessCompletedEvent event);
void afterProcessCompleted(ProcessCompletedEvent event);
void beforeNodeTriggered(ProcessNodeTriggeredEvent event);
void afterNodeTriggered(ProcessNodeTriggeredEvent event);
void beforeNodeLeft(ProcessNodeLeftEvent event);
void afterNodeLeft(ProcessNodeLeftEvent event);
void beforeVariableChanged(ProcessVariableChangedEvent event);
void afterVariableChanged(ProcessVariableChangedEvent event);
default void beforeSLAViolated(SLAViolatedEvent event){};
default void afterSLAViolated(SLAViolatedEvent event){};
```

ProcessEventListener：流程事项监听。它主要是用来针对流程与流程中元素节点的监听，在规则流章节中，介绍过规则流的 Java 代码是可以传变量的，该监听器还提供了变量的事项监听等。建议读者尝试一下这些监听器，既然 Drools 已经提供了这些功能，就无须自己再写一套 AOP 的功能了。

通过配置文件方式配置监听器，将上述 RuleRunTime 的相关代码复制到 8.1 章节 Spring+Drools 简单配置的测试用例中，如图 10-97 所示。

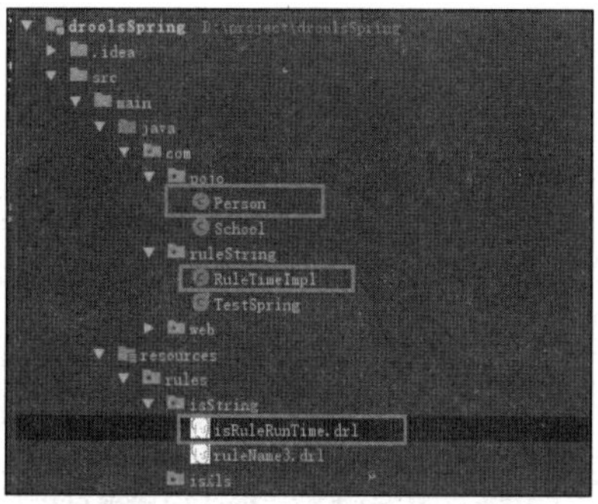

图10-97　配置文件中的事件监听

编辑 Spring.xml 文件，其内容为（注意加粗代码）：

```xml
<?xml version="1.0" encoding="UTF-8"?>
<beans xmlns="http://www.springframework.org/schema/beans"
       xmlns:xsi="http://www.w3.org/2001/XMLSchema-instance"
       xmlns:kie="http://drools.org/schema/kie-spring"
       xsi:schemaLocation="
       http://drools.org/schema/kie-spring
       http://drools.org/schema/kie-spring.xsd
       http://www.springframework.org/schema/beans
       http://www.springframework.org/schema/beans/spring-beans.xsd">
    <bean id="RuleTimeImpl" class="com.ruleString.RuleTimeImpl"/>
    <!-- 与kmodule.xml的配置是相似的 -->
    <kie:kmodule id="kmodule">
        <kie:kbase name="kbase" packages="rules.isString">
            <kie:ksession name="ksession"/>
        </kie:kbase>
        <kie:kbase name="kbaseTwo" packages="rules.isString">
            <kie:ksession name="ksessionRun">
                <kie:ruleRuntimeEventListener ref="RuleTimeImpl"/>
            </kie:ksession>
        </kie:kbase>
    </kie:kmodule>
    <bean id="kiePostProcessor" class="org.kie.spring.annotations.KModuleAnnotationPostProcessor"/>
</beans>
```

编辑 TestSpring.java 文件，添加一个新的注解变量与测试方法，其代码为：

```java
@KSession("ksessionRun")//注：这里的值与配置文件的值是一样的
KieSession ksessionRun;

@Test
public void ruleRunTime() {
    int count = ksessionRun.fireAllRules();
    System.out.println("总执行了" + count + "条规则");
    ksession.dispose();
}
```

执行 ruleRunTime() 方法，结果如图 10-98 所示。

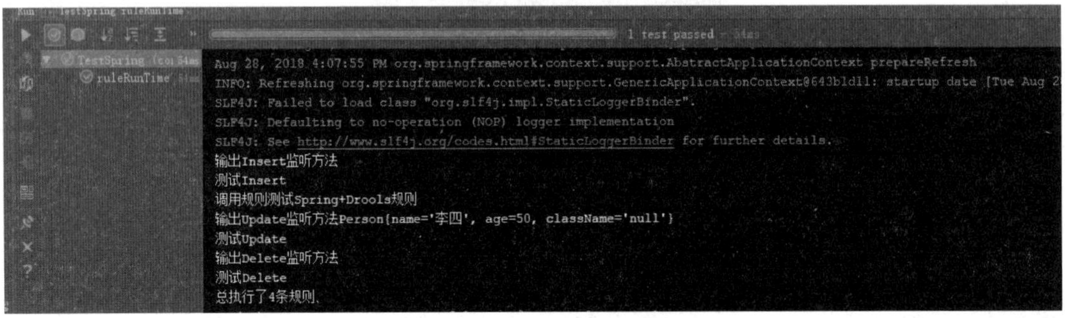

图10-98　使用事件监听后的效果

结果与之前通过 API 测试是相似的，还有一个写法与上面的差不多，但它有个并不友好的地方：必须要将 agendaEventListener、ruleRuntimeEventListener 和 processEventListener 一起实现，其代码为（注意加粗部分）：

```xml
<bean id="RuleTimeImpl" class="com.ruleString.RuleTimeImpl"/>
<!-- 与kmodule.xml的配置是相似的 -->
<kie:kmodule id="kmodule">
    <kie:kbase name="kbase" packages="rules.isString">
        <kie:ksession name="ksession"/>
    </kie:kbase>

    <kie:kbase name="kbaseTwo" packages="rules.isString">
        <kie:ksession name="ksessionRun">
            <kie:ruleRuntimeEventListener ref="RuleTimeImpl"/>
        </kie:ksession>
    </kie:kbase>

    <kie:kbase name="kbaseTwo2" packages="rules.isString">
        <kie:ksession name="ksessionRun2" listeners-ref="debugListeners">
            <kie:ruleRuntimeEventListener ref="RuleTimeImpl"/>
        </kie:ksession>
    </kie:kbase>
</kie:kmodule>
<kie:eventListeners id="debugListeners">
    <kie:ruleRuntimeEventListener ref="RuleTimeImpl"/>
    <kie:agendaEventListener ref="mock-agenda-listener"/>
    <kie:processEventListener ref="mock-process-listener"/>
</kie:eventListeners>
```

第11章 Workbench

11.1 Workbench

Workbench 是 KIE 组件中的元素，也称为 KIE-WB，是 Drools-WB 与 JBPM-WB 的结合体。它是一个 WEB-IDE，是一个可视化的规则编辑器。Workbench 功能十分强大，不仅提供了一系列的规则编辑器，还可以在提供的页面上编辑 JavaBean（数据对象）。它有插件扩展的功能，重要的功能之一是生成 kjar 包。Kie-Server 是一个服务，它可以对外提供接口调用，类似 Webserivce。

Workbench 是一个 War 包，它可以安装在 tomcat 或 wildfly 中。官方也提供了 Docker 的安装方式，本章将讲述 Workbench 的安装和使用。

本书使用的 Workbench 地址为 https://pan.baidu.com/s/1WGOHney0V5o2EIH8szOhFw。

11.2 Windows安装方式

介绍 Workbench 的安装前，先说明一下做测试的系统环境。

系统版本： Windows 10 64 位

JDK 版本： 1.8

Maven 版本： 3.5

Tomcat 版本： 8.5.27

为保证测试环境都可直接运行，下面的例子都是通过复制多份 Tomcat。

下载 Workbench 所需要的 war 包。在浏览器中输入"www.drools.org"找到下载菜单，如图 11-1 所示。

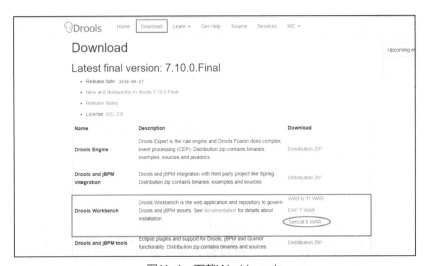

图11-1　下载Workbench

下载完成后的文件是一个 war 包，通过解压工具进行解压并打开，如图 11-2 所示。

图11-2　Workbench War包内容

打开 README.txt 文件，这个文件就是 Workbench 的安装说明，它的安装过程十分简单。根据内容进行配置。

（1）创建一个 setenv 文件，由于当前环境是 Windows，所以创建一个 setenv.bat 的批处理文件。放在 tomcat 的 bin 目录下。新建一个 setenv.bat 文件，其内容为：

```
set "CATALINA_OPTS=-Xmx512M
-Djava.security.auth.login.config=%CATALINA_HOME%\webapps\kie-drools-wb\
WEB-INF\classes\login.config -Dorg.jboss.logging.provider=jdk"
```

Windows 的变量与斜杠的用法是与 Linux 有所不同的，所以这里一定要注意。

（2）将下载 Workbench 的 war 包改名为 kie-drools-wb.war，并放在 Tomcat 的 webapps 目录下，如图 11-3 所示。

图11-3　改名后的war包

（3）为 Workbench 添加用户，修改 tomcat-users.xml 配置文件，在 \conf 目录下，添加如下内容。

```
<?xml version='1.0' encoding='utf-8'?>
<tomcat-users xmlns="http://tomcat.apache.org/xml"
              xmlns:xsi="http://www.w3.org/2001/XMLSchema-instance"
              xsi:schemaLocation="http://tomcat.apache.org/xml tomcat-users.xsd"
              version="1.0">
  <role rolename="admin"/>
  <user username="kie" password="kie" roles="admin"/>
</tomcat-users>
```

Workbench 的角色有很多种，但为了解 Workebch 的功能，这里使用 admin 权限来控制。

（4）下载 kie-tomcat-integration-7.10.0.Final.jar、javax.security.jacc-api-1.5.jar、slf4j-api-1.7.25.jar 这三个 jar 包，并放在 Tomcat/lib 目录下。

README.txt中的内容如下。

① 将`"kie-tomcat-integration"` JAR复制到TOMCAT_HOME/lib(org.kie:kie-tomcat-integration)。

② 将`"JACC"` JAR复制到TOMCAT_HOME/lib(javax.security.jacc:artifactId=javax.security.jacc-api in JBoss Maven Repository)。

③ 将`"slf4j-api"` JAR复制到TOMCAT_HOME/lib (org.slf4j:artifactId=slf4j-api in JBoss Maven Repository)。

（5）添加 KIE 的启动服务，修改 conf/server.xml 添加 Value 标签，内容为：

`<Valve className="org.kie.integration.tomcat.JACCValve" />`

（6）启动 Tomcat 时可能会报一些错误，这里先不要管，首先看一下 Workbench 的效果。在浏览器中输入"localhost:8080/kie-drools-wb"，会出现如图 11-4 所示的效果，证明 Workbench 配置成功了。

图11-4　成功启动的效果

输入事先配置好的用户名与密码，登录成功后，就会出现如图 11-5 所示的效果。左侧的红框是项目编辑，右侧的黄框是 Kie-Server。

图11-5　登录成功后的效果

Tomcat 启动时有一个问题，对于程序员来说只要有错误信息就应想办法纠正它，错误信息如

图 11-6 所示，找不到 org/codehaus/Jackson/Versioned 类。

```
no longer supported.  Use a servlet 3.0 container and the ResteasyServletInitializer
29-Aug-2018 14:53:20.670 WARN [localhost-startStop-1] org.jboss.resteasy.plugins.providers.RegisterBuiltin.registerProviders RESTEASY002145: NoClassDefFoundError: Una
e to load builtin provider org.jboss.resteasy.plugins.providers.jackson.ResteasyJacksonProvider from jar:file:/D:/apache-tomcat-8.5.27_Drools/webapps/kie-drools-wb/WE
INF/lib/resteasy-jackson-provider-3.0.24.Final.jar!/META-INF/services/javax.ws.rs.ext.Providers
 java.lang.NoClassDefFoundError: org/codehaus/jackson/Versioned
        at java.lang.ClassLoader.defineClass1(Native Method)
        at java.lang.ClassLoader.defineClass(ClassLoader.java:763)
        at java.security.SecureClassLoader.defineClass(SecureClassLoader.java:142)
        at org.apache.catalina.loader.WebappClassLoaderBase.findClassInternal(WebappClassLoaderBase.java:2283)
        at org.apache.catalina.loader.WebappClassLoaderBase.findClass(WebappClassLoaderBase.java:811)
        at org.apache.catalina.loader.WebappClassLoaderBase.loadClass(WebappClassLoaderBase.java:1260)
        at org.apache.catalina.loader.WebappClassLoaderBase.loadClass(WebappClassLoaderBase.java:1119)
        at java.lang.ClassLoader.defineClass1(Native Method)
        at java.lang.ClassLoader.defineClass(ClassLoader.java:763)
        at java.security.SecureClassLoader.defineClass(SecureClassLoader.java:142)
        at org.apache.catalina.loader.WebappClassLoaderBase.findClassInternal(WebappClassLoaderBase.java:2283)
        at org.apache.catalina.loader.WebappClassLoaderBase.findClass(WebappClassLoaderBase.java:811)
        at org.apache.catalina.loader.WebappClassLoaderBase.loadClass(WebappClassLoaderBase.java:1260)
        at org.apache.catalina.loader.WebappClassLoaderBase.loadClass(WebappClassLoaderBase.java:1119)
        at org.jboss.resteasy.plugins.providers.RegisterBuiltin.registerProviders(RegisterBuiltin.java:76)
        at org.jboss.resteasy.plugins.providers.RegisterBuiltin.register(RegisterBuiltin.java:37)
        at org.jboss.resteasy.spi.ResteasyDeployment.start(ResteasyDeployment.java:233)
        at org.jboss.resteasy.plugins.server.servlet.ResteasyBootstrap.contextInitialized(ResteasyBootstrap.java:28)
```

图11-6　Workbench启动时的异常信息

解决方案：下载 jackson-core-asl-1.9.13.jar、jackson-mapper-asl-1.9.13.jar，并将下载包放在 tomcat/lib 目录下。

讲述完 Workbench 7.10 的版本后，下面介绍 WIE-WB 6.4 版本的安装过程。

11.3 KIE-WB 6.4版本安装

1. 环境说明

（1）下载 linux 版本的 jdk 包，这里下载的是 jdk - 8u91 - linux - x64.tar.gz。

（2）在 /usr/local 下使用“mkdir java”创建 Java 目录，将 jdk 文件放入其中。

（3）通过“tar - zxvf ./jdk - 8u91 - linux - x64.tar.gz - C .”将其进行解压。

（4）配置环境变量，运行“sudo vi /etc/profile”，在最后插入要配置的内容：

```
export JAVA_HOME=/usr/local/java/jdk1.8.0_91/
    export PATH=$JAVA_HOME/bin:$PATH
    export CLASSPATH=.:$JAVA_HOME/lib/dt.jar:$JAVA_HOME/lib/tools.jar
```

（5）按 Esc 键，输入 (:wq 保存并退出)，再次运行“source /etc/profile”，使配置环境生效。

（6）运行“java –version”看是否生效，若出现 jdk 版本号，则表示安装并配置环境变量成功，如图 11-7 所示。

```
[root@zy49 /]# java -version
java version "1.8.0_91"
Java(TM) SE Runtime Environment (build 1.8.0_91-b14)
Java HotSpot(TM) 64-Bit Server VM (build 25.91-b14, mixed mode)
```

图11-7　jdk版本输出

下载 Tomcat，地址为 http://tomcat.apache.org/download - 70.cgi。

下载 Workbench 的 war 包，地址为 http://www.drools.org/download/download.html，如图 11-8 所示。

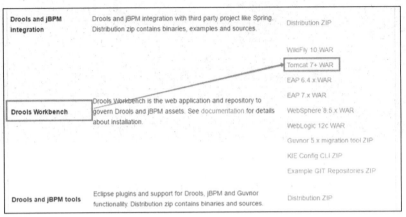

图11-8　Workbench 6.4版本下载

2. 安装说明

（1）下载成功后，将 war 包放在 tomcat-Workbench/webapps/ 目录下，并改名为 kie-wb.war。

（2）将下列 jar 复制到 TOMCAT_HOME/lib 中：

```
* btm-2.1.4.jar
* btm-tomcat55-lifecycle-2.1.4.jar
* h2-1.3.161.jar
* jta-1.1.jar
* slf4j-api-1.7.2.jar
* slf4j-jdk14-1.7.2.jar
* log4j-api-2.1.jar
* log4j-slf4j-impl-2.1.jar
```

log4j-api-2.1.jar 和 log4j-slf4j-impl-2.1.jar 是说明文件中没有提到的，但如果不添加，在 Tomcat 启动加载时会报错。

（3）在 TOMCAT_HOME/conf 目录下，新建 btm-config.properties(vim 命令新建)，其内容如下（注意，每个语句后面不能有空格）。

```
bitronix.tm.serverId=tomcat-btm-node0
bitronix.tm.journal.disk.logPart1Filename=${btm.root}/work/btm1.tlog
bitronix.tm.journal.disk.logPart2Filename=${btm.root}/work/btm2.tlog
bitronix.tm.resource.configuration=${btm.root}/conf/resources.properties
```

在 TOMCAT_HOME/conf 下，新建 resources.properties(vim 命令新建)，其内容如下（注意，每个语句后面不能有空格）。

```
resource.ds1.className=bitronix.tm.resource.jdbc.lrc.LrcXADataSource
resource.ds1.uniqueName=jdbc/jbpm
resource.ds1.minPoolSize=10
```

```
resource.ds1.maxPoolSize=20
resource.ds1.driverProperties.driverClassName=org.h2.Driver
resource.ds1.driverProperties.url=jdbc:h2:file:~/jbpm
resource.ds1.driverProperties.user=sa
resource.ds1.driverProperties.password=
resource.ds1.allowLocalTransactions=true
```

（4）在 TOMCAT_HOME/bin 目录下，新建 setenv.sh(vim 命令新建)，其内容如下。

```
CATALINA_OPTS="-Xmx512M -XX:MaxPermSize=512m -Dbtm.root=$CATALINA_HOME
    -Dbitronix.tm.configuration=$CATALINA_HOME/conf/btm-config.properties
    -Djbpm.tsr.jndi.lookup=java:comp/env/TransactionSynchronizationRegistry
    -Djava.security.auth.login.config=$CATALINA_HOME/webapps/kie-wb/WEB-INF/classes/login.config
    -Dorg.jboss.logging.provider=jdk
    -Dorg.jbpm.cdi.bm=java:comp/env/BeanManager
-Dorg.guvnor.m2repo.dir=/root/.m2/repository
    -Dorg.kie.demo=false
    -Dorg.kie.example=false"
```

（5）配置 JEE security。

将 "kie - tomcat - integration"JAR 复制到 TOMCAT_HOME/lib(org.kie:kie - tomcat - integration)。

将 "JACC"JAR 复制到 TOMCAT_HOME/lib(javax.security.jacc:artifactId=javax.security.jacc - api in JBoss Maven Repository)。

将 "slf4j - api"JAR 复制到 TOMCAT_HOME/lib (org.slf4j:artifactId=slf4j - api in JBoss Maven Repository)。

（6）在 TOMCAT_HOME/conf/server.xml 的 Host 节点最后添加如下内容。

```
<Valve className="org.kie.integration.tomcat.JACCValve" />
```

（7）编辑 TOMCAT_HOME/conf/tomcat-users.xml，添加 'analyst' 或 'admin' 角色，添加 KIE-WB 相应的用户，内容如下。

```
<role rolename="admin" />
<role rolename="analyst" />
<user username="kie" password="kie" roles="admin" />
<user username="kie-analyst" password="kie-analyst" roles="analyst" />
```

（8）启动 Tomcat，访问 http://liunx 的 IP 地址 "8080/kie - wb"，可以看到如图 11-9 所示的界面，证明搭建成功。

图11-9　Workbench 6.4版本页面效果

在无互联网环境下启动并运行 kie-web，如果不加 -Dorg.kie.demo=false 参数，kie-web 每次启动时会去默认的 GIT 地址中加载 kie-web 的 demo 参数；如果部署服务器为虚拟机，或者是无互联网环境时，它会因为建立 internet 连接超时而抛出一个疑似 memory leak 的 exception 而导致整个 war 工程加载失败。添加如下配置即可解决这个异常信息。

```
-Dorg.kie.demo=false
-Dorg.kie.example=false
```

当配置启动出现如下警告时：

```
Unable to instantiate EJB Asynchronous Bean. Falling back to Executors'
CachedThreadPool.javax.naming.NameNotFoundException: Name [module/
SimpleAsyncExecutorService] is not bound in this Context. Unable to find
[module].
```

在配置文件 setenv.sh 中添加：

```
-Dorg.uberfire.async.executor.safemode=true
```

该问题只是一个警告，不会影响到程序的正常运行。

Workbench 版本更新得快是有其道理的，通过版本对比，发现旧版本的配置比较麻烦，而且在 Windows 环境下，还会出现如下的问题。

```
在Windows环境下，启动时，可能存在一个严重的错误：
java.io.IOException: Cannot run program "bash" (in directory
"C:\Users\kangz"): CreateProcess error=2, 系统找不到指定的文件。
解决方案：下载Git-2.10.1-64-bit.exe 任意版本即可。
有这两个知识点，接下来让cmd支持UTF8就会变得容易。
①运行CMD。
②输入"CHCP"，按Enter键查看当前的编码。
③输入"CHCP 65001"，按Enter键。
④在窗体上右击，选择"属性"选项，设置字体；
⑤操作完上述步骤后，即使原来没有显示Lucida Console字体，现在也能看到了，选择它。如果原来就有，可以先选择它，出现问题再执行上述步骤（至少本机需要CHCP 65001）；
⑥选择"只应用到本窗体"选项，单击"确认"按钮。
```

11.4 Workbench操作手册

安装成功后，重要的就是如何使用 Workbench 了，Workbench 是一个 WEB-IDE，使用起来与 Eclipse 很相似。Workbench 是一个特别强大的页面开发应用工具，又结合了页面开发、Maven、服务器、数据库连接、检索、用户管理、发布、测试、规则编辑、任务等相关的功能，本节重点介绍如何更有效地使用 Workbench。

创建一个项目，单击"Add Project"按钮，如图 11-10 所示，会出现如图 11-11 所示的界面了。

图11-10　Workbench添加项目

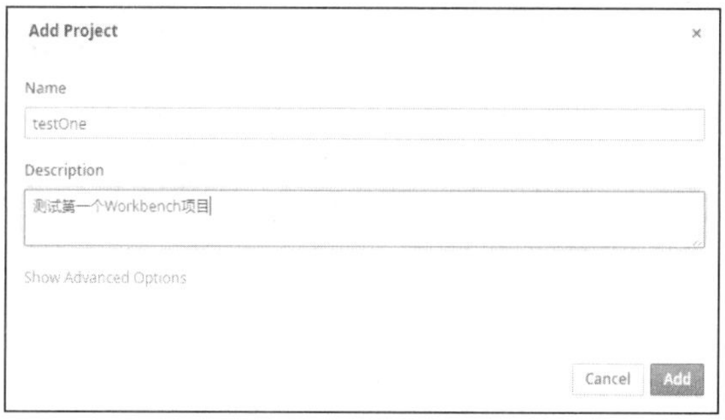

图11-11　项目添加页面

项目还有扩展选项的添加，单击"Show Advanced Options"链接，会出现如图 11-12 所示的界面，这是一个标准的 Maven 管理项目。

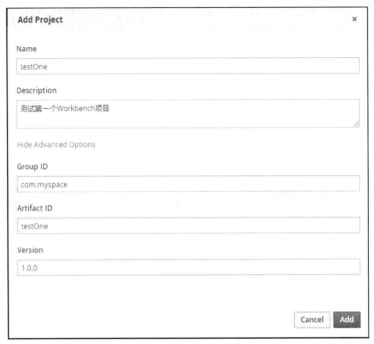

图11-12　项目扩展信息

有些用户在安装过程中会出现如下的问题，单击"Add"按钮后出现如图11-13所示的错误。

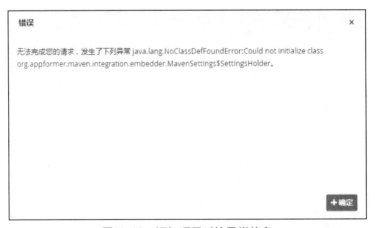

图11-13　添加项目时的异常信息

Workbench 的构建离不开 Maven，出现上面问题是因为 Workbench 没有找到默认的 .m2 目录，虽然当前环境安装了 Maven，并配置了环境变量，但服务并没有指到 .m2 路径下，这里需要运行 cmd，并执行 mvn help:system 命令，直到它构建完成后，重启 Tomcat。再次输入图 11-12 所示的内容，单击"Add"按钮后项目就创建成功了，如图 11-14 所示。

图11-14　添加项目成功

Workbench 是一个构建 kjar 的 WEB-IDE，它构建生成的文件与 jar 无异。所以学习 Workbench 之前，必须要先理解 Workbench 6.4 版本中的两个概念：组件单元和资料库。图 11-15 所示为 Workbench 6.4 版本的界面。

图11-15　Workbench 6.4版本界面

Workbench 理解起来很简单，读者可以把它当成项目的文件夹。组件单元是顶级目录，资料库是项目的子目录，一个组件单元可以有多个资料库。本节的重点并不是这些，之所以拿出来讲述一下，是因为组件单元与资料库在 7.x 版本中发生了变化，7.x 版本中的定义改成了空间，如图 11-16 所示。

图11-16　Workbench 7.10版本的目录管理

MySpace 是 Workbench 自带的，实际项目开发时，项目空间都是自定义的，具体操作如下。如图 11-17 所示，单击"Spaces"按钮，会出现图 11-18 所示的界面。

图11-17 组织目录

图11-18 创建后的项目管理

单击右边蓝色的"Add Space"按钮进行添加，如图 11-19 所示。输入空间名称后单击"Save"按钮即可，第二个参数 Contributors 可以不需要添加。默认为当前登录用户，如果是权限上的控制，权限低的用户是无法操作该功能的。

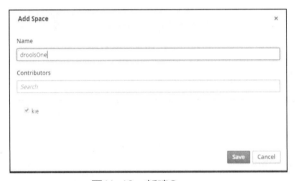

图11-19 新建Space

添加完成后的效果如图 11-20 所示。

图11-20 成功添加了Spaces

选择刚才创建的空间项目进行接下来的操作手册说明，重复上述讲到的创建项目过程，单击"Add Asset"按钮，如图 11-21 所示，右上角出现"取消"按钮。

第 11 章 Workbench

图11-21 项目可操作节点

从图 11-21 所示的界面中可以看出，功能基本包括了前面所讲到的知识点。下面先从操作 DRL 文件开始介绍。

1. DRL文件

单击"DRL 文件"图标，如图 11-22 所示。

图11-22 创建DRL文件

弹出如图 11-23 所示的界面，输入文件名，该文件名是必填项，软件包在规则相关文件中指的是资料目录，选择"com.droolsone.testdroolsone"选项是指把规则文件放在 resources/com/droolsone/testdroolsone 目录下。

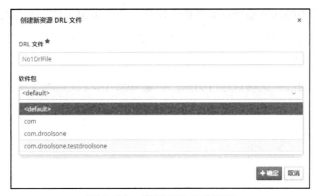

图11-23 编辑DRL文件界面

添加完成后，根据图 11-24 所示的效果进行规则的编辑，其内容如下。

295

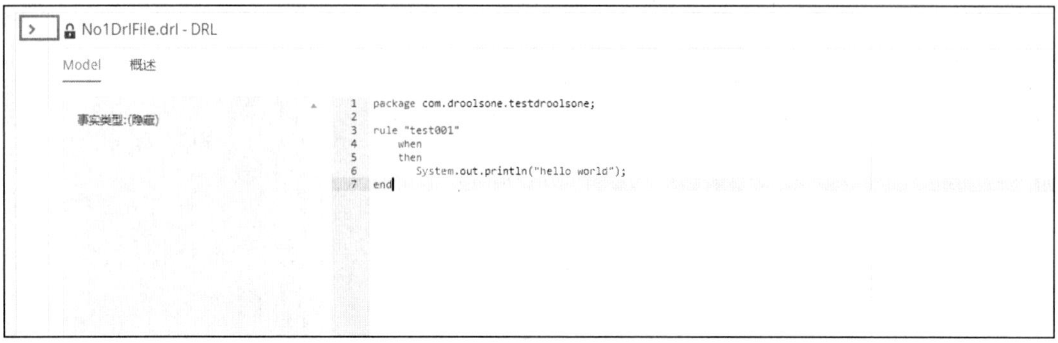

图11-24　编辑DRL文件

```
package com.droolsone.testdroolsone;
rule "test001"
    when
    then
        System.out.println("hello world");
end
```

选择图11-24中框出的地方，可以出现如图11-25所示的效果。

图11-25　界面扩展信息

当前操作界面共分为三个模块，左侧是项目文件管理模块、中间是项目文件编写模块，在项目文件编写模块中，右上角的部分是项目文件操作模块。单击编辑页面右上角的"保存"按钮，该按钮是规则的语法校验功能，编写模块左则的是项目文件目录管理模块，通过选择不同的目录，展现不同的项目文件。

规则编辑完成后，通过测试场景进行测试，创建 Add Asset 找到测试场景，如图11-26所示。

第 11 章 Workbench

图11-26　选择创建测试场景

单击"测试场景"图标出现如图 11-27 所示的界面，输出测试场景名称，软件包选择与规则文件一样的目录，并单击"确定"按钮会出现如图 11-28 所示的界面。

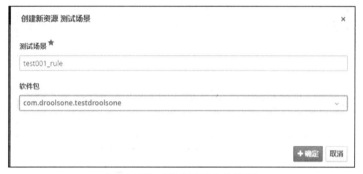

图11-27　测试场景文件编辑

图11-28　测试场景界面

单击"运行所有测试场景"按钮，通过 Tomcat 日志输出信息可以看到输出的结果，如图 11-29 所示。测试场景详细说明请看测试场景的相关章节。

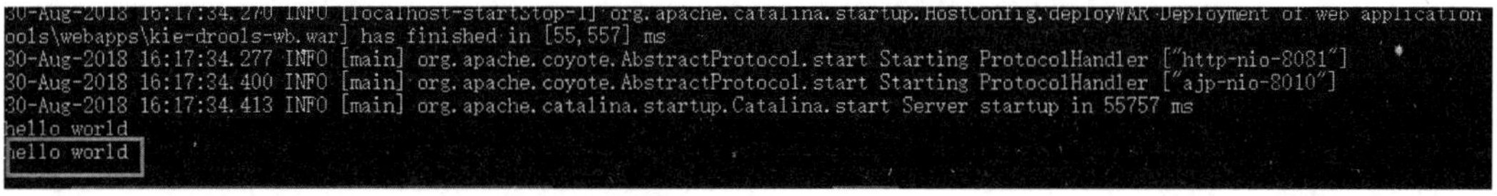

图11-29　输出的效果

2. 数据对象

单击"数据对象"图标，如图 11-30 所示。

图11-30　创建数据对象

弹出如图 11-31 所示的界面，输入文件名，该文件名是必填项，软件包在规则相关文件中指的是资料目录，选择"com.droolsone.testdroolsone"选项是指把规则文件放在 java/com/droolsone/testdroolsone 目录下，因为这是一个 java 文件。

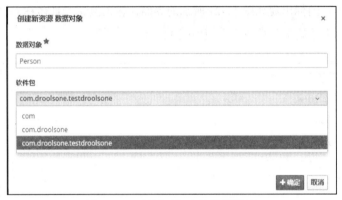

图11-31　数据对象文件编辑界面

添加完成后，根据图 11-32 所示的效果进行数据对象的操作。

图11-32　数据对象内容编辑界面

数据对象其实就是 JavaBean，在规则中也称为 Fact 事实对象。根据界面提供的"添加字段"按钮进行 JavaBean 属性的添加，效果如图 11-33 所示，其中 Id 是指 JavaBean 属性名，Id 与类型为必填项，"创建""创建并继续"就很好理解了。

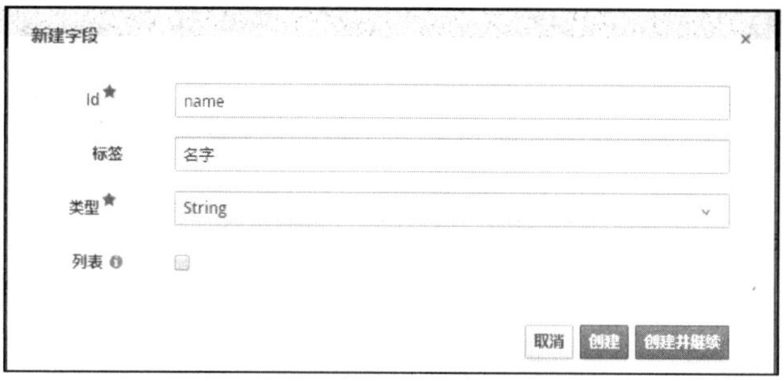

图11-33 数据对象属性编辑界面

创建完成后打开界面上的源代码，这个功能是非常有用的，可以直观地看到它生成代码后的内容，不只在数据对象上可以这样用，还可以在决策表、决策树上这样用。图 11-34 所示为一个标准的 JavaBean，并实现了 Get、Set 方法，重写 Equals、HashCode 方法。

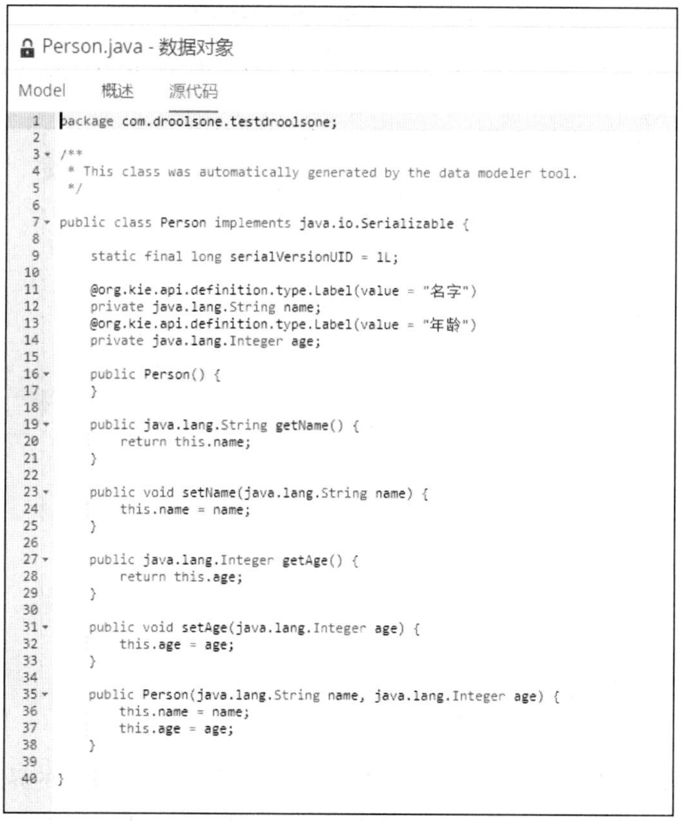


```
package com.droolsone.testdroolsone;

/**
 * This class was automatically generated by the data modeler tool.
 */

public class Person implements java.io.Serializable {

    static final long serialVersionUID = 1L;

    @org.kie.api.definition.type.Label(value = "名字")
    private java.lang.String name;
    @org.kie.api.definition.type.Label(value = "年龄")
    private java.lang.Integer age;

    public Person() {
    }

    public java.lang.String getName() {
        return this.name;
    }

    public void setName(java.lang.String name) {
        this.name = name;
    }

    public java.lang.Integer getAge() {
        return this.age;
    }

    public void setAge(java.lang.Integer age) {
        this.age = age;
    }

    public Person(java.lang.String name, java.lang.Integer age) {
        this.name = name;
        this.age = age;
    }

}
```

图11-34 数据对象源代码界面

单击“保存”按钮。Workbench 的功能是很强大的，对数据对象也是如此，在数据对象右侧有几个蓝色的按钮，如图 11-35 所示。

图11-35 数据对象扩展界面

当选中的是 Person 对象时,选择 Drools&jBPM 会出现如图 11-36 所示的界面。

图11-36 操作当前对象界面

选中 Person 的属性字段时,选择 Drools&jBPM 会出现如图 11-37 所示的界面。

不同的按钮有不同的功能,如果读者有兴趣可以对这几个按钮进行设置,完成后查看源代码会发生怎样的变化。

创建完成的数据对象有什么用呢?或者说如何在规则中使用呢?在添加之前创建的规则文件的编写规则区域多了一个事实类型,这就是刚刚创建的数据对象,如图 11-38 所示。

图11-37 操作数据对象属性界面

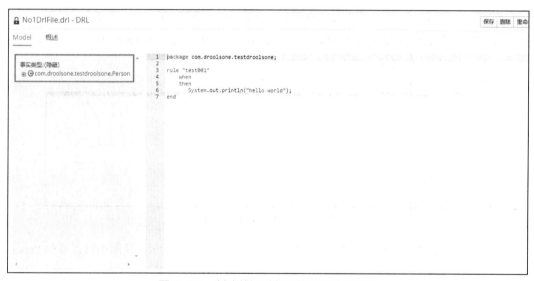

图11-38 创建数据对象对规则文件的影响

创建一个规则文件,即 ruleTestTwo.drl 文件,目录与 No1DrlFile.drl 文件一样,并在编写区域输入如下内容。

```
package com.droolsone.testdroolsone;
import com.droolsone.testdroolsone.Person;

rule "test002"
    when
     $p:Person(name=="张三")
    then
       System.out.println($p.getAge());
end
```

3.测试场景

测试场景是用来对规则文件进行测试的,在 Workbench 中编辑的规则最好通过测试场景进行验证，可以降低规则的出错率，提高规则的使用率。测试场景的功能也十分强大，如图11-39所示。

图11-39　测试场景界面

测试场景的核心在编写区，也就是GIVEN、调用方法、EXPECT。至于添加全局变量并不是一定的。GIVEN是添加输入数据和期望值，其实测试场景的功能与在Java调用Drools语法时的insert很相似，如图11-40所示。其中"事实名称"项是必添值。

图11-40　添加新事实对象

单击"添加"按钮后，测试场景界面会发生变化，如图11-41所示；从图中可以看到Person做了一个Facts事实对象插入规则中，并可以添加字段，如图11-42所示。

图11-41　添加事实对象后的界面

第 11 章　Workbench

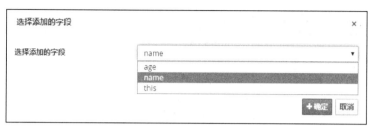

图11-42　操作事实对象属性界面

规则文件中 When 条件区是 name==" 张三 "，then 结果区是返回 age，所以需要添加两个字段。单击"插入 'Person'[p]"也可以进行字段的添加，如图 11-43 所示的用右框标记的地方。

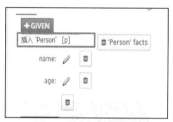

图11-43　操作事实对象界面

通过图 11-43 所示的 name 与 age 后面的笔图标进行操作，该图标是用来设置要进行比较的值。单击"编辑"按钮，选择字面值后，测试场景的界面就会变成如图 11-44 所示的效果，输入值并保存。

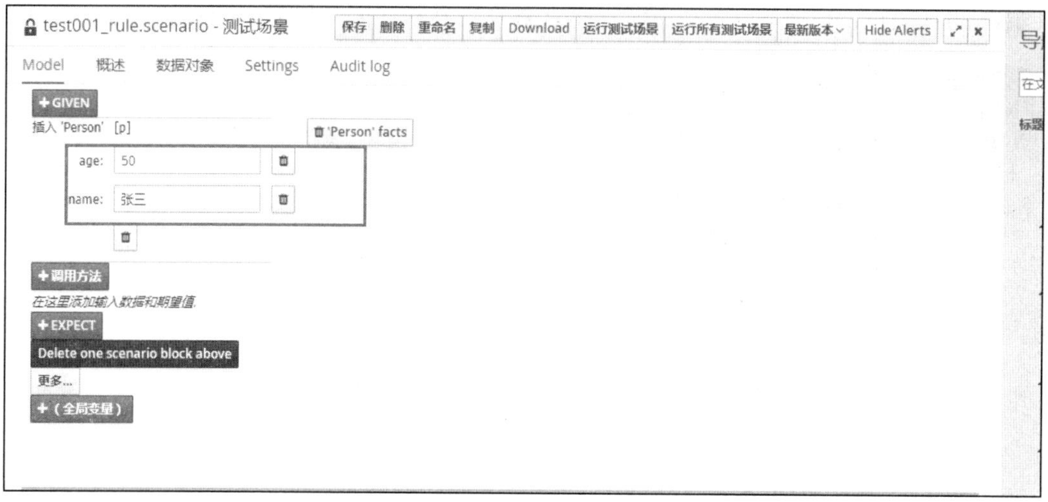

图11-44　添加事实对象insert的比较值

蓝色按钮 EXPECT 是指调用哪些规则，将 then 部分的参数是否设置为与测试场景的值相同。如图 11-45 所示，选择 test002 规则名，并单击"确定"按钮，测试场景就会出现如图 11-46 所示的效果，默认为"至少激活一次"选项，也可以选择其他选项。

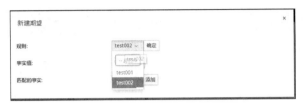

图11-45 选择要执行的规则

图11-46 选择活动模

单击"运行测试场景"按钮,在Tomcat的控制台就会出现如图11-47所示的输出结果。

图11-47 测试场景后的输出效果

为什么test001规则也输出了?因为当前的规则在同一个规则库中,而且test001规则条件永为True,所以在控制台的输出内容也自然包括了规则test001。

编辑ruleTestTwo.drl规则文件,添加test003规则,内容如下,效果如图11-48所示。

```
rule "test003"
    when
      $p:Person(name=="张三")
    then
        $p.setAge(30);
        $p.setName("李四")
        update($p)
        System.out.println($p.getAge());
end
```

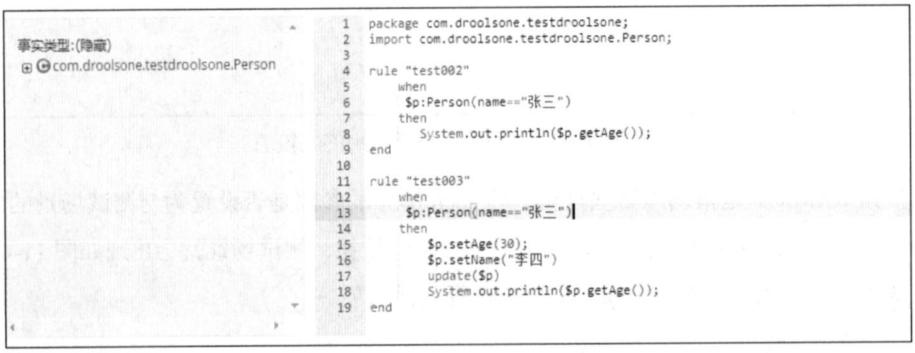

图11-48 添加test003规则

测试蓝色按钮 Expect 事实功能，单击"Expect"按钮选择事实对象，如图 11-49 所示。

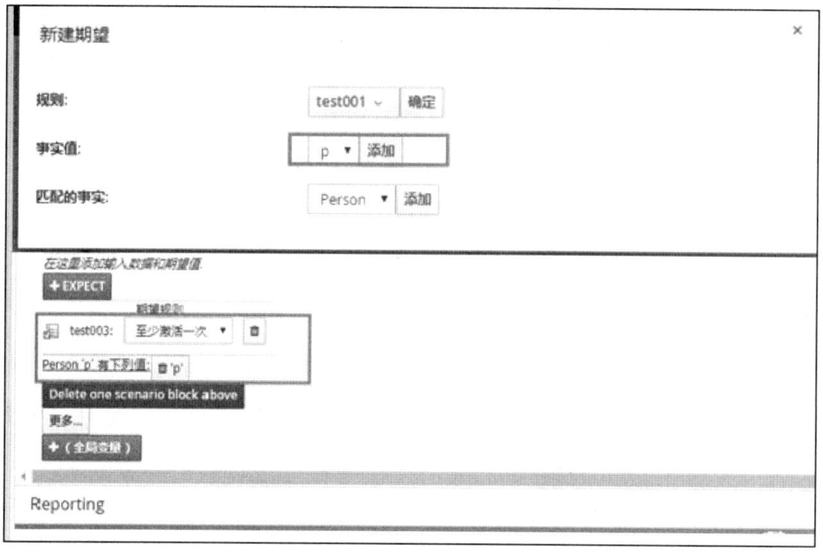

图11-49　编辑测试场景效果

单击框出部分，并输入规则 test003 的 then 所要求的值，如图 11-50 所示的效果。注意，本例中已经将新添加的规则 test003 进行测试了，所以一定要选择激活的规则。

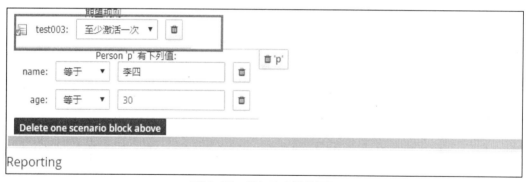

图11-50　添加事实操作界面

单击"运行测试场景"按钮，在 Tomcat 的控制台就会出现如图 11-51 所示的输出结果。

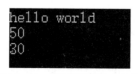

图11-51　测试场景输出结果

如果将 name 等于李四的值改为 name 等于李小四，再单击"运行测试场景"按钮就会出现如图 11-52 所示的效果，这表示没有期望值与结果是相同的。

305

图11-52 测试场景异常效果

创建测试场景时强调过一件事,就是选择的软件包要与规则目录保持一致,这样在测试场景默认得到的数据对象就是该路径下的所有数据对象,但有时数据对象与测试场景并不与数据对象在同一个目录下,就无法找到要添加的事实对象。解决方案是找到测试场景,并从其菜单中找数据对象,如图11-53所示。

图11-53 数据对象标签界面

单击"新建条目"按钮,在Import下拉列表中有自定义的数据对象,也有其他常用的,Workbench还可以进行jar包的导入,该jar包中的数据对象也是可以被引用的。

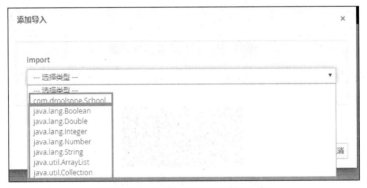

图11-54 测试场景引用数据对象

4. DSL定义

单击"DSL定义"图标,如图11-55所示。

图11-55　创建DSL文件

弹出如图 11-56 所示的界面，并输入文件名，该文件名是必填项，软件包在规则相关文件中指的是资料目录，选择"com.droolsone.testdroolsone"选项是指把规则文件放在 resources/com/droolsone/testdroolsone 目录下，因为这是一个 java 文件。

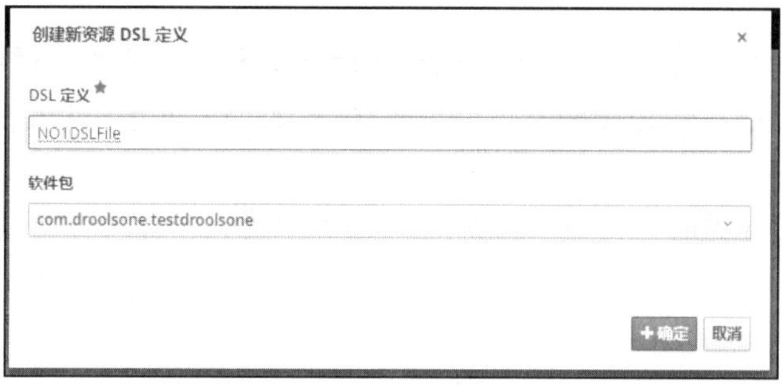

图11-56　DSL文件编辑界面

添加完成后，根据图 11-57 所示的效果进行规则编辑，其内容如下。

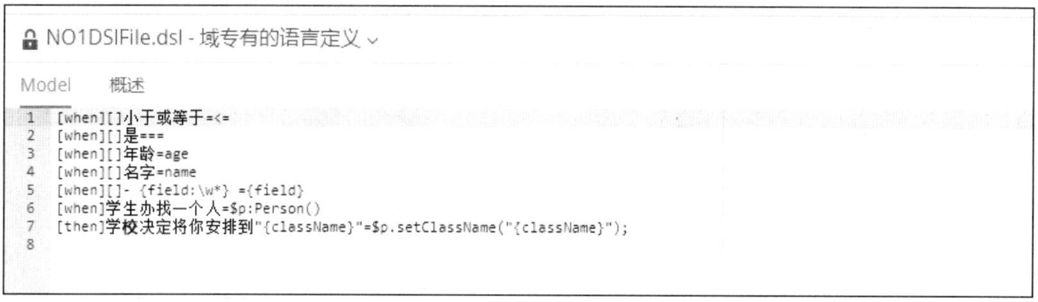

图11-57　添加DSL文件内容

```
[when][]小于或等于=<=
[when][]是===
[when][]年龄=age
[when][]名字=name
[when][]- {field:\w*} ={field}
[when]学生办找一个人=$p:Person()
[then]学校决定将你安排到"{className}"=$p.setClassName("{className}");
```

如何使用 DSL 文件？第 3 章 DSL 领域语言中介绍过 *.dslr 是领域语言的使用文件，在 Workbench 中也是一样的。在 DRL 文件中创建时有一个选择框，是将 *.drl 文件转换成 *.dslr 文件的复选框。创建一个 DRL 文件，输入 DRL 文件名，软件包选择与之前相同的路径，如图 11-58 所示。

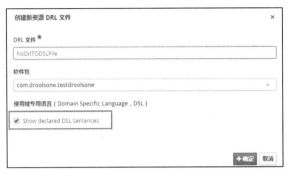

图11-58　创建DSL文件界面

单击"确定"按钮后，会出现如图11-59所示的界面，只要在规则文件编写区域正确地写上领域语言的内容即可。功能与第10章的DSL领域语言是一样的。通过测试场景进行测试就可以了。

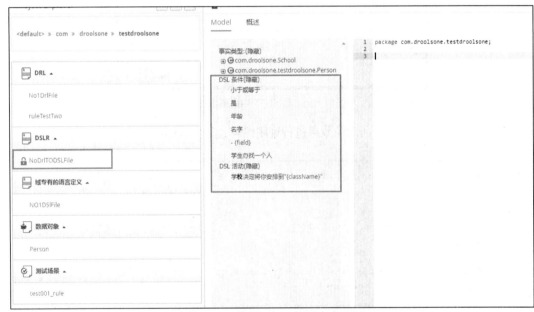

图11-59　编辑DSLR规则内容界面

5. 决策树

单击"向导型决策树"按钮，如图11-60所示。

图11-60　创建决策树文件

弹出如图11-61所示的界面，并输入文件名，该文件名是必填项，软件包在规则相关文件中指的是资料目录，选择"com.droolsone.testdroolsone"选项是指把规则文件放在resources/com/

droolsone/testdroolsone 目录下。

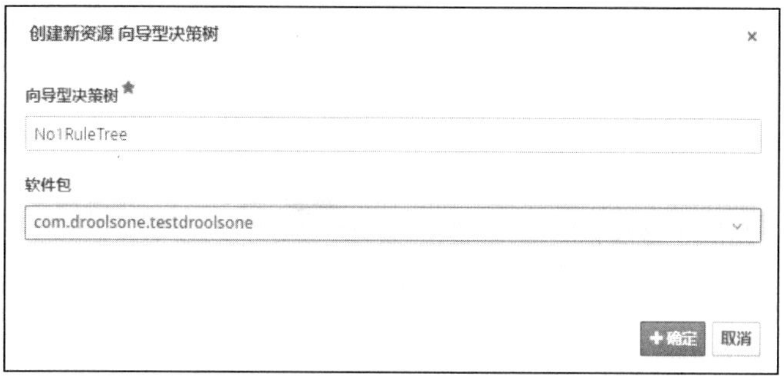

图11-61 决策树编辑界面

添加完成后，根据图 11-62 所示的效果进行决策树的操作。

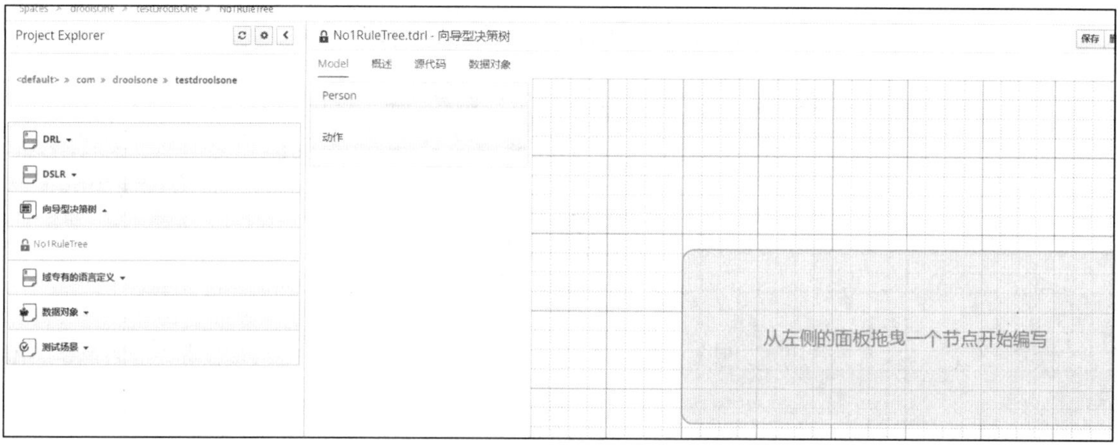

图11-62 决策树内容编辑界面

决策树的使用离不开数据对象，如果没有创建数据对象，或者是没有引用数据对象，只会出现动作选项，而且动作选项是无法直接拖曳到编辑区的。这是需要注意的，初学者很容易犯这个错误。

决策树是一个图形页面，可以用树状结构来展现规则，它的使用必须依赖数据对象，拖曳 Person 中的红色图标到右则的网格中，如图 11-63 所示。

红色的 Person 表示规则体中 when 的约束条件，即匹配模式。绿色图标是当前 Fact 事实对象的属性，没有 Person 或没有直接的关联关系的属性是不可以直接被拖曳到红色 Fact 事实对象下方的。

Workebch 在决策树中有一个特别容易出问题的地方，有时无法选中被拖曳后的 Fact 对象。解决方案有两种：第一种是重启 Workbench，第二种是单击右上角的类型药箱图标重建项目页面。

单击红色 Person 图标的编辑按钮，即笔＋纸的图标，注意，如图 11-64 所示的是正确操作决策树界面。

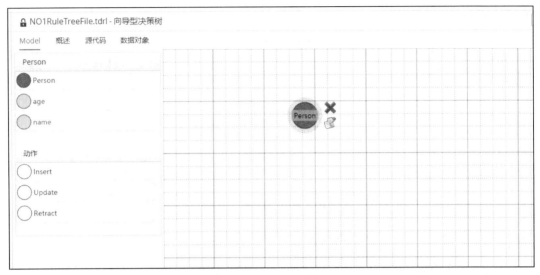

图11-63　拖曳Model数据对象的效果

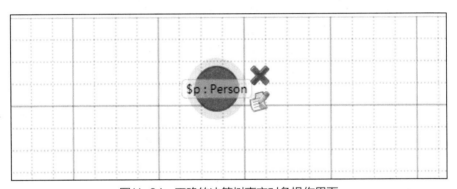

图11-64　正确的决策树事实对象操作界面

如何出现图11-64中的$p:Person？单击事实对象的操作按钮，这时会在页面上弹出如图11-65所示的效果，在文本框中添加"$p"。

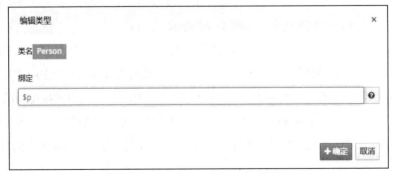

图11-65　添加事实对象绑定值界面

绑定$p的效果与在规则文件中写$p:Person()是一样的，而且这也是必须写的，因为这关系到动作的使用。如果不绑定变量，则动作无法拖曳到操作 Fact 对象的下方，这也是决策树离不开数据对象的原因。

通过对 Fact 事实对象属性的拖曳与动作的拖曳完成如图 11-66 所示的效果。

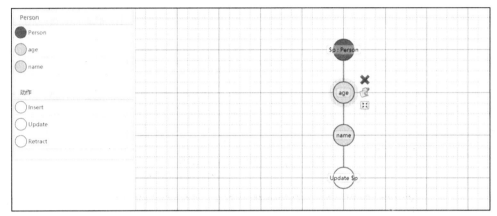

图11-66　拖曳事实对象属性和动作后的效果

分别对 age、name 进行参数设置。单击 age 的编辑图标，弹出如图 11-67 所示的效果。

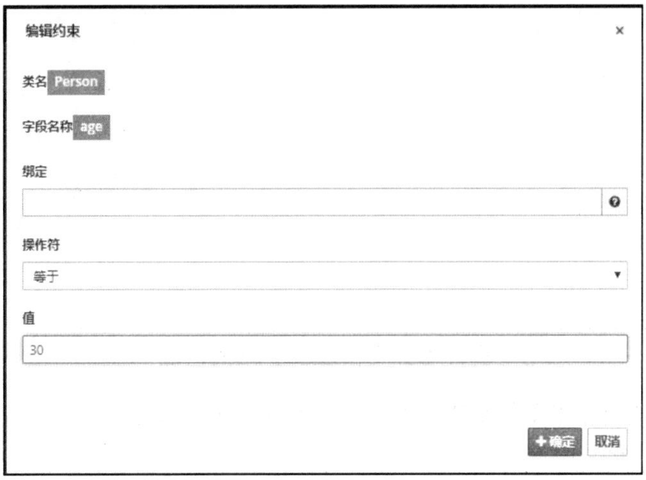

图11-67　事实属性约束界面

绑定变量与 Fact 是一样的，如果写绑定值 $age，则效果就如同在规则中设置 $p:Person($age:age)。操作符有很多种，本例中选择"等于"选项，如果是值一定要注意与数据对象属性的类型一致，然后再设置 name。设置完成后如图 11-68 所示。

选择"Update"动作，该动作是用来操作 Fact 对象的修改功能的，图中有一个复选框，选中该复选框则会修改 Fact 事实对象，用的是 modify 功能。如果不选中，则代表是用 set 进行操作。设置字段值需要单击下方的"添加"按钮，该值为 then 部分的设置，如图 11-69 所示。

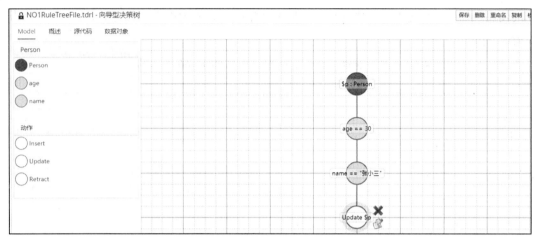

图11-68　完整的决策树效果

图11-69　Update动作操作界面

取消选中"用修改更新引擎"复选框时决策树生成的规则源码，如图11-70所示。

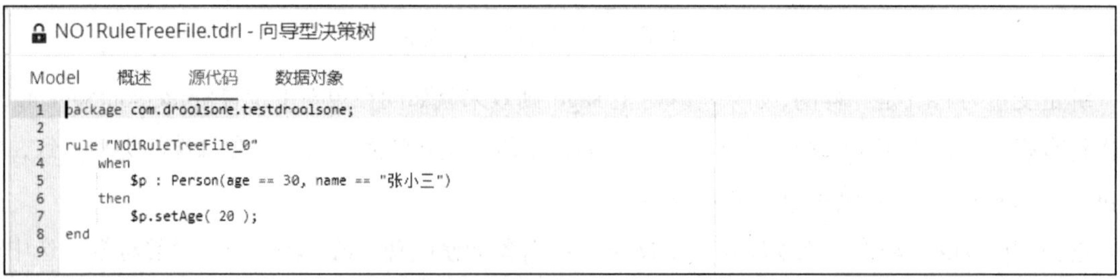

图11-70　取消选中"用修改更新引擎"复选框时的决策树源码

选中"用修改更新引擎"复选框时决策树生成的规则源码，如图11-71所示。

第 11 章 Workbench

```
NO1RuleTreeFile.tdrl - 向导型决策树
Model  概述  源代码  数据对象
1  package com.droolsone.testdroolsone;
2
3  rule "NO1RuleTreeFile_0"
4    when
5      $p : Person(age == 30, name == "张小三")
6    then
7      modify( $p ) {
8        setAge( 20 )
9      }
10  end
11
```

图 11-71　选中"用修改更新引擎"复选框时的决策树源码

想要测试编写的决策树是否正确，可以通过新建条目中的测试场景来操作。通过测试场景的应用得出，决策树是一个单独的文件（.tdrl 文件），不能进行相互引用，有关决策树需要注意如下几点。

（1）树必须有一个根数据对象。

（2）一棵树只能有一个根。

（3）数据对象可以是字段属性或其他数据对象，其他数据对象必须是根数据对象中的子类或是成员属性。

（4）字段的字段约束必须是相同的数据对象作为父节点。

（5）编辑决策树时，要注意死循环，因为决策树不可添加属性。

6.Guided 决策表

单击"Guided 决策表"图标，如图 11-72 所示。

图 11-72　创建决策文件

弹出如图 11-73 所示的界面，输入文件名，该文件名是必填项，软件包在规则相关文件中指的是资料目录，选择"com.droolsone.testdroolsone"选项是指把规则文件放在 resources/com/droolsone/testdroolsone 目录下。

创建 Guided 决策表有扩展的条目与限制的条目两种方式，这两种方式对决策表的使用都有所影响。

（1）扩展的条目，值在表体里定义。选中该单选按钮，则可使用向导，通过使用向导更容易理解 Guided 决策表的创建与应用。

添加名称并选择软件包之后单击"确定"按钮会出现如图 11-74 所示的对话框。

图11-73 决策表文件编辑界面

图11-74 扩展的条目文件配置界面

单击"下一步"按钮,会出现如图 11-75 所示的对话框,其功能是让决策表导入所需要的数据对象。

第 11 章 Workbench

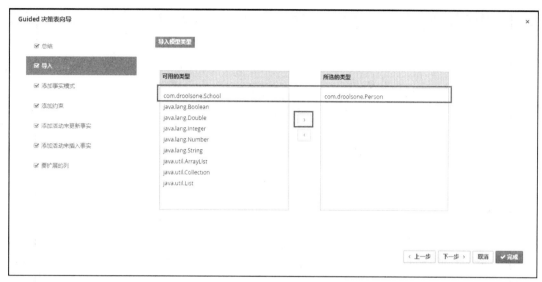

图11-75 决策表导入数据对象

单击"下一步"按钮，会出现如图 11-76 所示的对话框，其功能是绑定数据对象，如在规则文件中的 $p:Person() 效果，只有选择了数据对象才会出现该信息。

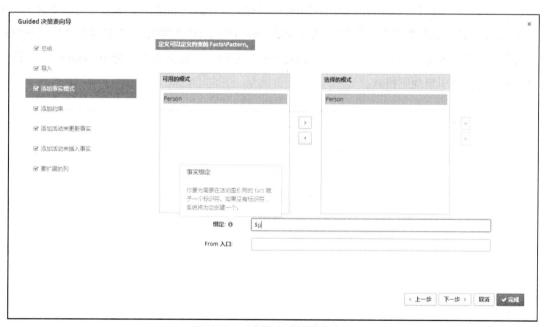

图11-76 决策表设置绑定变量

单击"下一步"按钮，会出现如图 11-77 所示的对话框，其功能是设置约束条件，本例中类型设置了 name，转换成规则文件后就是 $p:Person(name==' 张三 ')。

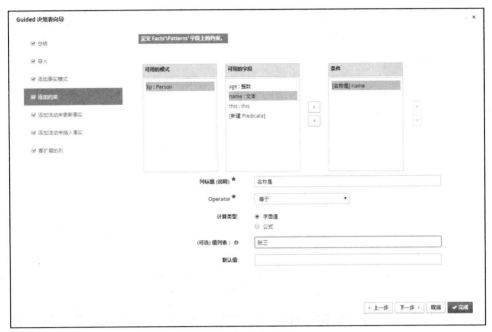

图11-77 决策表设置约束条件

单击"下一步"按钮，会出现如图 11-78 所示的对话框，其功能是设置 then 结果，本例中类型设置了 name，转换成规则文件后就是类似 modify($p){setName(' 李四 ');}，当然这是选中"用修改更新引擎"复选框才有这样的效果，如果取消选中该复选框，则就是直接的 $p.setName("李四")，与决策树设置更新引擎规则相似。

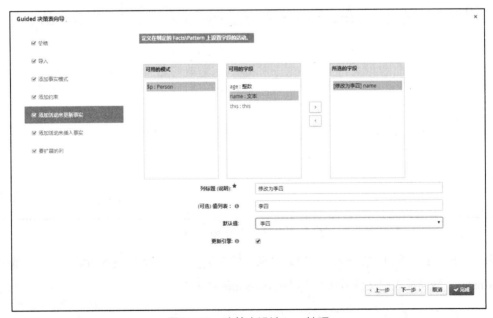

图11-78 决策表设计then处理

单击"下一步"按钮，会出现如图 11-79 所示的对话框，其功能是设置 insert 的规则操作，本例中暂不需要。

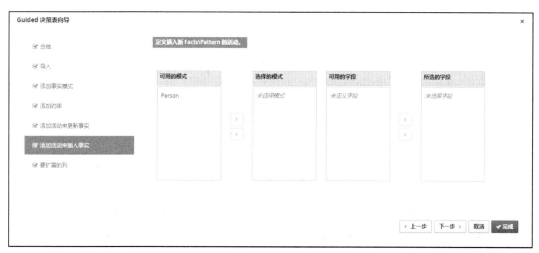

图11-79　决策表数据初始化

单击"下一步"按钮，会出现如图 11-80 所示的对话框。

图11-80　决策表配置完成

单击"完成"按钮，会在 resources/com/droolsone/testdroolsone 目录创建一个如图 11-81 所示的决策表，注意画框的地方，Workbench 7.x 版本与 Workbench 之前的 6.4 版本对决策表的操作有所不同。

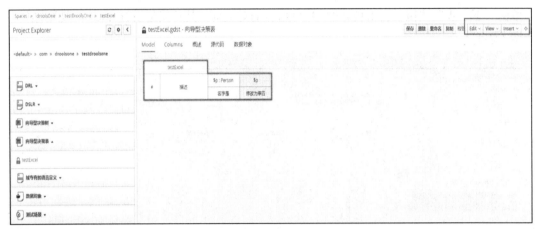

图11-81　决策表创建完成后的页面

既然已经创建好了决策表，如何生成规则，这是一个问题，先来看看Workbench 6.4版本是如何操作的，如图11-82所示，通过添加行来操作。

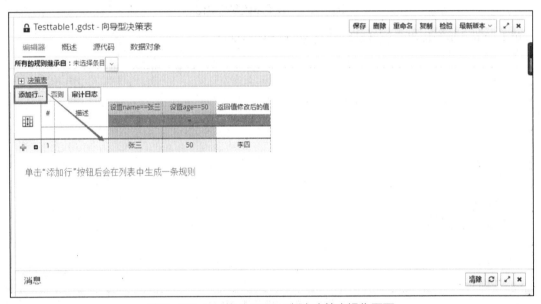

图11-82　Workbench 6.4版本决策表操作页面

既然Workbench 6.4版本提供了可以添加行的功能，7.10版本也应该提供。通过图11-81所示的画框部分，可以看出新版本Workbench的决策表有一个很好用的功能，就是它的决策表格是可以拖曳的，但这并不是重点，还是先看一下它是如何添加行的。拖曳表格到insert按钮下。如图11-83所示，选择"Append row"选项，描述列与之前介绍的决策表功能一样，是用来添加注释的。

第 11 章 Workbench

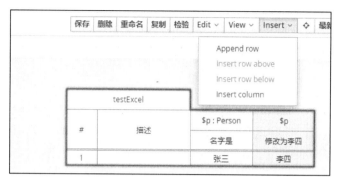

图11-83　决策表添加数据

单击"保存"按钮后，通过查看源代码，可以看到生成的规则文件，如图 11-84 所示。

图11-84　决策表编辑完成后生成的规则代码

（2）限制的条目，值在列里定义。选中该单选按钮，则可使用向导，通过使用向导更容易理解 Guided 决策表的创建与应用。

添加名称并选择软件包之后单击"确定"按钮会出现如图 11-85 所示的对话框。

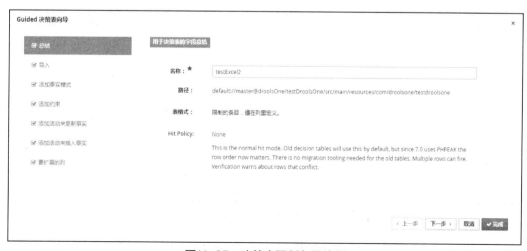

图11-85　决策表限制条目编辑

添加过程基本相同，在添加约束条件与添加事实活动更新事实时有不一样的地方，如图 11-86

319

和图 11-87 所示。

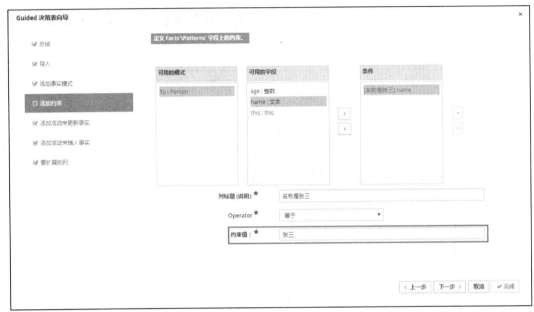

图11-86　决策表限制条目添加约束条件

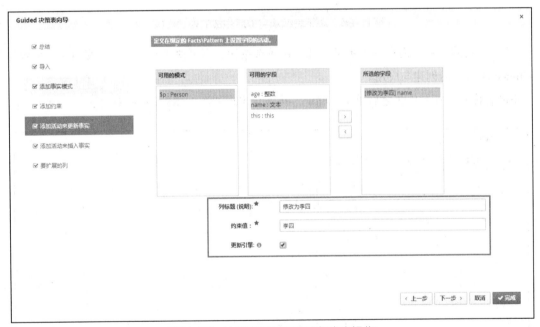

图11-87　决策表限制条目更新事实操作

单击"完成"按钮，会在 resources/com/droolsone/testdroolsone 目录创建一个如图 11-88 所示的决策表，与扩展的条目有所不同，每一行是可以进行选中复选框操作的。

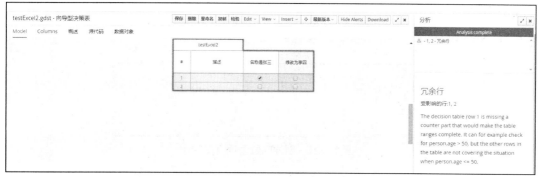

图11-88　决策表限制条目界面

可以选中第一行"修改为李四"复选框，查询源代码，如图11-89所示，而且在决策表的右侧可以很清晰地看出决策表哪里出了问题。

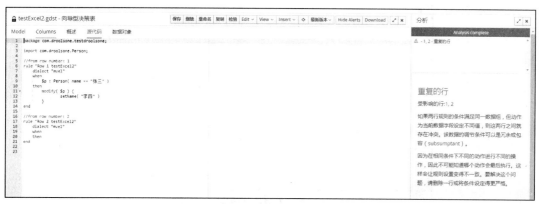

图11-89　决策表限制条目生成规则源码

添加行与扩展的条目相似，当选中的复选框出现如图11-90所示的效果，右侧将不再出现提示，说明这个决策表是符合基本规则设计的，即使全选中也不会有问题。

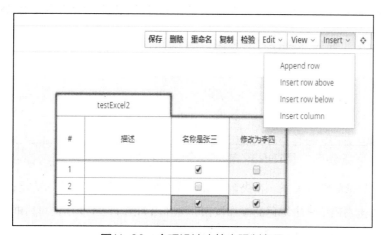

图11-90　合理设计决策表限制条目

7. Guided 规则

单击"Guided 规则"图标，如图 11-91 所示。

图11-91　创建向导型规则文件

弹出如图 11-92 所示的界面，输入文件名，该文件名为必填项，软件包在规则相关文件中指的是资料目录，选择"com.droolsone.testdroolsone"选项是指把规则文件放在 resources/com/droolsone/testdroolsone 目录下。

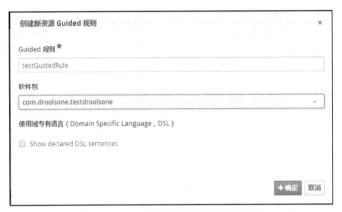

图11-92　向导型规则文件编辑页面

创建 Guided 规则与创建规则文件一样，也有两种选择，一种是使用 DSL，另一种是不使用 DSL。单击"确定"按钮后，在 Workbench 的界面上就会看到如图 11-93 所示的效果。

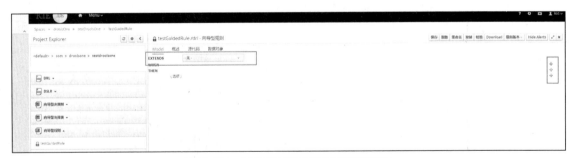

图11-93　向导型规则文件内容编辑页面

EXTENDS 是规则的继承关系。在决策表或是决策树中其实都有这样的功能，具体的设计还要切合实际业务。单击右侧第一个绿色的添加按钮，会弹出如图 11-94 所示的对话框。选择"Person"选项，相当于在规则文件中的 LHS 部分引用了约束条件的 Fact 对象是 Person，这类 Guided 引导型规则文件还支持自定义格式的 DRL，但需要编辑者编写的规则符合语法。

图11-94　向导型规则文件添加事实对象

这时 Guided 规则界面上会出现如图 11-95 所示的效果。

图11-95　向导型规则文件配置事实对象后的效果

其实在设置 WHEN 或是 THEN 时就是在设置规则体中的 LHS 与 RHS 部分，单击"存在 Person"按钮后会弹出如图 11-96 所示的对话框。

可以在引导型规则中设置添加字段限制、多字段线束、表达式编辑器等，选择添加字段限制，这个字段限制是针对某一个字段的，选择 Person 的 name 属性后，Guided 规则界面就会出现如图 11-97 所示的效果。变量名的设置与决策树中的效果很相似，设置了变量名就可以在设置 THEN 时设置删除或修改。

图11-96　向导型规则文件添加约束条件

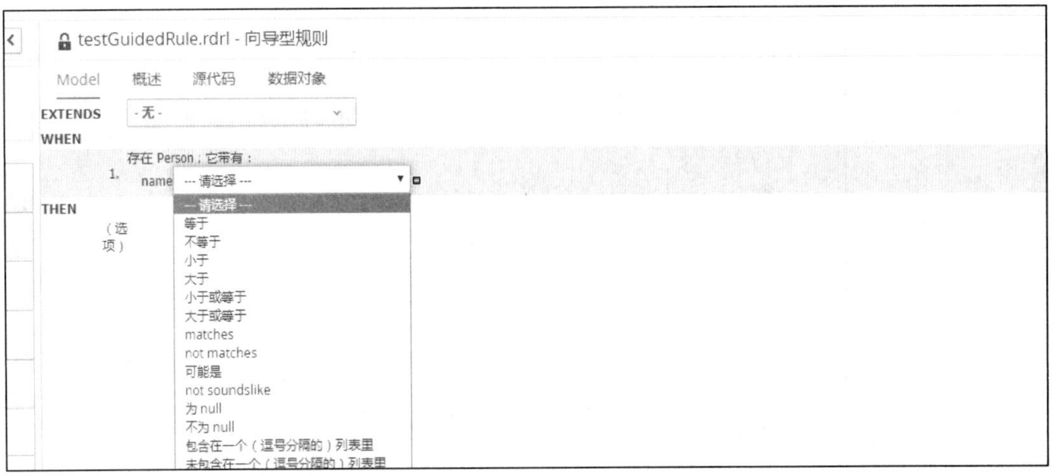

图11-97　向导型规则事实属性编辑约束

通过下拉列表框选择 name 的约束条件，选择"等于"约束条件后，单击旁边的铅笔状的图标，如图 11-98 所示，设置字面值与之前讲到的测试场景相似，是用来设置规则的约束值的，单击"字面值"按钮后会返回 Guided 规则界面，在它的输入框中输入想要比较的参数就可以了。

图11-98　向导型规则事实属性编辑约束值

第二个绿色的添加图标是用来设置 THEN 部分的，设置方式与 WHEN 相似，第三个绿色的添加图标是用来设置规则属性的，单击该图标会弹出如图 11-99 所示的对话框。

图11-99　向导型规则设计属性

它还能识别其他属性的设置，如已经在规则中选择了 no-loop 与 salience，所以在添加规则属性时就不会再出现，如图 11-100 所示。在图 11-99 中就没有这两个属性了，证明一个属性是不能被设置多次的。

图11-100　向导型规则设计属性后的效果

属性的参数不同，所展示出来的方式也就不相同，salience 是 int 类型，所以它是文本框，要注意其类型不能添写错；no-loop 是 Boolean 类型，所以它是通过选中复选框来控制它是 true 还是 false 的。

设置完成后保存，查询源代码，效果如图 11-101 所示。

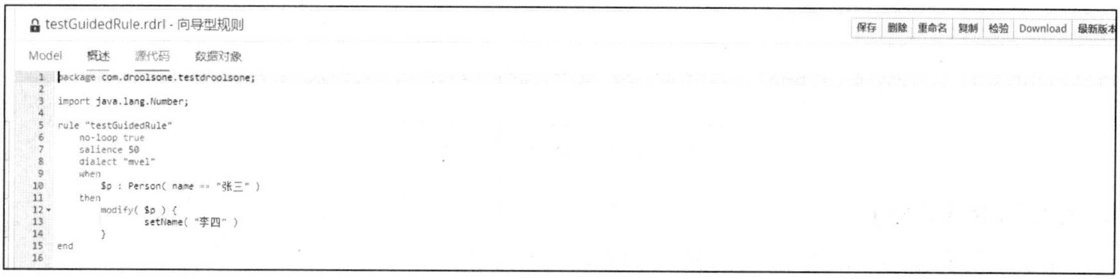

图11-101　向导型规则配置完成后生成规则源代码的效果

8. 软件包

单击"软件包"图标，如图 11-102 所示。

图11-102　创建软件包

弹出如图 11-103 所示的界面，输入文件名，该文件名为必填项。软件包是指创建了一个目录，选择软件包"com"选项，单击"确定"按钮，表示在 com 包路径下创建一个 droolsTwo 的目录。

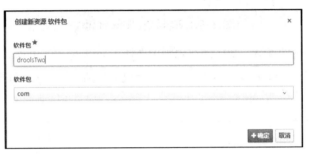

图11-103　软件包编辑界面

单击"确定"按钮后，在左侧的项目管理中会出现创建后的软件包，如图 11-104 所示。

图11-104　软件包显示效果

9. 决策表（电子表格）

单击"决策表（电子表格）"图标，如图 11-105 所示。

第 11 章 Workbench

图11-105　创建电子表格

弹出如图 11-106 所示的界面，输入文件名，该文件名为必填项，软件包在规则相关文件中指的是资料目录，选择"com.droolsone.testdroolsone"选项是指把规则文件放在 resources/com/droolsone/testdroolsone 目录下。

图11-106　决策表（电子表格）文件编辑

这个功能很容易理解，就是将 Excel 文件上传到该目录下，并修改文件名，但需要注意的是，选择正确的文件扩展名，并且上传的 Excel 文件必须符合 Drools 决策表的语法规定，特别需要关注的是其 import 事实对象的包路径。单击"确定"按钮后会出现如图 11-107 所示的效果，它提供再次上传与下载的功能。

图11-107　上传与下载的操作

10. 业务流程

业务流程有两种，在目前的 Workbench 中有一种比较原始的在之前版本就存在的业务流程创建插件，还有一种新的业务流程插件，如图 11-108 所示。

图11-108 创建业务流程文件

单击"Business Process(legacy)"图标,弹出如图 11-109 所示的界面,输入文件名,该文件名为必填项,软件包在规则相关文件中指的是资料目录,选择"com.droolsone.testdroolsone"选项是指把规则文件放在 resources/com/droolsone/testdroolsone 目录下。无论是老版本还是新版本的业务流程文件,创建过程都是相似的。

图11-109 业务流程文件编辑

单击"确定"按钮后,会出现如图 11-110 所示的老版本业务流程文件内容编辑的界面,通过方框中的菜单折叠按钮来控制操作业务规则元素节点的选择和设置值。

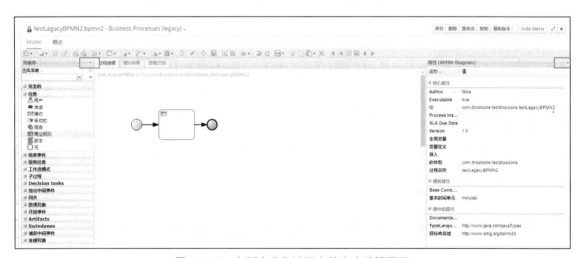

图11-110 老版本业务流程文件内容编辑界面

图 11-111 所示的是新版本业务流程文件内容编辑的界面,通过方框中的菜单折叠按钮来控制业务规则元素节点的内容,通过左侧的圆圈等图标来选择元素节点。页面相比之前老版本来说更加简单。

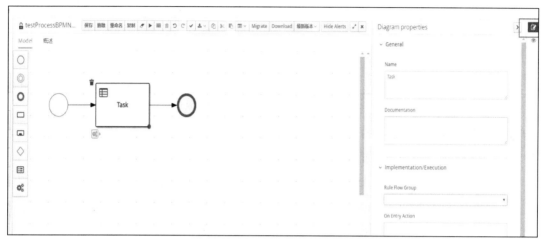

图11-111　新版本业务流程文件内容编辑界面

业务流程的操作整体与 Eclipse 相似，可以通过左右两侧的操作进行业务流程的设置。有规则体设置了规则流程的属性，就会在规则任务元素节点的设置值中设置了指定规则流程属性的规则。

11. 全局变量

单击"Global Variable(s)"图标，如图 11-112 所示。

图11-112　创建全局变量

弹出如图 11-113 所示的界面，输入文件名，该文件名为必填项，软件包在规则相关文件中指的是资料目录，选择"com.droolsone.testdroolsone"选项是指把规则文件放在 java/com/droolsone/testdroolsone 目录下。

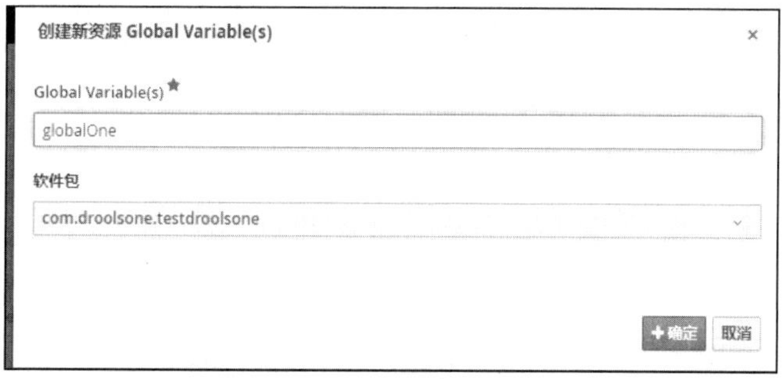

图11-113　全局变量文件编辑

单击"确定"按钮后，会出现如图 11-114 所示的界面。

图11-114　全局变量文件内容编辑

单击"Add"按钮，弹出如图11-115所示的对话框，设置Alias值为全局变量的别名，在规则文件中是指它的名称，下面是设置全局变量的名称。

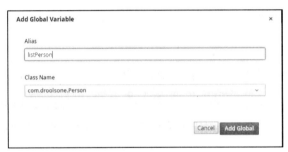

图11-115　设置全局变量名称

11.5　Workbench与Java交互

Workbench常用的操作在前面有了较为详细的说明，但Workbench如何在实际中应用是一个问题，Workbench只在本地服务上创建了规则，其中的规则又如何被业务端调用呢？下面就来介绍Workbench与业务端（Java端）的两种交互方式。与Java交互前先要做一些准备工作。

第一步：创建一个新的空间及项目，如图11-116所示。

图11-116　创建Workbench项目

第二步：在javaIsWorkbench中创建一个Person数据对象和一个决策树，如图11-117所示。

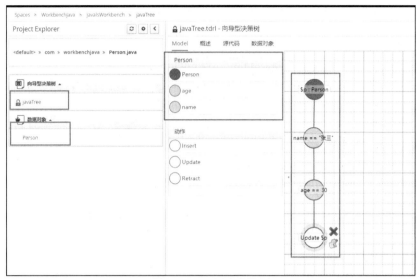

图11-117 创建决策树

单击决策树源代码标签，生成的规则内容如图 11-118 所示。

javaTree.tdrl - 向导型决策树

Model 概述 源代码 数据对象

```
package com.workbenchjava.javaisworkbench;

rule "javaTree_0"
    when
        $p : Person(name == "张三", age == 30)
    then
        modify( $p ) {
            setName( "李四" )
        }
end

```

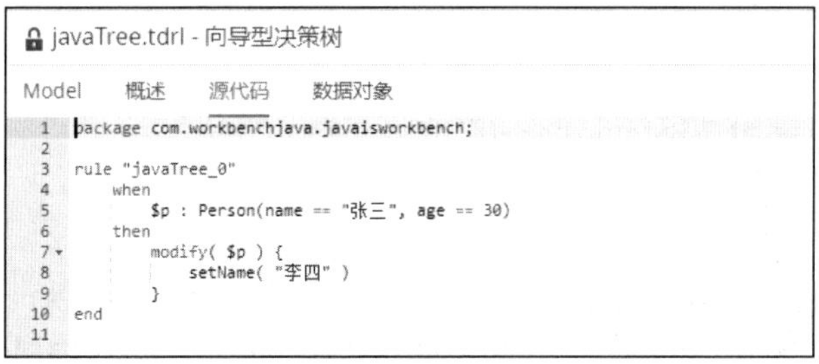

图11-118 决策树源代码

第三步：通过测试场景进行测试，证明决策树是正确的，虽然这个决策树相当简单，但还是要养成良好的习惯，对规则文件进行测试。

第四步：构建项目，界面如图 11-119 所示。

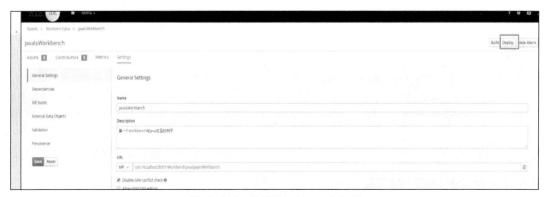

图11-119 构建Workbench项目

第五步：创建一个新的 droolsWorkbench 项目。修改 Pom.xml 文件，其内容如下。

```xml
<?xml version="1.0" encoding="UTF-8"?>
<project xmlns="http://maven.apache.org/POM/4.0.0"
         xmlns:xsi="http://www.w3.org/2001/XMLSchema-instance"
         xsi:schemaLocation="http://maven.apache.org/POM/4.0.0 http://maven.apache.org/xsd/maven-4.0.0.xsd">
    <modelVersion>4.0.0</modelVersion>

    <groupId>com</groupId>
    <artifactId>droolsWorkbench</artifactId>
    <version>1.0-SNAPSHOT</version>
    <properties>
        <drools.version>7.10.0.Final</drools.version>
        <log4j2.version>2.5</log4j2.version>
    </properties>
    <dependencies>
        <dependency>
            <groupId>org.drools</groupId>
            <artifactId>drools-compiler</artifactId>
            <version>${drools.version}</version>
        </dependency>
        <dependency>
            <groupId>junit</groupId>
            <artifactId>junit</artifactId>
            <version>4.11</version>
        </dependency>
        <dependency>
            <groupId>org.drools</groupId>
            <artifactId>drools-decisiontables</artifactId>
            <version>RELEASE</version>
        </dependency>
    </dependencies>
    <build>
        <plugins>
            <plugin>
                <groupId>org.apache.maven.plugins</groupId>
                <artifactId>maven-compiler-plugin</artifactId>
                <configuration>
                    <source>1.8</source>
                    <target>1.8</target>
                </configuration>
            </plugin>
        </plugins>
        <finalName>drools_test_test</finalName>
    </build>
</project>
```

1. Java与Workbench的一般交互

Workbench 与 Java 的简单方式交互需要在 Java 项目中创建一个与 Workbench 项目中的 Fact 事

实对象一样的 JavaBean 对象,该对象要与 Workbench 的数据对象路径一致,属性或方法可以相同但不能少于 Workbench 中数据对象的属性或方法。图 11-120 中的 Workbench 数据对象的路径是 com.Workbenchjava.javaisWorkbench。

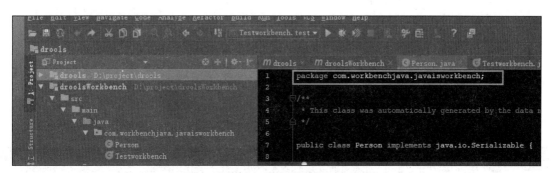

图11-120　Workbench数据对象源代码

而在 Java 项目中的 Person 包路径也必须是 com.Workbenchjava.javaisWorkbench,如图 11-121 所示。

图11-121　Java项目的数据对象目录

创建 Workbench 与 Java 交互的测试类 TestWorkbench.java 文件,在 com.Workbenchjava.javais-Workbench 目录下,其内容如下(注意加粗部分)。

```
package com.Workbenchjava.javaisWorkbench;

import org.drools.core.io.impl.UrlResource;
import org.junit.Test;
import org.kie.api.KieServices;
import org.kie.api.builder.KieModule;
import org.kie.api.builder.KieRepository;
import org.kie.api.runtime.KieContainer;
import org.kie.api.runtime.KieSession;

import java.io.IOException;
import java.io.InputStream;

public class TestWorkbench {
```

```java
@Test
public void test() throws IOException {
    String url = "http://localhost:8081/kie-drools-wb/maven2/com/Workbenchjava/javaIsWorkbench/1.0.0/javaIsWorkbench-1.0.0.jar";
    KieServices ks = KieServices.Factory.get();
    KieRepository kr = ks.getRepository();
    UrlResource urlResource = (UrlResource) ks.getResources().newUrlResource(url);
    urlResource.setUsername("kie");
    urlResource.setPassword("kie");
    urlResource.setBasicAuthentication("enabled");
    InputStream is = urlResource.getInputStream();
    KieModule kModule = kr.addKieModule(ks.getResources().newInputStreamResource(is));
    KieContainer kContainer = ks.newKieContainer(kModule.getReleaseId());
    KieSession kieSession = kContainer.newKieSession();
    Person p = new Person();
    p.setAge(30);
    p.setName("张三");
    kieSession.insert(p);
    int i = kieSession.fireAllRules();
    System.out.print("Java调用Workbench,共执行了" + i + "条规则");
    System.out.print("修改后的结果" + p.getName());
}
```

执行 test() 方法，结果如图 11-122 所示。

图11-122　Java项目与Workbench交互成功

通过图 11-122 所示，证明调用 Workbench 是成功的。在前面讲到过，Workbench 是很多功能的结合体，构建成功的项目，其实是生成了一个 jar 包，至于与 Java 交互，相信读者也已经明白了，就是为了要引用这个 jar 包。

分析一下上面代码的功能，String url 是一个很重要的代码。它的功能主要是读取 Workbench 对外提供的 jar 包下载地址，内容如下：

```
String url = "http://localhost:8081/kie-drools-wb/maven2/com/Workbenchjava/javaIsWorkbench/1.0.0/javaIsWorkbench-1.0.0.jar";
```

怎么找到这个下载地方呢？返回 Workbench 界面，选择右上角的管理员按钮，如图 11-123 所示。

图11-123　管理员按钮

单击"Artifacts"图标，如图 11-124 所示。

图11-124　Artifacts图标

页面跳转到如图 11-125 所示的界面。

图11-125　下载Workbench资源目录

选择构建的项目，这里选择 javaIsWorkbench-1.0.0.jar，并单击"下载"按钮，就会弹出如图 11-126 所示的页面，在浏览器地址栏中找到该 jar 包的下载地址就可以了。

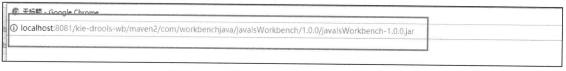

图11-126　获取Workbench生成的jar包地址

若要想找到新版本的 Workbench 中项目构建生成的 jar 包地址并不容易，与 Workbench 6.4 版本相比较，如图 11-127 所示，找到项目构建生成的 jar 包只需要在其资料库中找到即可。

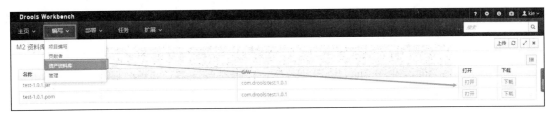

图11-127　Workbench 6.4版本下载jar包

单击"打开"按钮就可找到 Workbench 对外提供的下载地址，虽然这是当前项目 pom.xml 配置文件中的信息，但提供了更多的便捷操作。通过构建项目的 pom.xml 就可以找到相关的信息了，如图 11-128 所示。

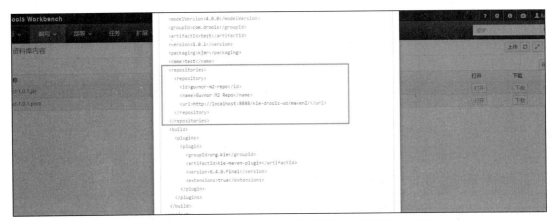

图11-128　查询Workbench构建生成的jar包配置文件内容

Java 项目获取 Workbench 的登录信息，其内容如下。

```
UrlResource urlResource = (UrlResource) ks.getResources().
newUrlResource(url);
urlResource.setUsername("kie");
urlResource.setPassword("kie");
urlResource.setBasicAuthentication("enabled");
InputStream is = urlResource.getInputStream();
```

上述的代码就相当于在 IE 中打开 KIE-WB 的 Web 地址，然后输入用户名、密码，并单击"登录"按钮的操作。

```
KieRepository kr = ks.getRepository();
KieModule kModule = kr.addKieModule(ks.getResources().
newInputStreamResource(is));
KieContainer kContainer = ks.newKieContainer(kModule.getReleaseId());
KieSession kieSession = kContainer.newKieSession();
```

上述代码是访问并执行规则。写法与前面讲的例子有一些区别，因为代码中没有指定 KieSession 的名称，理论上是应该指定的，但是因为在 Workbench 中没有设置，所以这里为默认值，可以不写，如果一定要写上，则要去 Workbench 项目代码的 kmodule.xml 中查找。

Java 项目与 Workbench 交互还有一种写法，其效果与上述代码的效果是一样的。虽然这种写法非常简单，但也有一个小小的弊端，如果本地环境中没有 jar 包，则无法进行调用。

编辑 TestWorkbench.java 文件，添加 runRules2() 方法，其内容如下。

```
@Test
public void runRules2() {
    KieServices kieServices = KieServices.Factory.get();
    ReleaseId releaseId = kieServices.newReleaseId("com.Workbenchjava",
"javaIsWorkbench", "1.0.0");
    KieContainer kContainer = kieServices.newKieContainer(releaseId);
    KieSession kSession = kContainer.newKieSession();
    Person p = new Person();
    p.setAge(30);
    p.setName("张三");
    kSession.insert(p);
    int i = kSession.fireAllRules();
    System.out.print("Java调用Workbench,共执行了" + i + "条规则");
    System.out.print("修改后的结果" + p.getName());
}
```

执行 runRules2() 方法，结果与图 11-122 所示的效果是一样的。第二种方式一般应用在自动扫描上，如果本地 Maven 与 Workbench 提供的下载地址可以交互，也可以通过这种方式进行，毕竟 Workbench 使用时会部署在其他机器上，不会是的本地计算机。

2. Java与Workbench的自动扫描

自动扫描的原理与一般方式交互是相似的，所不同的是自动扫描用到了 kie-ci 的包，而且它支持 Workbench 重新构建项目后 jar 包的读取，是动态规则实现方式之一。

Workbench 实现与 Java 的自动扫描功能具体方式有两种：第一种是 ci-api 的形式，第二种是 Spring 整合。Workbeanch 与 Maven 相似，这里的组 ID、构建 ID、版本号就相当重要了，做自动扫描前，要有几个前置条件，与一般方式交互的前置条件是一样的，但一定要注意 Java 项目与 Workbench 的版本问题，其中，Workbench 使用的版本是 7.10，而 Java 项目使用的是 7.9 版本，这里通过自动扫描方式是无法获取到新的规则值的。新版本的自动扫描比较简单，先做一些前置工作。修改 pom.xml 配置文件，添加 kie-ci 的引用，其内容如下。

```
<dependency>
    <groupId>org.kie</groupId>
    <artifactId>kie-ci</artifactId>
    <version>${drools.version}</version>
</dependency>
```

编辑 TestWorkbench.java 文件，添加 runkScanner() 方法，其内容如下。

```
@Test
public void runkScanner() {
    KieServices kieServices = KieServices.Factory.get();
    ReleaseId releaseId = kieServices.newReleaseId("com.Workbenchjava",
```

```
"javaIsWorkbench", "1.0.0");
    KieContainer kContainer = kieServices.newKieContainer(releaseId);
    KieScanner kScanner = kieServices.newKieScanner(kContainer);
    // 启动KieScanner轮询Maven存储库每10秒
    kScanner.start(10000L);
    while (true) {
        try {
            KieSession kSession = kContainer.newKieSession();
            Person p = new Person();
            p.setAge(30);
            p.setName("张三");
            kSession.insert(p);
            int i = kSession.fireAllRules();
            System.out.print("Java调用Workbench, 自动扫描 共执行了" + i + "条规则");
            System.out.println("修改后的结果" + p.getName());
            Thread.sleep(5000);
        } catch (InterruptedException e) {
            e.printStackTrace();
        }
    }
}
```

执行 runkScanner() 代码，由于之前并没有修改过 Workbench 中决策树的 update 节点，因此运行的结果与一般方式交互是一样的，如图 11-129 所示。

图11-129　自动扫描的效果1

返回 Workbench 的决策树页面，并修改 update 节点，将 name 的字段值由"李四"修改为"王五"并保存，如图 11-130 所示。

图11-130　编辑决策树

找到项目构建界面，单击"Deploy"按钮，提示构建成功后，返回本机 IDE 中，查询控制台的输出，如图 11-131 所示。

图11-131　自动扫描的效果2

除了上述的 ci-api 调用方式外，还有一种通过整合 Spring 配置文件的方式也可以实现自动扫描的交互。

整合 Spring，通过配置文件的方式需要修改 pom.xml，引用 Spring 的相关 jar 包。编辑 pom.xml 文件，添加 Spring 引用，其内容如下。

```xml
<spring.version>4.2.6.RELEASE</spring.version>
<dependency>
    <groupId>org.kie</groupId>
    <artifactId>kie-spring</artifactId>
    <version>${drools.version}</version>
</dependency>
<dependency>
    <groupId>org.springframework</groupId>
    <artifactId>spring-core</artifactId>
    <version>${spring.version}</version>
</dependency>
<dependency>
    <groupId>org.springframework</groupId>
    <artifactId>spring-beans</artifactId>
    <version>${spring.version}</version>
</dependency>
<dependency>
    <groupId>org.springframework</groupId>
    <artifactId>spring-context</artifactId>
    <version>${spring.version}</version>
</dependency>
<dependency>
    <groupId>org.springframework</groupId>
    <artifactId>spring-test</artifactId>
    <version>${spring.version}</version>
</dependency>
<dependency>
    <groupId>org.springframework</groupId>
    <artifactId>spring-context-support</artifactId>
```

```xml
        <version>${spring.version}</version>
</dependency>
<dependency>
    <groupId>org.springframework</groupId>
    <artifactId>spring-tx</artifactId>
    <version>${spring.version}</version>
</dependency>
```

创建 Srping.xml 配置文件，在 resources 目录中，其内容如下。

```xml
<?xml version="1.0" encoding="UTF-8"?>
<beans xmlns="http://www.springframework.org/schema/beans"
      xmlns:xsi="http://www.w3.org/2001/XMLSchema-instance"
      xmlns:kie="http://drools.org/schema/kie-spring"
      xsi:schemaLocation="
      http://drools.org/schema/kie-spring
      http://drools.org/schema/kie-spring.xsd
      http://www.springframework.org/schema/beans
      http://www.springframework.org/schema/beans/spring-beans.xsd">

    <kie:import releaseId-ref="sptringweb" enableScanner="true" scannerInterval="12000"/>
    <kie:releaseId id="sptringweb" groupId="com.Workbenchjava" artifactId="javaIsWorkbench" version="1.0.0"/>
    <bean id="kiePostProcessor" class="org.kie.spring.annotations.KModuleAnnotationPostProcessor"/>
</beans>
```

创建 TestSpring.java 文件，在 com/Workbenchjava/javaisWorkbench 目录中，其内容如下。

```java
package com.Workbenchjava.javaisWorkbench;

import org.junit.Test;
import org.junit.runner.RunWith;
import org.kie.api.KieBase;
import org.kie.api.cdi.KBase;
import org.kie.api.cdi.KSession;
import org.kie.api.runtime.KieSession;
import org.springframework.test.context.ContextConfiguration;
import org.springframework.test.context.junit4.SpringJUnit4ClassRunner;

@RunWith(SpringJUnit4ClassRunner.class)
@ContextConfiguration({"classpath:Spring.xml"})
public class TestSpring {
    @KSession("defaultKieSession")
    KieSession kSession;

    @Test
    public void runRules() {
        while (true) {
            try {
```

```
                Person p = new Person();
                p.setAge(30);
                p.setName("张三");
                kSession.insert(p);
                int i = kSession.fireAllRules();
                System.out.print("Java调用Workbench 自动扫描，共执行了" + i
 + "条规则");
                System.out.println("修改后的结果" + p.getName());
                Thread.sleep(5000);
            } catch (InterruptedException e) {
                e.printStackTrace();
            }
        }
    }
}
```

执行 runRules() 方法，由于在 ci-api 交互时，已经将决策树的 update 节点修改成了"王五"，通过执行该方法输出 Person name 的值是"王五"也是正常的，如图 11-132 所示。

图11-132　Spring自动扫描效果1

返回 Workbench 的决策树页面，并修改 update 节点，将 name 的字段值由"王五"修改为"赵六"并保存，如图 11-133 所示。

图11-133　编辑决策树

找到项目构建界面，单击"Deploy"按钮，提示构建成功后，返回本机 IDE 中，查询控制台的输出，如图 11-134 所示。

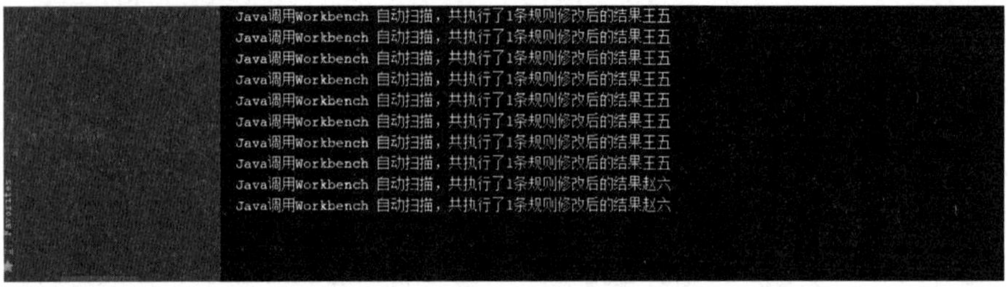

图11-134　Spring自动扫描效果2

Workbench 与 Java 交互是 KIE-WB 学习的高级应用，它可以让使用者对 Workbench 有一个更好的认知，在介绍 Workbench 的过程中一直都强调一个核心点就是 Maven，Workbench 部署离不开 Maven，它的创建、构建都离不开 Maven 相关的规范。在本章中都是通过 Windows 10（本地）+ Maven（本地）+ Java 项目（本地）+ Tomcat（本地）测试的。所以看似一切都是很有顺序的，因为 Workbench 构建生成的 kjar 包是直接放在本地 .m2 目录下的，所以配置上没有任何有关项目与远程服务器 Maven 关联的配置。如果 Workbench 部署到其他服务器上，这时就需要让项目或是本地 Maven 与 Workbench 所在的 Maven 进行关联。

通过一般方式交互中的阐述，Workbench 构建后会生成一个 kjar 包，并提供下载地址，也可以理解成 Workbench 向外提供了一个下载地址用来让 Java 项目或其他基于 JVM 项目进行交互。这就需要修改本地的 Maven 配置或 Java 项目 pom.xml 配置文件来进行关联。

在简单方式交互中，如果 Workbench 在其他服务器上，就需要修改 Java 项目的 pom.xml 配置文件。添加与 Workbench 交互的配置，内容如下（通过该代码可以将远程 Workbench 上构建的项目下载到本地 .m2 目录中）。

```
<repositories>
    <repository>
      <id>guvnor-m2-repo</id>
      <name>Guvnor M2 Repo</name>
      <url>http://localhost:8888/kie-drools-wb/maven2/</url>
    </repository>
  </repositories>
```

在自动扫描交互中，修改的就是本地环境变量中的 maven settings.xml 文件，其内容如下。

```
<?xml version="1.0" encoding="UTF-8"?>
<settings xmlns="http://maven.apache.org/SETTINGS/1.0.0"
        xmlns:xsi="http://www.w3.org/2001/XMLSchema-instance"
        xsi:schemaLocation="http://maven.apache.org/SETTINGS/1.0.0 http://maven.apache.org/xsd/settings-1.0.0.xsd">
<servers>
    <server>
```

```xml
      <id>guvnor-m2-repo</id>
      <username>kie</username>       <!-- 这里是登录时的账号密码，如果不设置将没有
权限会报错 -->
      <password>kie</password>
      <configuration>
        <wagonProvider>httpclient</wagonProvider>
        <httpConfiguration>
          <all>
            <usePreemptive>true</usePreemptive>
          </all>
        </httpConfiguration>
      </configuration>
    </server>
</servers>

<profiles>
    <profile>
      <id>guvnor-m2-repo</id>
      <repositories>
        <repository>
            <id>guvnor-m2-repo</id>
            <name>Guvnor M2 Repo</name>
            <url>http://localhost:8888/kie-drools-wb/maven2/</url>
          <layout>default</layout>
          <releases>
            <enabled>true</enabled>
            <updatePolicy>always</updatePolicy>
          </releases>
          <snapshots>
            <enabled>true</enabled>
            <updatePolicy>always</updatePolicy>    <!--更新策略，常常 -->
          </snapshots>
        </repository>
      </repositories>
    <activation>
      <activeByDefault>true</activeByDefault>       <!--这里要设置成true -->
    </activation>
    </profile>
  </profiles>
  <activeProfiles>
    <activeProfile>guvnor-m2-repo</activeProfile>    <!-- 这个设置也是必须要
有的 -->
  </activeProfiles>
</settings>
```

总结：Workbench 与 Java 的交互离不开 Maven 对 jar 包与版本的管理，Workbench 的主要核心之一是构建 kjar 包，kjar 包是一个 jar 包文件，Workbench 对该 jar 提供了一个对外下载的地址，同时也添加权限限制。当 Workbench 与 Java 项目部署不在同一台机器上时，则需要两台机器都有

Maven，并且需要 Java 项目配置与 Workbench 提供的下载接口进行通信，可以这样理解，Java 服务通过配置将 Workbench 构建生成的 kjar 下载到本地并进行读取和操作。

11.6 构建项目的版本控制

Workbench 项目中版本是非常重要的，它直接影响 Workbench 与 Java 项目的交互，并且还会影响 Kie-Server 的使用。关于版本的问题，其实写法和 Mavne 是大同小异的，这里推荐两种方式：第一种是将版本设置为 LATEST，第二种在版本后添加 SNAPSHOT。

第一种写法：将设置的 version 写成 LATEST 即可，简单来说就是将设置的版本号写成 LATEST，如 JavaSpring 项目与 Workbench 交互时，其代码如下。

```
<kie:releaseId id="sptringweb" groupId="com.dools.web.test" artifactId="droos_web_class" version="LATEST"/>
```

如果将 Workbench 中的项目版本号设置成非固定值，构建项目则仅对已部署的 kjar 重新获取并进行更改。固定版本不会在运行时自动更新。

在 Workbeanch 中设置版本号时就必须写成 x.y.z，如 1.0.0。但如果是用这个属性，下一次升级时，就要将版本号进行一个累加。举例说明，当前版本号是 1.0.0，下一次升级的值必须大于 1.0.0，可以修改成 1.0.1，如图 11-135 所示。

图11-135　设置Workbench项目版本号

第二种写法：将设置的 Version 写成 1.0.0-SNAPSHOT，这样的写法有一个好处就是可以将当前版本进行覆盖，如图 11-136 所示。

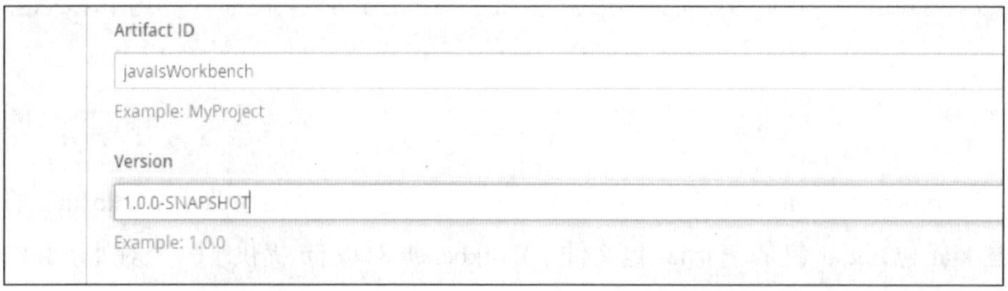

图11-136　设置Workbench版本添加"1.0.0-SNAPSHOT"

11.7 Workbench上传文件与添加依赖关系

1. 上传文件

Workbench 是支持上传的，在 Workbench 7.10 版本中上传文件很简单，单击 "Import Asset" 按钮即可，如图 11-137 所示。

图11-137　Workbench上传文件添加页面

弹出如图 11-138 所示的对话框。输入文件名，如果上传文件名与输入的名称不同，则会以输入的名称为准，也可以选择软件包等。

图11-138　Workbench上传文件编辑页面

上传文件的功能能解决由本地编辑的一些文件无法直接移植到 Workbench 中的尴尬场景。

2. 依赖关系

设想一下，在具体的开发中，就拿 Spring 来说，想要使用 Spring 的框架则必须通过引用 jar 包或是通过 pom.xml 进行配置，介绍了 Workbench 的功能后，是否 Workbench 也有提供这种功能呢？毕竟 Workbench 的核心之一是 Maven，而且它有 pom.xml 配置文件，答案是肯定的。Workbench 项目与项目之间或是引用其他 jar 包该如何建立依赖关系呢？下面就来介绍一下在 Workbench 中是如何操作的。

首先读者要明白 Workbench 是可以将项目打成 jar 包的，所以可以通过这一点来进行依赖操作，找到 Workbench 项目中的 Settings 菜单，如图 11-139 所示。

图11-139　Workbench项目编辑页面

单击左侧的"Dependencies"按钮，右侧会出现出现两种添加依赖关系的方式。第一种是直接通过引用 Group ID、Artifact ID、Version 的方式。单击"Add Dependency"链接，会弹出如图 11-140 所示的对话框，添加相关的信息就可以添加依赖关系了。

图11-140　编辑依赖页面

第二种是通过本地 Artifacts 的方式，直接选择 jar 包进行上传。单击右侧的"Add from Repository"链接，就会弹出如图 11-141 所示的对话框，选择需要上传的 jar 包就可以添加依赖关系了。但这种方式只能选择 Workbench 中已构建的项目或是通过 M2 存储库界面存在的项目。

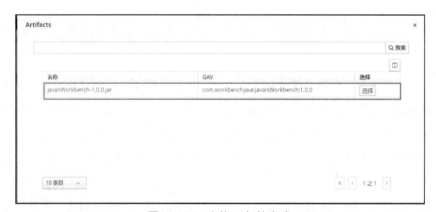

图11-141　上传jar包的方式

M2 存储库界面是指 Artifacts 页面，管理员按钮中设置的 Artifacts 选项中，如图 11-142 所示（在介绍 Workbench 与 Java 简单方式交互中提到过），页面中有上传和刷新的按钮。

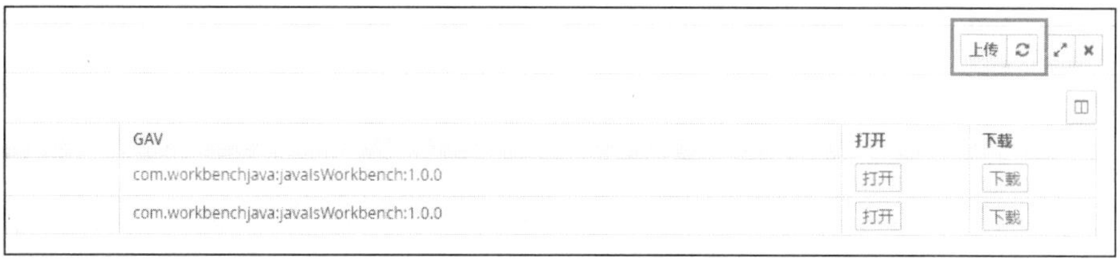

图11-142　上传与刷新

上传功能的使用说明。通过本机 IDE 创建一个 Maven 项目，如图 11-143 所示，并将该项目打成 jar 包。

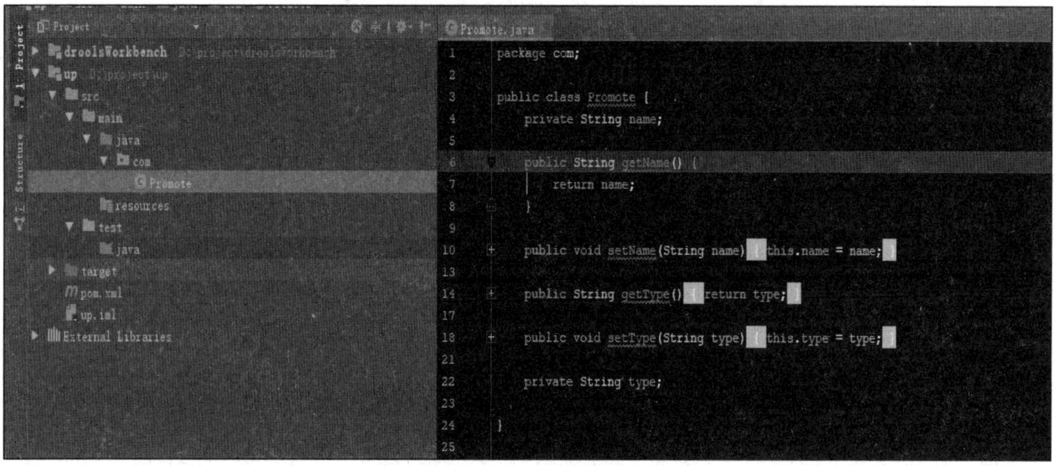

图11-143　创建一个Maven项目

选择相应的 jar 包文件，单击"上传"按钮即可，其效果如图 11-144 所示，在存储库列表中多了一个上传的 jar 包。

图11-144　成功上传Maven项目

上传的 jar 包是全局性的，是指只要在 Workbench 中创建的项目都可以使用，但要使用上传的 jar 包中的文件，则需要在项目添加依赖关系中进行关联，这与之前讲到的在项目中添加依赖关系是一样的，想想看一个项目如果没有添加引用又怎么能使用这个引用 jar 包呢。

通过第二种方式添加依赖关系将上传的 jar 包进行引用，单击"Add from Repository"链接，会弹出如图 11-145 所示的对话框。

图11-145　选择上传的jar包

找到刚刚上传的项目，如果项目特别多可以通过上方的搜索框进行查找，单击"选择"按钮后，就会返回到 Dependencies 界面。界面会发生变化，如图 11-146 所示。

图11-146　Workbench项目添加引用

"Package white list"是可选的，如果是引用全部，则选中"All"单选按钮即可。单击左侧的"保存"按钮后，返回决策树的编辑页面，通过数据对象栏目新建条目，发现这时已经上传 jar 包中的 Promote 类了，如图 11-147 所示。

图11-147　决策树引用外部jar包数据对象

决策树编辑页面会出现 Promote 的数据对象，如图 11-148 所示。

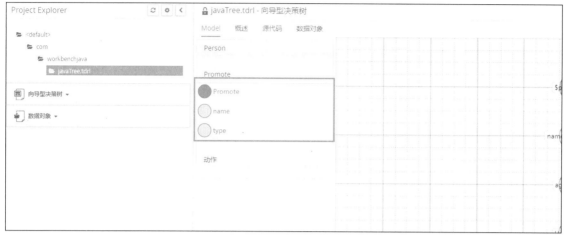

图11-148 决策树引用外部jar数据对象后的效果

11.8 Workbench中设置Kbase+KieSession

KieBase 与 KieSession 的重要性相信读者已经很清楚了，在 Workbench 与 Java 交互中自动扫描的第二种方式 Java+Spring 配置文件中使用了 KieSession 的注解方式进行的调用，如图 11-149 所示。这是因为没有在项目中设置指定 KieSession 的 name。而 Workbench 构建项目时自动为当前 jar 添加一个默认参数。但这里有一个约束，就是只能通过默认值来编辑自动扫描。

```
@RunWith(SpringJUnit4ClassRunner.class)
@ContextConfiguration({"classpath:Spring.xml"})
public class TestSpring {
    @KSession("defaultKieSession")//注: 这里的值与配置文件中的值是一样的
    KieSession kSession;

    @Test
    public void runRules() {
        while (true) {
            try {
                Person p = new Person();
                p.setAge(30);
                p.setName("张三");
                kSession.insert(p);
                int i = kSession.fireAllRules();
                System.out.print("Java调用Workbench 自动扫描，共执行了" + i + "条规则");
                System.out.println("修改后的结果" + p.getName());
```

图11-149 Java交互默认的KieSession

新版本的 Workbench 创建 KieBase 在项目的 Settings 菜单中，单击左侧 KIE bases 选项，右侧就会出现如图 11-150 所示的界面效果。

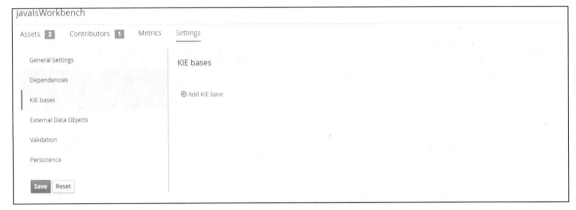

图11-150　Workbench项目编辑操作KieBase

单击"Add KIE base"链接，会弹出如图 11-151 所示的对话框。

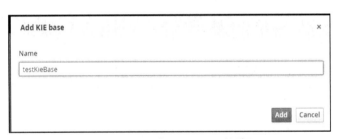

图11-151　添加KieBase对话框

输入"KieBsae"的名称，这个名称是指设置 kmodule.xml 中的 KieBase 标签的名称，所以一定要符合 kmodule.xml 的配置格式。单击"Add"按钮后会返回到 KieBases 界面，如图 11-152 所示。

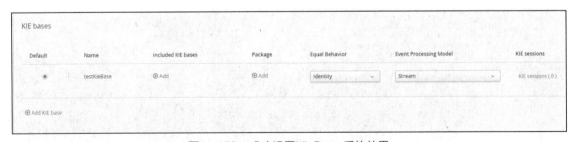

图11-152　成功设置KieBase后的效果

它可以设置包含 KieBase、路径地址、行为、事件处理和 KieSession 功能。这些功能与中级篇中讲到的 kmodule 配置中是一样的。这里关系 Package 的配置与 KieSession 的配置。

单击 Package 下方的"Add"按钮，并输入决策树源代码中的 Package 参数，如图 11-153 所示。

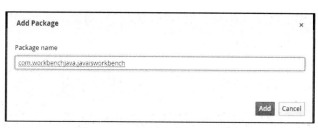

图11-153　添加Package对话框

输入正确的参数后单击"Add"按钮，出现如图 11-154 所示的效果。

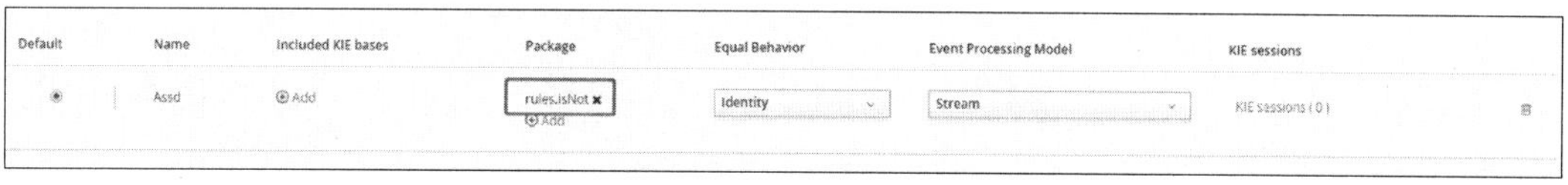

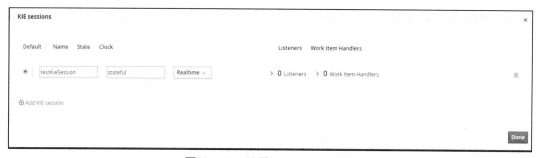

图11-154　添加package后的效果

单击"KIE sessions"链接，出现如图 11-155 所示的界面。

图11-155　设置KieSession界面

选中"Add KIE session"单选按钮，并输入 KieSession 的名称，默认值是无状态的，这里改为有状态的，添加完成后，关闭对话框并单击左侧的"保存"按钮。

查看 KieBase 和 KieSession 是否添加成功有以下两种方式。

第一种方式：通过项目的 kmodule.xml 配置文件。通过项目自定义视图按钮选择"资料库视图"选项。找到资料文件下的 META-INF 目录，单击 kmodule.xml 配置文件，如图 11-156 所示的自定义视图可以将项目展示为不同的效果。

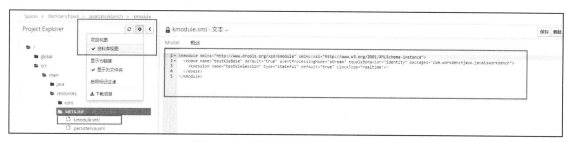

图11-156　查看Workbench的文件

第二种方式：重新构建项目。通过自动扫描的方式进行测试，在droolsWorkbench项目中修改TestSpring.java文件中的@KieSession参数即可，运行结果如图11-157所示。

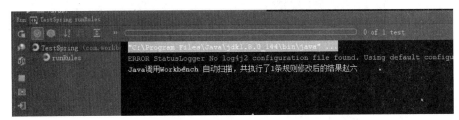

图11-157　设置KieSession指定名称并执行规则

11.9　Workbench构建jar包到Maven私服

Workbench构建jar包到Maven私服的一般方式有两种，因Workbench构建出来的jar包是一个标准的Maven项目。

第一种方式：通过Maven的命名进行上传。

第二种方式：通过操作Workbench项目中的pom.xml配置进行上传。

Workbench构建时生成的jar包直接放到公司所在的私服，其实这个不只是针对Workbench的，更准确地说是关于Maven的配置，在pom.xml中加入配置，其内容如下。

```
<distributionManagement>
<repository>
<id>thirdparty</id>
<name>Nexus Release Repository</name>
<url>http://localhost:8081/nexus/content/repositories/thirdparty/</url>
</repository>
</distributionManagement>
```

第12章
Kie-Server

Kie-Server 是一个全新的知识概念，从字面上可以看出，这是一个 Kie 的服务，既然是服务，就要进行配置，Kie-Server 可以做很多的事情，为了更容易理解，将通过以下几个概念进行说明。Kie-Server 与自动扫描大同小异，根据 Workbench 构建 jar 并通过对外提供接口服务的方式完成启动与访问，一个类似 Webservice 的 Web 访问服务器，对外提供访问接口。Kie-Server 是可以独立部署的应用服务器，也可以部署成集群应用，通过 Nginx 做负载均衡。

Kie-Server 可以理解为自动扫描的升级配置，它更重要的是提供服务，根据所描述的概念，下面就对 Kie-Server 进行安装部署。部署 Kie-Server 有整合部署、分离部署及集群部署 3 种方式，且从简单到复杂。使用 Kie-Server 时需要先下载相关的 war 包。下载界面与 Workbench 页面一样，在下载列表中找到 KIE Execution Server，如图 12-1 所示。

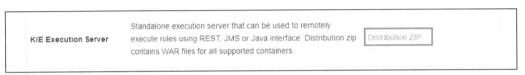

图12-1　下载Kie-Server war安装包

下载完成后将压缩包中的 kie-server-7.10.0.Final-webc.war 包文件解压到新 Tomcat 的 WebApps 目录下，并改名为 Kie-Server.war，如图 12-2 所示。

图12-2　解压包的内容

12.1　整合部署

整合部署是指将 Workbench 与 Kie-Server 这两个 War 放在同一台 Tomcat 服务器中，添加用户信息，为 Kie-Server 添加登录信息。编辑 tomcat-user.xml，添加如下内容。

```
<role rolename="Kie-Server" />
<user username="kieserver" password="kieserver1!" roles="Kie-Server" />
```

值是 Kie-Server 的默认密码，但必须要进行配置，如图 12-3 所示。

| org.kie.server.controller.user | string (default is "kieserver") | Username used to connect to the controller REST api | Yes when using a controller |
| org.kie.server.controller.pwd | string (default is "kieserver1!") | Password used to connect to the controller REST api | Yes when using a controller |

图12-3　Kie-Server的配置说明

这个值是可以自定义的，如果想要使用自定义的用户名或密码则需要在 setenv.sh（setenv.bat）中进行修改。

```
<user username="helloword" password="helloword" roles="Kie-Server"/>
只要在setenv.sh中设置两个属性：
org.kie.server.controller.user=helloword
org.kie.server.controller.pwd=helloword
```

如果在 setenv.sh 中并未设置上面两个属性，则必须要在 tomcat.user.xml 中设置默认的用户名和密码，否则就会报 kie-Server 401 等没有权限的错误。如果在脚本文件 setenv.sh 中设置了该属性，则在 tomcat.user.xml 中也可配置自定义的用户名和密码。在 Windows 环境下，该配置是在同一行的。编辑 setenv.bat 文件，添加 Kie-Server 相关的配置，其内容如下。

```
-Dorg.kie.server.id=kieserver
-Dorg.kie.server.location=http://localhost:8081/Kie-Server/services/rest/server
-Dorg.kie.server.controller=http://localhost:8081/kie-drools-wb/rest/controller
-Dorg.kie.server.persistence.ds=java:comp/env/jdbc/jbpm
-Dorg.kie.server.persistence.tm=org.hibernate.service.jta.platform.internal.BitronixJtaPlatform
-Dorg.jbpm.cdi.bm=java:comp/env/BeanManager
-Dorg.jbpm.server.ext.disabled=true
-Dfile.encoding=UTF-8
```

这里要注意的是 kie.server.id，这个值是与服务器模板挂钩的。在 Workbench 提供的操作服务器页面中手动添加服务器模板是没有效果的，配置时必须是 kie.server.id 的值才会生效。那如何进行手动添加呢？

```
-Dorg.kie.server.location=http://ID:端口/Kie-Server/services/rest/server \
-Dorg.kie.server.controller=http://ID:端口/kie-drools-wb/rest/controller \
IP地址是根据本地的IP地址配置的
```

配置完成后重启 Tomcat，等待启动成功（等待的过程如果很慢可以适当加大内存）。启动成功后进入登录页面，单击 Workbench 主页面 Deploy 下的 "servers" 链接，如图 12-4 所示。

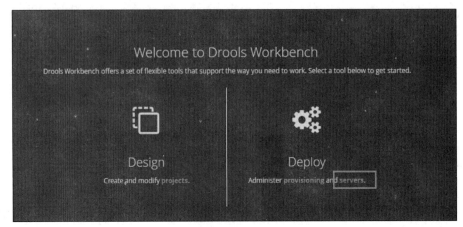

图12-4 单击"servers"链接

单击"servers"链接后会跳转到如图 12-5 所示的界面。

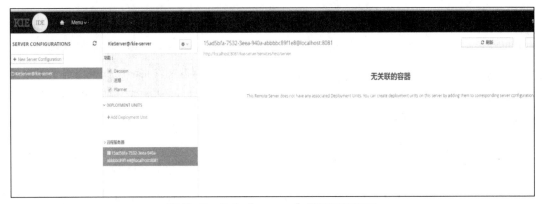

图12-5 单击"servers"链接后的界面

当左侧出现如图 12-5 所示的效果，则表示 Kie-Server 配置成功，当前 Workbench 所关系的 Kie-Server 服务器 ID，也就是在 sh 或 bat 文件中配置的 kie.server.id 值。同一个 Kie-Server 中的 Kie-Server 是不可重复的。中间栏是当前选中的 Kie-Server 的配置管理控制台，它包含了 Kie-Server 的功能，添加容器和远程服务器管理。单击远程服务器后，Kie-Server 右侧界面出现如图 12-6 所示的效果。

图12-6 Kie-Server链接地址

单击页面上的 URL 地址，会在浏览器中出现需要进行身份验证的对话框，如图 12-7 所示。

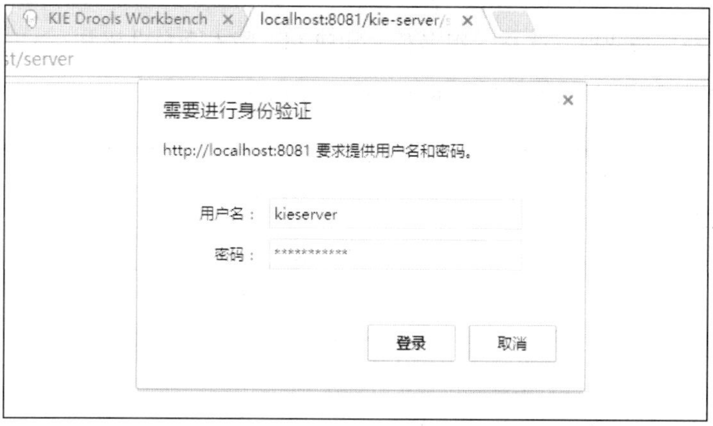

图12-7　Kie-Server登录页面

输入在 Tomcat-user.xml 中配置的 Kie-Server 的用户名与密码，并单击"登录"按钮，如浏览器界面出现如图 12-8 所示的页面，就证明 Kie-Server 配置成功了。

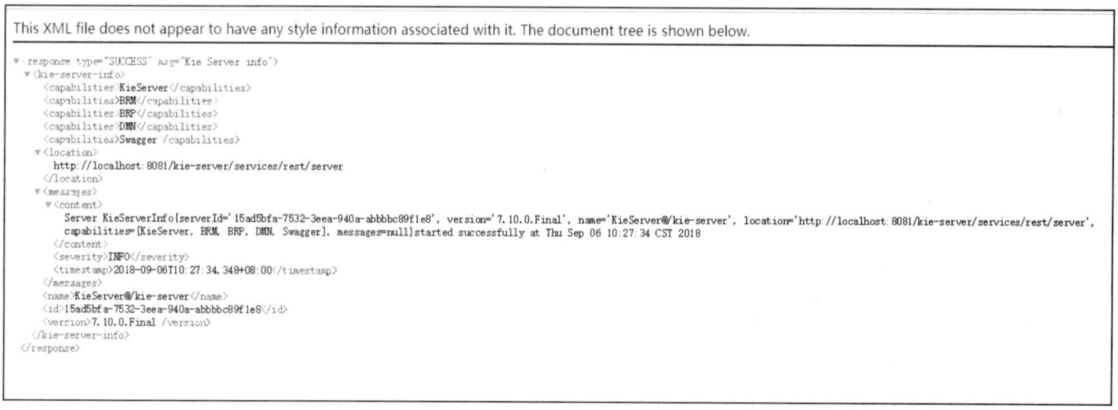

图12-8　Kie-Server登录成功页面

添加容器模板，单击"Add Deployment Unit"链接，进行添加模板容器操作，如图 12-9 所示。

图12-9　添加模板容器操作

Workbench 会弹出如图 12-10 所示的对话框，在该对话框中发现有之前构建的项目和之前上传的项目，选择 WorkbenchJava 项目，选择项目也有一个说明，要选择设置过 KieBase 与 KieSession 的项目，单击"选择"按钮后，名称、组名称、构建 Id、版本会自动添加。Alias 是项目的别名，这个参数是需要特别注意的，该值关系到 Kie-Server 与 Java 的交互。

图12-10　为Kie-Server添加操作项目

单击"完成"按钮。在 Kie-Server 界面的中间栏与右侧栏中会出现刚刚添加的信息，如图 12-11 所示，其实这时就可以单击右侧的"启动"按钮。但在启动之前，需要对右侧栏中的 3 个菜单做一个说明。

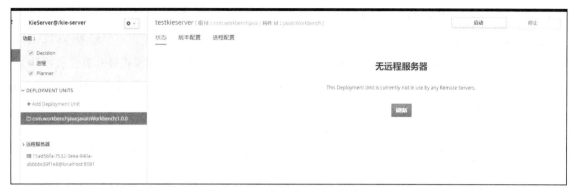

图12-11　添加操作项目后的效果

“状态”菜单：在单击“启动”按钮后，就会出现 Kie-Server 所代理的对外提供的远程服务器，如图 12-12 所示。

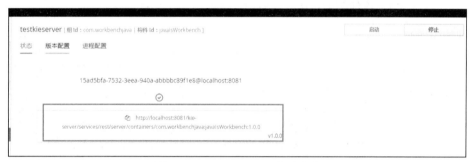

图12-12　Kie-Server启动成功

单击远程服务器的访问地址，会在浏览器中出现如图 12-13 所示的界面，这也表示远程服务器添加操作项目启动成功了。

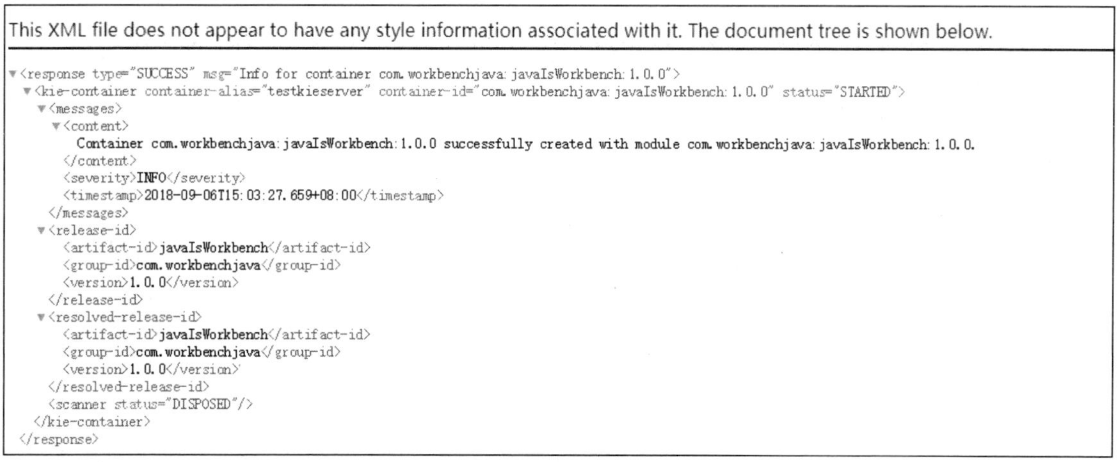


```
<response type="SUCCESS" msg="Info for container com.workbenchjava:javaIsWorkbench:1.0.0">
  <kie-container container-alias="testkieserver" container-id="com.workbenchjava:javaIsWorkbench:1.0.0" status="STARTED">
    <messages>
      <content>
        Container com.workbenchjava:javaIsWorkbench:1.0.0 successfully created with module com.workbenchjava:javaIsWorkbench:1.0.0.
      </content>
      <severity>INFO</severity>
      <timestamp>2018-09-06T15:03:27.659+08:00</timestamp>
    </messages>
    <release-id>
      <artifact-id>javaIsWorkbench</artifact-id>
      <group-id>com.workbenchjava</group-id>
      <version>1.0.0</version>
    </release-id>
    <resolved-release-id>
      <artifact-id>javaIsWorkbench</artifact-id>
      <group-id>com.workbenchjava</group-id>
      <version>1.0.0</version>
    </resolved-release-id>
    <scanner status="DISPOSED"/>
  </kie-container>
</response>
```


图12-13　Kie-Server添加操作项目启动成功

版本配置：该功能与之前介绍的自动扫描是有相似之处的。版本的升级只对最新版本进行扫描，有且只升级一次，操作界面如图 12-14 所示。例如，操作添加容器之前的版本是 1.0.1，升级时输入的版本还是 1.0.1，这个版本进行的升级是不起作用的，但当 Workbench 构建一个新版本时，则日志中就会提示引用成功，所以升级时，新版本是不会出现不生效情况的（建议选择升级比较可靠）；现在就扫描则是对当前 Jar 再次进行扫描更新，但有时可能不起作用。启动扫描程序则是根据前面的间隔值自动扫描（可能会很占服务器空间）；配置成功后就可以通过工具访问或代码的形式访问服务器上的规则。

部署 Kie-Server 是动态规则的一种，在自动扫描中讲到过，自动扫描是动态规则的一种方式，而 Kie-Server 与自动扫描是相似的。一般交互或自动扫描其实是通过 Java 与 Workbench 进行交互，从而操作 Workbench 构建生成的规则 kjar 包。而 Kie-Server 是做了构建 KieBase 的过程，同时抛出了一个接口，供 Java 远程调用。Kie-Server 最大的好处是可以部署成集群，做负载，但要注意的是做负载时 KieSession 是不能共享的，原因为 KieBase 是基于内存的，JVM 是不能进行共享的。使用的过程中，遇到高并发多线程的调用时，使用的也都是无状态的 KieSession。

图12-14　Kie-Server操作项目的版本页面

进程配置：该功能很强大，可以设置特定的 KieBase、KieSession 等，还可以执行多种登录方式、合并模式等，如图 12-15 所示。

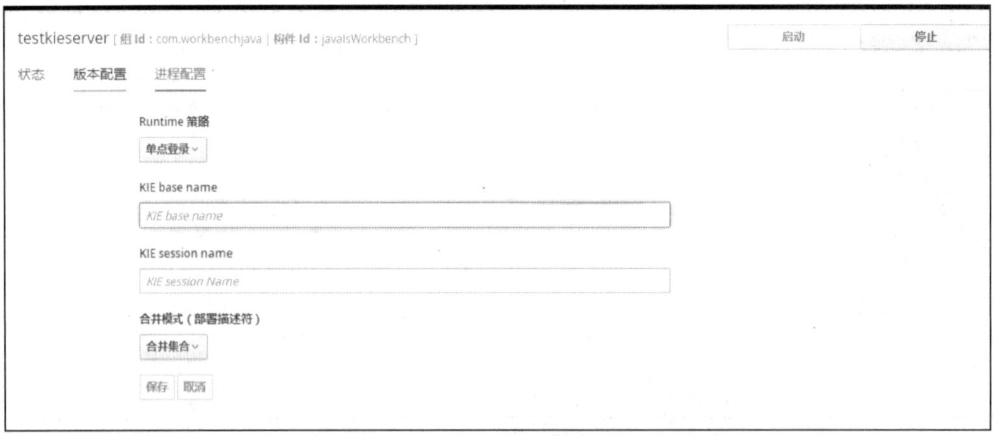

图12-15　Kie-Server的高级配置

启动成功后，单击中间栏的远程服务器也可以清楚地看到当前服务器所支持的 kjar 对外的服务，如图 12-16 所示。

图12-16　添加项目成功后的效果

部署过程中可能会出现一些问题，下面是比较常见的问题。

```
Error when initializing server extension of type OptaPlanner KIE Server
extension due to null
```

解决方法：如果 Kie-Drools-wb 与 Kie-Server-wb 设置在同一台机器上则至少需要两个处理器。

```
-Dorg.optaplanner.server.ext.disabled=true
```

脚本配置的说明，页面中有些功能配置，如规则、进程、规划，其实为默认选中的，但这几个功能说明是可配置的。编辑 setenv.bat(setenv.sh) 文件，添加如下内容，就可以对该功能进行修改了。

```
-Dorg.jbpm.server.ext.disabled=true        进程禁用
-Dorg.drools.server.ext.disabled=true      规则、规划
```

虽然这些值是可配置的，但并不建议这样做，否则规则、进程、规划所相关的信息将无法访问，尤其会影响规则流和规则的正常使用。这是非常重要的，切记不要盲目设置。

在 Tomcat 或任何其他 Web 容器上运行时，无法利用 JMS 界面。WAR 的 Web 容器版本仅包含 REST 接口。

特别注意：如果在 Windows 系统中安装则要修改 tomcat/conf/server.xml。

```
Realm className="org.apache.catalina.realm.LockOutRealm" lockOutTime="1"
failureCount="999999">
        <Realm className="org.apache.catalina.realm.UserDatabaseRealm"
            resourceName="UserDatabase"/>
</Realm>
```

因为 Kie-Server 一旦启动后，它会带着用户在浏览器的 Cookie 中不断地与 KIE-WB 同步，而有些用户的机器上装有 360 杀毒软件，过一阵就会把用户的 Cookie 清除掉，一旦 Kie-Server 与 KIE-WB 同步，用户名和密码失效，会造成 Tomcat 的默认用户 5 次登录失败后被锁 5 分钟。

12.2 分离部署

分离部署是指 Workbench 与 Kie-Server 不在同一个 Tomcat 中，注意这里所讲到的是使用 Tomcat 进行配置。为保证 Tomcat 可直接执行，这里复制了两个 Tomcat，将一个 Tomcat 用来部署 Workbench 的 Tomcat 端口号改成 8080，以区别 Kie-Server 的分离部署。其他配置与 Windows 安装方式是一样的。但 Tomcat-user.xml 需要添加 Kie-Server 的用户名与密码，否则会一直在 Tomcat 控制中输出如图 12-17 所示的尝试验证 Kie-Server。

图12-17　尝试验证Kie-Server

第二个 Tomcat 是用来部署 Kie-Server 的，将 Kie-Server.war 包复制到 Tomcat/WebApps 目录下，并将端口号改为 8081。创建 setenv.bat（setenv.sh）文件，在目录 tomcat/bin 中，内容如下（加粗部分相比在整合部署中区别是 Workbench 的端口或者说它的地址有所不同，其他都是一样的）。

```
set "CATALINA_OPTS=-Xmx2048M
-Djava.security.auth.login.config=%CATALINA_HOME%\webapps\kie-drools-wb\
WEB-INF\classes\login.config
-Dorg.jboss.logging.provider=jdk
-Dorg.kie.server.location=http://localhost:8081/Kie-Server/services/
rest/server
-Dorg.kie.server.controller=http://localhost:8080/kie-drools-wb/rest/
controller
-Dorg.kie.server.persistence.ds=java:comp/env/jdbc/jbpm
-Dorg.kie.server.persistence.tm=org.hibernate.service.jta.platform.
internal.BitronixJtaPlatform
-Dorg.jbpm.cdi.bm=java:comp/env/BeanManager
-Dorg.jbpm.server.ext.disabled=true -Dfile.encoding=UTF-8 "
```

分别启动两个 Tomcat，成功后，重复整合部署的操作 Kie-Server，会出现如图 12-18 所示的页面，注意浏览器地址栏的端口与 Kie-Server 界面的端口号。

第 12 章 Kie-Server

图12-18 分离部署的效果

特别强调：这里要设置访问 Workbench 的权限和 Jar 下载的指定路径。原因：分离部署是在不同机器进行的操作，Workbench 负责 Jar 包，Kie-Server 负责服务。但服务本身是找不到 Workbench 所有的 Jar 包的，这里就要进行指定，在 setenv.bat（setenv.sh）文件中添加如下配置。

-Dkie.maven.settings.custom=/root/.m2/settings.xml

配置完成后，添加 settings.xml 在 /root/.m2/ 下。

```xml
<?xml version="1.0" encoding="UTF-8"?>
<settings>
<servers>
    <server>
      <id>guvnor-m2-repo</id>
      <username>Workbench</username>
      <password>admin</password>
      <configuration>
        <wagonProvider>httpclient</wagonProvider>
        <httpConfiguration>
          <all>
            <usePreemptive>true</usePreemptive>
          </all>
        </httpConfiguration>
      </configuration>
    </server>
</servers>
<profiles>
<profile>
      <id>guvnor-m2-repo</id>
      <repositories>
        <repository>
          <id>guvnor-m2-repo</id>
          <name>Guvnor M2 Repo</name>
          <url>http://Workbench:端口/kie-wb/maven2/</url>
          <layout>default</layout>
```

```xml
            <releases>
                <enabled>true</enabled>
                <updatePolicy>always</updatePolicy>
            </releases>
            <snapshots>
                <enabled>true</enabled>
                <updatePolicy>always</updatePolicy>
            </snapshots>
        </repository>
      </repositories>
  <activation>
       <activeByDefault>true</activeByDefault>
    </activation>
    </profile>
  </profiles>
  <activeProfiles>
    <activeProfile>guvnor-m2-repo</activeProfile>
  </activeProfiles>
</settings>
```

这点特别符合从开始就强调的事情，Workbench 是一个构建项目成 kjar 的过程，kjar 包是包含有 kmodule.xml 配置文件 jar 的。不管是 Workbench 与 Java 的两种交互，还是 Kie-Server 与 Workbench 的整合，其关键点是让 Java 所在的服务器或 Kie-Server 所在服务器的 Maven 与 Workbench 所在的 Maven 进行一个连接。保证 Kie-Server 或 Java 的 Maven 可以访问到 Workbench 所管理的 kjar。

12.3 集群部署

集群部署有两种，一种是针对 Kie-Server 的集群部署，另一种是 Workbench+Kie-Server，它们都是多台机器的集群部署。但它们所使用的服务器容器有所不同，图 12-19 所示是 Kie-Server 的集群部署。

Kie-Server 集群并不难理解，它们都为一台 Workbench 服务。将已经配置好的分离部署方案中的 Kie-Server 服务器进行复制，放在不同的机器上或修改 Tomcat 的端口号。它们的区别是 location 的配置，内容如下。

```
-Dorg.kie.server.id=kieserver
-Dorg.kie.server.location=http://localhost:8081/Kie-Server/services/rest/server
```

复制多台 Tomcat，并通过 Nginx 做负载，就可以实现集群了，但需要注意，Kie-Server 在 Workbench 中的用户名与密码是必需的，如果是多台 Kie-Server，建议使用同一套用户名和密码，除非有特别要求可以进行分离，但必须在 Workbench 所在的 Tomcat-user.xml 中有所体现，否则会出现尝试连接分离部署中的问题。

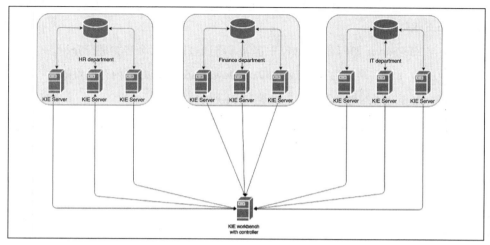

图12-19　Kie-Server的集群部署

还有一点需要强调的是 KieSession 的状态问题，在 3.5 节中讲到，KieSession 为有状态时会保留其状态，并会出现产生迭代方式插入（笛卡儿积），而且在 KieBase 内存中，服务器与服务器之间无法共享。因此，强烈建议使用无状态的 KieSession 进行对 Kie-Server 的操作。

Workbench 也是可以实现集群的，但在 Tomcat 中是无法使用的。Workbench 是一个 WEB-IDE，其一定会产生文件，所有 Workbench 的集群就是表示两台服务器中文件的同步。这就要用到 Zookeeper 了。但通过多方面的尝试，最终还是失败了，通过与国外知名的 Drools 专家进行交流后，发现他们也并没有实现通过 Tomcat 进行 Workbench 的集群方式，如图 12-20 所示为交流内容。

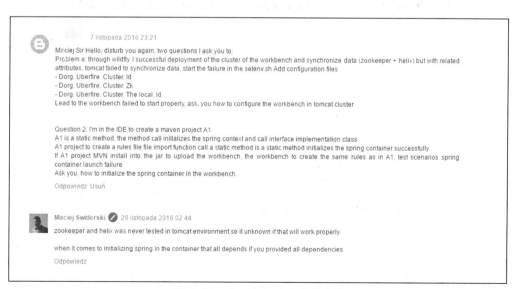

图12-20　与国外知名专家的交流

1. Wildfly服务器的配置及Kie-Server+Workbench

开始介绍 Workbench 时，提到过 Workbench 或 Kie-Server 是可以配置在 Wildfly 服务器上的，并且 Wildfly 是支持通过 Zookeeper 做 Workbench 集群的，做集群之前需要了解一下如何在 Wildfly 上配置 Workbench+Kie-Server。

Workbench 与 Kie-Server 如何在 Wildfly 中部署，将通过以下几个步骤进行介绍。

第一步：Wildfly 下载。网址为 http://wildfly.org/downloads/，最好将内存设置的大一些。

其版本众多，这里选用的是 8.2 版本，如图 12-21 所示。

8.2.0.Final	2014-11-20	Java EE7 Full & Web Distribution	LGPL	126 MB	ZIP
				113 MB	TGZ
		Update Existing 8.1.0.Final Install	LGPL	62 MB	ZIP
		Application Server Source Code	LGPL	36 MB	ZIP
				22 MB	TGZ
		Release Notes			Notes

图12-21　下载Wildfy

第二步：下载完成后放到 Liunx 上，这里存放在 /usr/local 路径下，主机 IP 为 192.168.80.10。

安装过程：tar -zxvf wildfly-8.2.0.Final.tar.gz 并将 wildfly-8.2.0.Final 改名为 wildfly8，如图 12-22 所示。

```
[root@localhost local]# pwd
/usr/local
[root@localhost local]# ls
jdk  jdk-7u79-linux-x64.tar.gz  wildfly8  wildfly-8.2.0.Final.tar.gz
```

图12-22　解压文件并进行文件夹名称的修改

第三步：配置 wildfly 的用户名及密码为 /usr/local/wildfly8/bin，执行 ./add-user.sh 命令。

```
./add-user.sh        Enter
What type of user do you wish to add?
 a) Management User (mgmt-users.properties)
 b) Application User (application-users.properties)
(a): a        Enter
Enter the details of the new user to add.
Using realm 'ManagementRealm' as discovered from the existing property
files.
Username :admin        Enter  这个可自定义  但建议使用admin

Password recommendations are listed below. To modify these restrictions
edit the add-user.properties configuration file.
 - The password should not be one of the following restricted values
{root, admin, administrator}
```

```
- The password should contain at least 8 characters, 1 alphabetic
character(s), 1 digit(s), 1 non-alphanumeric symbol(s)
- The password should be different from the username
Password :    这里输入密码    Enter

JBAS015267: Password must have at least 1 non-alphanumeric symbol.
Are you sure you want to use the password entered yes/no?    yes    Enter

Re-enter Password :    这里是重复密码    Enter

What groups do you want this user to belong to? (Please enter a comma
separated list, or leave blank for none)[ ]: admin    Enter    设置用户组

About to add user 'admin' for realm 'ManagementRealm'
Is this correct yes/no?  yes    Enter

Added user 'admin' to file
'/usr/local/wildfly8/standalone/configuration/mgmt-users.properties'
Added user 'admin' to file '/usr/local/wildfly8/domain/configuration/
mgmt-users.properties'
Added user 'admin' with groups admin to file '/usr/local/wildfly8/
standalone/configuration/mgmt-groups.properties'
Added user 'admin' with groups admin to file '/usr/local/wildfly8/
domain/configuration/mgmt-groups.properties'
Is this new user going to be used for one AS process to connect to
another AS process?
e.g. for a slave host controller connecting to the master or for a
Remoting connection for server to server EJB calls.
yes/no? yes    Enter

To represent the user add the following to the server-identities defini-
tion <secret value="YWRtaW4xMjM=" />    这个值是用来做集群的
```

第四步：新建用户完成之后，编辑其他配置文件，执行 cd/usr/local/wildfly8/domain/configuration 命令，进入这个目录下并编辑 host.xml，其内容如下。

```xml
<?xml version='1.0' encoding='UTF-8'?>

<host name="master" xmlns="urn:jboss:domain:2.2">

    <management>
        <security-realms>
            <security-realm name="ManagementRealm">
                <authentication>
                    <local default-user="$local" skip-group-loading="true" />
                    <properties path="mgmt-users.properties" relative-to="jboss.domain.config.dir"/>
                </authentication>
                <authorization map-groups-to-roles="false">
```

```xml
                <properties path="mgmt-groups.properties" relative-to="jboss.domain.config.dir"/>
            </authorization>
        </security-realm>
        <security-realm name="ApplicationRealm">
            <authentication>
                <local default-user="$local" allowed-users="*" skip-group-loading="true" />
                <properties path="application-users.properties" relative-to="jboss.domain.config.dir" />
            </authentication>
            <authorization>
                <properties path="application-roles.properties" relative-to="jboss.domain.config.dir"/>
            </authorization>
        </security-realm>
    </security-realms>
    <audit-log>
        <formatters>
            <json-formatter name="json-formatter"/>
        </formatters>
        <handlers>
            <file-handler name="host-file" formatter="json-formatter" relative-to="jboss.domain.data.dir" path="audit-log.log"/>
            <file-handler name="server-file" formatter="json-formatter" relative-to="jboss.server.data.dir" path="audit-log.log"/>
        </handlers>
        <logger log-boot="true" log-read-only="false" enabled="false">
            <handlers>
                <handler name="host-file"/>
            </handlers>
        </logger>
        <server-logger log-boot="true" log-read-only="false" enabled="false">
            <handlers>
                <handler name="server-file"/>
            </handlers>
        </server-logger>
    </audit-log>
    <management-interfaces>
        <native-interface security-realm="ManagementRealm">
            <socket interface="management" port="${jboss.management.native.port:9999}"/>
        </native-interface>
        <http-interface security-realm="ManagementRealm" http-upgrade-enabled="true">
            <socket interface="management" port="${jboss.management.http.port:9990}"/>
```

```xml
            </http-interface>
        </management-interfaces>
    </management>

    <domain-controller>
        <local/>
        <!-- Alternative remote domain controller configuration with a host and port -->
        <!-- <remote host="${jboss.domain.master.address}" port="${jboss.domain.master.port:9999}" security-realm="ManagementRealm"/> -->
    </domain-controller>

    <interfaces>
        <interface name="management">
            <inet-address value="${jboss.bind.address.management:192.168.80.10}"/>
        </interface>
        <interface name="public">
            <inet-address value="${jboss.bind.address:192.168.80.10}"/>
        </interface>
        <interface name="unsecure">
            <!-- Used for IIOP sockets in the standard configuration.
                 To secure JacORB you need to setup SSL -->
            <inet-address value="${jboss.bind.address.unsecure:192.168.80.10}"/>
        </interface>
    </interfaces>

    <jvms>
        <jvm name="default">
            <heap size="64m" max-size="256m"/>
            <permgen size="256m" max-size="256m"/>
            <jvm-options>
                <option value="-server"/>
            </jvm-options>
        </jvm>
    </jvms>

    <servers>
        <server name="server-one" group="main-server-group">
        </server>
    </servers>
</host>
```

第五步：配置服务，执行 vim /etc/default/wildfly.conf 命令，并添加配置信息，其内容如下。

```
## jdk的安装路径
JAVA_HOME="/usr/local/jdk"
## JBOSS_HOME是Wildfly的安装根目录 之前为什么要进行改名 就是要用在这里
JBOSS_HOME="/usr/local/wildfly8"
```

```
## 这里需要改为执行当前登录Liunx系统的用户名
JBOSS_USER=root
## 指定运行模式为domain
JBOSS_MODE=domain
## 指定domain的配置文件为domain.xml，slave的配置文件为host.xml
JBOSS_DOMAIN_CONFIG=domain.xml
JBOSS_HOST_CONFIG=host.xml
```

第六步：为系统配置服务。

```
##将.sh启动命令放到系统目录下，执行命令
cp wildfly8/bin/init.d/wildfly-init-redhat.sh /etc/init.d/wildfly
##增加执行权限
chmod +x /etc/init.d/wildfly
##增加系统服务
chkconfig --add wildfly
##设置开机启动
chkconfig wildfly on
##启动wildfly，记得先启动master
service wildfly start
```

第七步：进行验证，在浏览器中输入 http://192.168.80.10:9990/console/App.html，如图 12-23 所示，登录成功后，浏览器会出现如图 12-24 所示的界面。

图12-23　Wildfly启动后的效果

第八步：配置 Kie-Server+wornbench。

下载 Workbench 与 Kie-Server 所用的两个 war 包，下载地址为 http://www.drools.org/down load/download.html，如图 12-25 所示。

图12-24　成功配置Wildfly的效果

图12-25　下载wildfly war包

第九步：上传 war 包。

与之前一样将下载好的 war 包进行改名操作，并根据下面步骤完成上传文件的操作。

选择"Deployments"菜单，如图 12-26 所示。

图12-26　选择菜单

单击"Add"按钮，如图 12-27 所示。

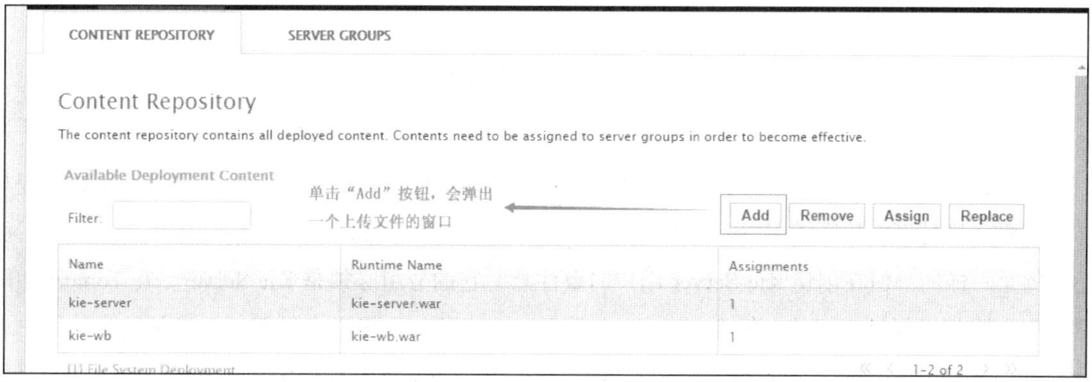

图12-27　单击"Add"按钮

上传文件后单击"下一步"按钮，如图 12-28 所示。

图12-28　上传kie-wb.war包

第十步：启动 war 包。

单击"完成"按钮后，选中列表中的行，单击界面上的"Assign"按钮。发布成功后列表会发生变化，如图 12-29 所示。

图12-29　发布war包

第十一步：添加配置属性。

选择如图 12-30 模块完成配置文件的设置，这里的配置与 setenv.sh 类似，配置后效果如图 12-31 所示。

图12-30　添加配置属性

第十二步：创建用户信息。

配置完成后，还要去服务器上创建新的用户，即 Workbench 和 Kie-Server 的用户。配置过程与前面一样。只是之前是系统用户，现在要创建应用用户，将之前的"a"改成"b"用户类型，后面的步骤是一样的。但在创建 Kie-Server 用户时要注意，它的分组必须是 Kie-Server。在 Tomcat 中用户名和密码都必须是默认的，但在 Wildfly 中是可以改变的，如果没有设置，就必须是默认用户名及密码。完成所有步骤后，重启服务，与之前介绍的 Workbench 一样直接打开就可以了，在浏览器中输入地址 http://192.168.80.10:8080/kie-wb，页面正常打开即可成功，如果页面不能正常访问，则可能是其中某一步骤出现了问题，重新配置即可。

java.net.preferIPv4Stack	true 这个是默认的	true
org.kie.demo	false	true
org.kie.example	false	true
org.kie.server.controller	http://192.168.80.10:8080/kie-wb/rest/controller	true
org.kie.server.controller.pwd	kieserver	true
org.kie.server.controller.user	kieserver	true
org.kie.server.id	wildfly-kieserver	true
org.kie.server.location	http://192.168.80.10:8080/kie-server/services/rest/server	true
org.kie.server.persistence.dialect	org.hibernate.dialect.H2Dialect	true
org.kie.server.persistence.ds	java:jboss/datasources/ExampleDS	true
org.kie.server.persistence.tm	org.hibernate.service.jta.platform.internal.JBossAppServerJtaPlatform	true

图12-31 设置完成后的配置属性

2. Wildfly分离部署

Wildfly 分离部署与 Tomcat 相似，过程与 Wildfly 整合部署相似，部署的集群还是 Kie-Server 的集群，具体的安装部署如下。

第一步：首先需要启动两台 Liunx，下载并安装 Wildfly。

第二步：下载 Kie-wb 的 War 包及 Kie-Server 的 War 包。

第三步：安装 Wildfly 后，添加系统用户名及密码，然后启动 Wildfly 服务并将 Kie-wb 与 Kie-Server 的 War 包上传到 Wildfly 上，这里要注意的是，一定要分别上传，且在每台机器上只上传一个 War 包，并部署 War 包。

第四步：完成上面三个步骤后，就是最为关键的配置了，配置 Workbench 所在机器的 Wildfly 服务器，即在 $WILDFLY_HOST/domain/configuration 中配置 host.xml 文件。

第五步：前面已有一些简单地说明，这里强调 Server 的配置，IP 地址一定要与本地的 IP 相对应才可以。

```
<server name="server-one" group="main-server-group">
<!--配置属性 -->
<system-properties>
 <property name="jboss.node.name" value="nodeOne" boot-time="true"/>
 <property name="org.uberfire.nio.git.dir" value="/tmp/kie/nodeone" boot-time="true"/>
 <property name="org.uberfire.metadata.index.dir" value="/tmp/kie/nodeone" boot-time="true"/>
 <property name="org.kie.demo" value="false" boot-time="true" />
 <property name="org.kie.example" value="false" boot-time="true"/>
 <property name="org.uberfire.nio.git.daemon.enabled" value="true" boot-time="true"/>
```

```xml
 <property name="org.uberfire.nio.git.daemon.host" value="192.168.80.31" boot-time="true" />
 <property name="org.uberfire.nio.git.daemon.port" value="9418" boot-time="true"/>
 <property name="org.kie.server.user" value="kieserver" boot-time="true" />
 <property name="org.kie.server.pwd" value="kieserver" boot-time="true" />
 <property name="org.guvnor.m2repo.dir" value="/root/.m2/repository" boot-time="true" />
 </system-properties>
</server>
```

第六步：配置 domin.xml 文件，添加内容如下。

```xml
<security-domain name="kie-ide" cache-type="default">
 <authentication>
  <login-module code="Remoting" flag="optional">
    <module-option name="password-stacking" value="useFirstPass"/>
  </login-module>
  <login-module code="RealmDirect" flag="required">
    <module-option name="password-stacking" value="useFirstPass"/>
  </login-module>
 </authentication>
</security-domain>
```

配置 Kie-Server 所在机器的 Wildfly 服务器。

第七步：在 $WILDFLY_HOST/domain/configuration 中配置 host.xml 文件。

```xml
<server name="server-one" group="main-server-group">
    <system-properties>
        <!--配置本机地址-->
        <property name="org.kie.server.location" value="http://192.168.80.33:8080/Kie-Server/services/rest/server" boot-time="false"/>
        <property name="org.kie.server.id" value="wildfly-kieserver" boot-time="false"/>
        <property name="org.kie.server.controller.user" value="kieserver" boot-time="false"/>
        <property name="org.kie.server.controller.pwd" value="kieserver" boot-time="false"/>
        <!--配置kie-web地址 failover策略-->
        <property name="org.kie.server.controller" value="http://192.168.80.31:8080/kie-wb/rest/controller" boot-time="false"/>
        <property name="org.kie.server.persistence.dialect" value="org.hibernate.dialect.H2Dialect" boot-time="false"/>
        <property name="org.kie.server.persistence.ds" value="java:jboss/datasources/ExampleDS" boot-time="false"/>
        <property name="org.kie.server.persistence.tm" value="org.hibernate.service.jta.platform.internal.JBossAppServerJtaPlatform" boot-
```

```
time="false"/>
        <property name="kie.maven.settings.custom" value="/root/.m2/settings.xml" boot-time="false" />
    </system-properties>
</server>
```

这里有两个 IP 地址，org.kie.controller 的值为另一台 Workbench 服务器上的 IP 地址，从配置上来看，与之前写的 $TOMCAT_HOST/bin/setenv.sh 很相似，需要指定 settings.xml 配置，其修改内容如下。

```
<?xml version="1.0" encoding="UTF-8"?>
<settings>
    <localRepository>/root/.m2/repository</localRepository>
    <proxies>
    </proxies>
    <servers>
        <server>
            <id>kie-Workbench</id>
            <username>drools</username>
            <password>drools123</password>
            <configuration>
                <wagonProvider>httpclient</wagonProvider>
                <httpConfiguration>
                    <all>
                        <usePreemptive>true</usePreemptive>
                    </all>
                </httpConfiguration>
            </configuration>
        </server>
    </servers>
    <mirrors>
    </mirrors>
    <profiles>
        <profile>
            <id>kie</id>
            <properties>
            </properties>
            <repositories>
                <repository>
                    <id>kie-Workbench</id>
                    <name>JBoss BRMS Guvnor M2 Repository</name>
                    <url>http://192.168.80.31:8080/kie-wb/maven2/</url>
                    <activation>
                        <activeByDefault>true</activeByDefault>
                    </activation>
                    <layout>default</layout>
                    <releases>
                        <enabled>true</enabled>
                        <updatePolicy>always</updatePolicy>
```

```xml
                </releases>
                <snapshots>
                    <enabled>true</enabled>
                    <updatePolicy>always</updatePolicy>
                </snapshots>
            </repository>
        </repositories>
        <pluginRepositories>
            <pluginRepository>
                <id>kie-Workbench</id>
                <name>JBoss BRMS Guvnor M2 Repository</name>
                <url>http://192.168.80.31:8080/kie-wb/maven2/</url>
                <layout>default</layout>
                <releases>
                    <enabled>true</enabled>
                    <updatePolicy>always</updatePolicy>
                </releases>
                <snapshots>
                    <enabled>true</enabled>
                    <updatePolicy>always</updatePolicy>
                </snapshots>
            </pluginRepository>
        </pluginRepositories>
    </profile>
</profiles>
<activeProfiles>
    <activeProfile>kie</activeProfile>
</activeProfiles>
</settings>
```

3. Wildfly集群部署

Wildfly服务器的集群特点，重点在这个集群方面，首先来看一下集群的结构，如图12-32所示。

第一步：准备5台服务器。

安装过程中配置了5台机器，IP地址分别为：

```
10.0.5.213(kie-wb1——Workbench)服务器
10.0.5.119(kie-wb2——Workbench)服务器
10.0.5.207(Kie-Server)服务器
10.0.5.208(Kie-Server)服务器
10.0.5.84 (zookeeper+Helix)服务器
```

第二步：安装Zookeeper，安装过程使用的是Zookeeper 3.4.6版本。

第三步：安装Helix，安装过程使用的是Helix 0.6.5版本。

第四步：将Zookeeper-3.4.6/conf/ 的 zoo_sample.cfg 改名为zoo.cfg，不需要进行其他设置。

第五步：启动Zookeeper ./zkServer.sh start，在 $ZOOKEEPER_HOME/bin 目录下执行。

第六步：配置Helix 服务器到指定的目录下，执行 "cd $HELIX_HOME/bin" 命令。

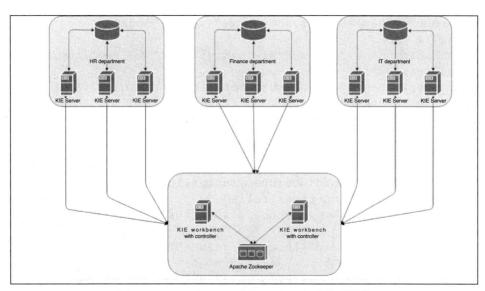

图12-32　Workbench与Kie-Server的集群结构

创建集群：

```
:./helix-admin.sh --zkSvr localhost:2181 --addCluster kie-cluster
```

Zookeeper 服务器使用的 zkSvr 值必须匹配。集群名称 (Kie-cluster) 可以根据需要改变。

将节点添加到集群：

```
# Node 1 $ ./helix-admin.sh --zkSvr localhost:2181 --addNode kie-cluster nodeOne:12345 # Node 2 $ ./helix-admin.sh --zkSvr localhost:2181 --addNode kie-cluster nodeTwo:12346 ...
```

这个添加只要在 Zookeeper 的机器上做就可以了，节点的值是唯一的。

添加资料到集群：

```
./helix-admin.sh --zkSvr localhost:2181 --addResource kie-cluster vfs-repo 1 LeaderStandby AUTO_REBALANCE
```

平衡集群的初始化：

```
./helix-admin.sh --zkSvr localhost:2181 --rebalance kie-cluster vfs-repo 2
```

开始 Helix 控制器管理集群：

```
./run-helix-controller.sh --zkSvr localhost:2181 --cluster kie-cluster 2>&1 > /tmp/control
```

第七步：配置 Workbench 所在服务器。

在机器 1 上配置：

```
<server name="server-one" group="main-server-group">
   <!--配置属性 -->
   <system-properties>
```

```xml
        <property name="jboss.node.name" value="nodeOne" boot-time="false"/>
        <property name="org.uberfire.nio.git.dir" value="/tmp/kie/nodeone" boot-time="false"/>
        <property name="org.uberfire.metadata.index.dir" value="/tmp/kie/nodeone" boot-time="false"/>
        <property name="org.uberfire.cluster.id" value="kie-cluster" boot-time="false"/>
        <property name="org.uberfire.cluster.zk" value="10.0.5.84:2181" boot-time="false"/>
        <property name="org.uberfire.cluster.local.id" value="nodeOne_12345" boot-time="false"/>
        <property name="org.uberfire.cluster.vfs.lock" value="vfs-repo" boot-time="false"/>
        <property name="org.kie.demo" value="false" boot-time="false"/>
        <property name="org.kie.example" value="false" boot-time="false"/>
        <property name="org.uberfire.nio.git.daemon.enabled" value="true" boot-time="false"/>
        <property name="org.uberfire.nio.git.daemon.host" value="10.0.5.213" boot-time="false" />
        <property name="org.uberfire.nio.git.daemon.port" value="9418" boot-time="false"/>
        <property name="org.kie.server.user" value="kieserver" boot-time="false" />
        <property name="org.kie.server.pwd" value="kieserver" boot-time="false" />
        <property name="org.guvnor.m2repo.dir" value="/root/.m2/repository" boot-time="false" />
    </system-properties>
</server>
```

配置过程中发现，其实只是操作 org.uberfire.cluster.xxx 的配置，但需要注意的是，有些地方还是要进行修改的，这里只是简单地说明一下：将机器 1 上的配置，完全复制到机器 2 上，并将 org.uberfire.nio.git.daemon.host 的值设置为机器 2 上的地址，org.uberfire.cluster.local.id 的值设置为 nodeTwo_12346。并将其余的 One or one 全部改为 Two or tow，就算完成了第 2 台服务器的配置。

第八步：配置 Kie-Server 所在的服务器，仍是对 host.xml 进行配置。

```xml
<server name="server-one" group="main-server-group">
    <system-properties>
        <!--配置本机地址-->
        <property name="org.kie.server.location" value="http://10.0.5.207:8080/Kie-Server/services/rest/server" boot-time="false"/>
        <property name="org.kie.server.id" value="wildfly-kieserver1" boot-time="false"/>
        <property name="org.kie.server.controller.user"
```

```xml
value="kieserver" boot-time="false"/>
        <property name="org.kie.server.controller.pwd" value="kieserver" boot-time="false"/>
        <!--配置kie-web地址 failover策略-->

        <property name="org.kie.server.controller" value="http://10.0.5.213:8080/kie-wb/rest/controller,http://10.0.5.119:8080/kie-wb/rest/controller" boot-time="false"/>

        <property name="org.kie.server.persistence.dialect" value="org.hibernate.dialect.H2Dialect" boot-time="false"/>
        <property name="org.kie.server.persistence.ds" value="java:jboss/datasources/ExampleDS" boot-time="false"/>
        <property name="org.kie.server.persistence.tm" value="org.hibernate.service.jta.platform.internal.JBossAppServerJtaPlatform" boot-time="false"/>
        <property name="kie.maven.settings.custom" value="/root/.m2/settings.xml" boot-time="false" />
    </system-properties>
</server>
```

简单地说明一下：org.kie.server.id 的值是可变的，在 Kie-Server 机器上如果设置 value="xxx1" 则也可能在 Kie-Server 机器 2 上设置 value="xxx2"。这样的设置，其实是将 Kie-Server 模板进行了分组，并且每个服务器都是独立的，还有一个区别在于 org.kie.server.controller 的设置，这个属性是用来做 Kie-wb 请求的，如果这里设置了多个有效的 Kie-wbURL，则在页面上的体现如图 12-33 所示。

图12-33 集群部署效果

总结：Wildfly 集群是可以为 Workbench 做集群服务的，Workbench 最大的用途之一在于创建或编辑规则文件，而 Workbench 的集群所处理的核心部分在于文件同步。使用 Zookeeper 是为了做文件同步。Kie-Server 的集群也是一样的，配置多台服务器，使用第三方工具做负载均衡，完成集群的配置。这样，发现所谓的 Workbench 集群、Kie-Server 集群就是使用 Web 容器启动的一些 war 包，进行文件同步、接口负载。

12.4 Kie-Server与Java交互

Kie-Server 是对外提供接口访问的,既然有接口访问,那么如何请求 Kie-Server 是一个问题,通过大量的查阅资料和实践得出,常用的有两种请求方式,一种是 JSON 方式,另一种是 XML 方式。

使用 Kie-Server 与 Java 交互时,需要先引用 Kie-Server 的 jar 包编辑 DroolsWorkbench 项目的 pom.xml,添加 Kie-Server 引用,其内容如下。

```xml
<dependency>
<groupId>org.kie.server</groupId>
<artifactId>Kie-Server-api</artifactId>
<version>${drools.version}</version>
</dependency>
<dependency>
<groupId>org.kie.server</groupId>
<artifactId>Kie-Server-client</artifactId>
<version>${drools.version}</version>
</dependency>
```

在操作 Workbench 自动扫描或是与项目的一般交互中,提到过,插入的 Fact 因子要与 Workbench 项目中生成的 jar 因子(数据对象)路径一样,所以有以下两种方案。

第一种:手动将这些 jar 包的数据对象添加到项目中。

使用 Workbench 下载 jar 包进行复制,自定义编写方式。

第二种:将 Workbench 生成的 jar 包放在公司 Maven 的私服库中,通过 Maven 的引用实现。

创建 TestKieWBServer.java 文件,目录在 com.Workbenchjava.javaisWorkbench 中。

1. JSON请求方式

添加 testJson() 测试方法,其内容如下。

```java
@Test
public void testJson(){
    Person person=new Person();
    person.setAge(30);
    person.setName("张三");
    String url = "http://localhost:8081/Kie-Server/services/rest/server";
    String username = "kieserver";
    String password = "kieserver1!";

    KieServicesConfiguration config = KieServicesFactory.newRestConfiguration(url, username, password);
    config.setMarshallingFormat(MarshallingFormat.JSON);//请求方式
    config.setTimeout(30000L);//如果请求失败,再次请求的间隔时间

    KieServicesClient client = KieServicesFactory.newKieServicesClient(config);//创建Kie-Server客户端
```

```
        RuleServicesClient rules = client.getServicesClient(RuleServicesCli
ent.class);//创建访问规则的客户端

        KieCommands cmdFactory = KieServices.Factory.get().getCommands();

        List<Command<?>> commands = new LinkedList<>();
        commands.add(cmdFactory.newInsert(person, "person"));//输入事务,在获
取请求时,与web请求获取一样
        commands.add(cmdFactory.newFireAllRules());
        ServiceResponse<ExecutionResults> response = rules.executeCommandsW
ithResults("testkieserver",cmdFactory.newBatchExecution(commands,
"testKieSession"));
        //第一个参数,容器名称;第二个参数,将传放的值放到容器中;testKieSession  表
示KieSession 第二个参数可有可无,如果没有,则使用的是kiesession默认值
        System.out.println(response.getMsg());
        ExecutionResults result = response.getResult(); //获取请求
        ServiceResponse.ResponseType type = response.getType();    //请求状态
        System.out.println(type.name());
        person = (Person) result.getValue("person");  //与web 获取前端传值很像
        System.out.println(person.getName());
}
```

引用的类,其内容如下。

```
import org.kie.api.KieServices;
import org.kie.api.command.Command;
import org.kie.api.command.KieCommands;
import org.kie.api.runtime.ExecutionResults;
import org.kie.server.api.marshalling.MarshallingFormat;
import org.kie.server.api.model.ServiceResponse;
import org.kie.server.client.KieServicesClient;
import org.kie.server.client.KieServicesConfiguration;
import org.kie.server.client.KieServicesFactory;
import org.kie.server.client.RuleServicesClient;
```

执行 testJson() 方法,结果如图 12-34 所示。

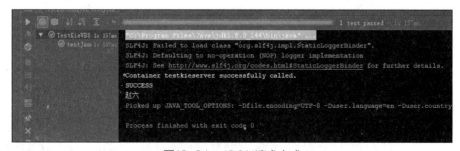

图12-34 JSON请求方式

2. XML请求方式

添加 esttssxml () 测试方法,其内容如下。

```java
@Test
public void esttssxml(){
    Person person=new Person();
    person.setAge(30);
    person.setName("张三");
    String url = "http://localhost:8081/Kie-Server/services/rest/server";
    String username = "kieserver";
    String password = "kieserver1!";
    InsertObjectCommand insertObjectCommand1 = new InsertObjectCommand(person, "person");//person输入事务,在获取请求时,与web请求获取一样
    GetObjectsCommand getObjectsCommand = new GetObjectsCommand();
    getObjectsCommand.setOutIdentifier("objects");  //输出标识符
    FireAllRulesCommand fireAllRulesCommand = new FireAllRulesCommand("RunAllRules");
    List commands = new ArrayList<>();
    commands.add(insertObjectCommand1);
    commands.add(fireAllRulesCommand);
    commands.add(getObjectsCommand);
    BatchExecutionCommand command = new BatchExecutionCommandImpl(commands);
    String xStreamXml = BatchExecutionHelper.newXStreamMarshaller().toXML(command);//将请求内容设置成为XML
    KieServicesConfiguration config = KieServicesFactory.newRestConfiguration(url, username, password);//登录服务器
    config.setMarshallingFormat(MarshallingFormat.**XSTREAM**);//请求方式
    KieServicesClient client = KieServicesFactory.newKieServicesClient(config);//获取请求
    String containerId = "testkieserver";//请求容器的名称
    Kie-ServerCommand call = new CallContainerCommand(containerId, xStreamXml);//拼接Kie-Server命名
    List<Kie-ServerCommand> cmds = Arrays.asList(call);//命名集合
    CommandScript script = new CommandScript(cmds);
    ServiceResponsesList reply = client.executeScript(script);//服务响应列表 请求服务
    for (ServiceResponse<? extends Object> r : reply.getResponses()) {
        System.out.println(r.getResult());
        if (r.getResult() != null) {
            ExecutionResultImpl result = (ExecutionResultImpl) BatchExecutionHelper.newXStreamMarshaller().fromXML((String)r.getResult());
            person = (Person) result.getResults().get("person");  //与web results功能是一样的
            // Objects From insert(fact0) in rule. The problem is that they are staying and multiplying there in Drools, don't know yet how to manage it. ToDo.
            ArrayList<Object> objects = (ArrayList<Object>) result.getResults().get("objects");//与web results功能是一样的
```

```
            System.out.println(objects+ person.getName());
        }
        else
            System.out.println("Empty result...?");
    }
}
```

执行 esttssxml () 方法，结果如图 12-35 所示。

图12-35　XML请求方式

下面总结一下注意点，使用事实对象时，与自动扫描是一样的，实体类型与 package 路径必须保持一致，请求方式可以是集合，也可以是单个实体请求，体现在命名请求。针对 Workbench 菜单中 Kie-Server 页面的说明，每次启动就是一个新的 KieSession。但是没有地方去释放，除非是重新启动。

3.扩展两个新概念

（1）判断容器是否存在，如果容器不存在，就没有执行的必要。

```
public void listContainers() {
    KieContainerResourceList containersList = kieServicesClient.listContainers().getResult();
    List<KieContainerResource> kieContainers = containersList.getContainers();
    System.out.println("Available containers: ");
    for (KieContainerResource container : kieContainers) {
        System.out.println("\t" + container.getContainerId() + " (" + container.getReleaseId() + ")");
    }
}
```

container.getContainerId() 是获取容器名称的，既然能拿到容器名称，就能很容易进行判断了。

（2）JSON 中 JavaBean 的集合请求，其实写法也特别简单，List 中可以放各种各样的数据，所以在"commands.add(cmdFactory.newInsert(obj));"之前对 newInsert 的值进行一个封装，通过遍历的方式将值放在命名中，如图 12-36 所示，这样就能容易地进行操作了。

```
//待执行命令列表
//commands.add(commandsFactory.newInsertElements(listObjs));
List<Command<?>> commands = new ArrayList<Command<?>>();
for(Object o : listObjs){
    commands.add(commandsFactory.newInsert(o));
}
for(Map.Entry<String, ?> entry : mapObjs.entrySet()){
    commands.add(commandsFactory.newInsert(entry.getValue(), entry.getKey()));
}
commands.add(commandsFactory.newFireAllRules());
```

图12-36　批量操作规则

第13章 动态规则

所谓动态规则，是指在不重启服务器的前提下使业务规则发生变化，且不影响服务器的正常使用，从而实现动态业务变化的规则。这也是规则引擎的魅力所在。

规则引擎要实现动态规则并不容易，动态规则可分为以下 7 种方式。

（1）字符串方式，拼接规则语法，形成完成的规则内容，通过 string 方式调用规则。

（2）通过规则模板方式对规则进行修改。

（3）Workbench 自动扫描。

（4）Workbench 整合 Kie-Server。

（5）指定文件的方式，通过规则文件执行规则。

（6）通过代码生成 kjar。

（7）模仿官方文档方式。

比较熟悉的应该是（2）~（4）这 3 种方案，规则模板的使用及用途已经在第 3 章规则引擎中讲述过，这里就不再重复了。

1. 字符串方式

什么是字符串方式？理解起来很简单，规则文件是一个纯文本文件，规则内容也就是文本中的内容。所以可以理解为，将规则文件中的规则内容以字符串方式进行加载并执行。

在 Drools 项目中创建 StringRule.java 文件，目录为 comTwo.sf，其内容如下。

```java
package comTwo.sf;

import org.junit.Test;
import org.kie.api.io.ResourceType;
import org.kie.api.runtime.KieSession;
import org.kie.internal.utils.KieHelper;

public class StringRule {
    @Test
    public void ruleDaseTestString() throws Exception {
        String myRule = "package rules rule \"Hello World 2\" when eval(true) then System.out.println(\"Test, Drools!\"); end";
        System.out.println(myRule);
        KieHelper helper = new KieHelper();
        helper.addContent(myRule, ResourceType.DRL);
        KieSession ksession = helper.build().newKieSession();
        ksession.fireAllRules();
        ksession.dispose();
    }
}
```

执行 ruleDaseTestString() 方法，结果如图 13-1 所示。

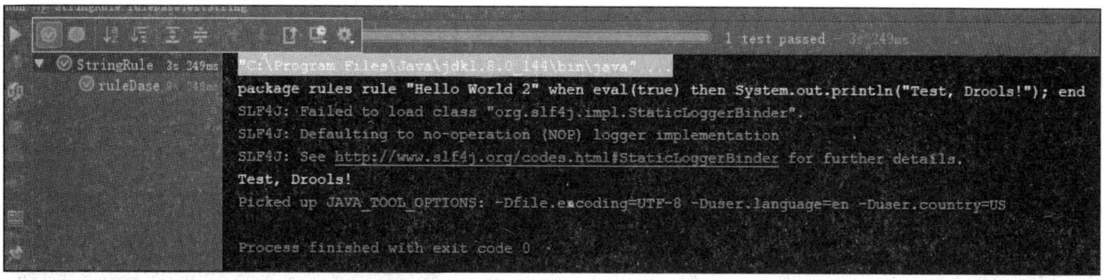

图13-1 字符串方式的动态规则

变量 myRule 内容比较简单，转成 Rule 文件就是输出一个 Test，Drools 在控制台，KieHelper 是核心，它是 Kie 提供的工具类，是自己封装的。但在官方教程中并没有提到这个类的使用，在本书的第 5 章中已详细说明。

其实这种字符串方式比较简单，既然可以通过字符串的格式使用规则，那是否可以把内容放到数据库管理中呢？字符串的方式只是通过变量做了一个引子，在实战应用中，规则内容比较少时可以通过数据库字段读取，量特别大时，需要通过写文件的方式进行读取。

拼接规则内容也是基于字符串方式动态规则的一个难点，在拼接的过程中要注意以下几点。

（1）转义问题。

（2）基本语法的正确性。

（3）效率问题。

2.指定文件的方式

指定文件的方式在前面提过，所谓指定文件方式，就是通过 Kie 提供的 API 获取特定目录下的规则文件并构建执行规则。

创建 TestResource.java 文件，目录在 comTwo.sf，其内容如下（注意加粗部分）。

```java
package comTwo.sf;

import org.junit.Test;
import org.kie.api.io.Resource;
import org.kie.api.io.ResourceType;
import org.kie.api.runtime.KieSession;
import org.kie.internal.io.ResourceFactory;
import org.kie.internal.utils.KieHelper;

public class TestResource {
    @Test
    public void  testPathClasses() throws Exception {
        Resource drl = ResourceFactory.newClassPathResource("rules/isSalience/salience.drl");
        KieHelper helper = new KieHelper();
        helper.addResource(drl, ResourceType.DRL);
        KieSession session = helper.build().newKieSession();
        session.fireAllRules();
```

```
        session.dispose();
    }
}
```

执行 testPathClasses() 方法，结果如图 13-2 所示。

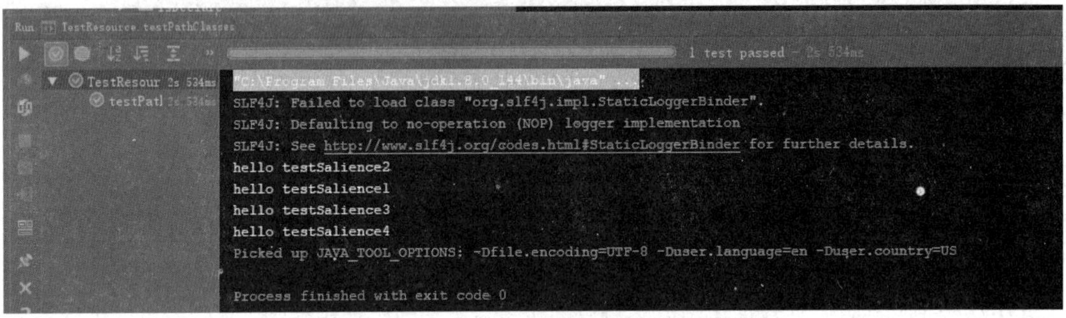

图13-2 指定规则文件相对路径

还有一种指定规则文件绝对路径的方式，编辑 TestResource.java 文件，添加 testFilePath() 方法，其内容如下（注意加粗部分）。

```
@Test
public void testFilePath() throws IOException {
    Resource dis = ResourceFactory.newFileResource( "D:\\project\\drools\\src\\main\\resources\\rules\\isSalience\\salience.drl" );
    KieHelper helper = new KieHelper();
    helper.addResource(dis,ResourceType.DRL);
    KieSession ksession = helper.build().newKieSession();
    int i = ksession.fireAllRules();
    System.out.println( "        " + i + "次");
    ksession.dispose();
}
```

执行 testFilePath () 方法，结果如图 13-3 所示。

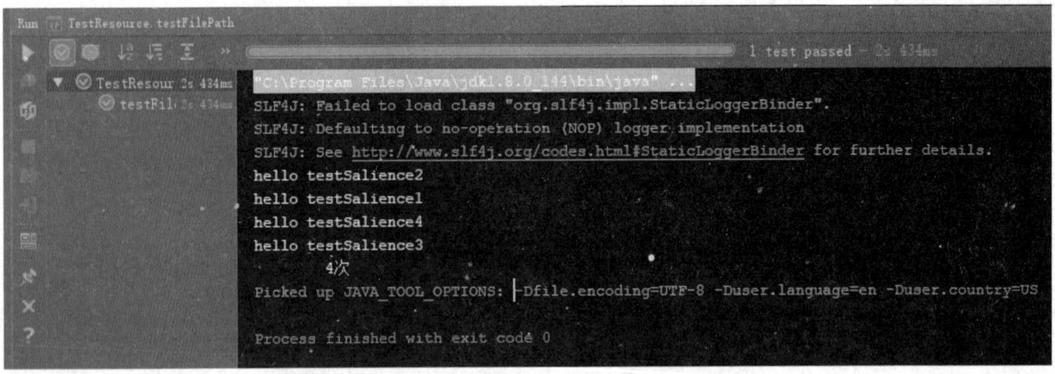

图13-3 指定规则文件的绝对路径

另外，还有一些指定 URL pach 字节的方式，本章就不一一说明了，有兴趣的读者可以自己尝试一下。既然已经能通过指定规则文件的方式来调用规则，也就可以通过 IO 进行创建规则文件，然后获取路径并调用了，其风险及注意事项与指定字符串方式是一样的。

上述例子介绍的都是一个文件的使用，能不能一次性多加载几个合法的规则文件呢？答案是一定的，原理与指定一个文件是一样的，都是通过 IO 读取。

创建 TestResourceList.java 类。目录为 comTwo.sf，其内容如下。

```java
package comTwo.sf;

import org.kie.api.KieServices;
import org.kie.api.builder.KieBuilder;
import org.kie.api.builder.KieFileSystem;
import org.kie.api.builder.Message;
import org.kie.api.builder.ReleaseId;
import org.kie.api.builder.model.KieModuleModel;
import org.kie.api.runtime.KieContainer;
import org.kie.api.runtime.KieSession;
import org.kie.internal.io.ResourceFactory;

import java.io.File;
import java.io.IOException;
import java.util.ArrayList;
import java.util.List;

public class TestResourceList {
    private static final String RULES_PATH = "rules";
    private static List<File> getRuleFiles() throws IOException {
        List<File> list = new ArrayList<File>();
        String filePath = Thread.currentThread().getContextClassLoader().getResource("").getPath();
        File rootDir = new File(filePath);
        File[] files = rootDir.listFiles();
        for (File itemFile : files) {
            if (itemFile.isDirectory() && itemFile.getName().equals(RULES_PATH)) {
                for (File f : itemFile.listFiles()) {
                    if (f.getName().endsWith(".drl")) {
                        list.add(f);
                    }
                }
            }
        }
        return list;
    }
    public static void main(String[] args) throws IOException {
        // TODO Auto-generated method stub
        KieServices ks = KieServices.Factory.get();
        KieFileSystem kfs = ks.newKieFileSystem();
        ReleaseId rid = ks.newReleaseId("org.drools",
```

```
"kiemodulemodel","1.0");
        kfs.generateAndWritePomXML(rid);
        KieModuleModel kModuleModel = ks.newKieModuleModel();
        kModuleModel.newKieBaseModel("kiemodulemodel")
            .newKieSessionModel("ksession");
        kfs.writeKModuleXML(kModuleModel.toXML());
        System.out.println(kModuleModel.toXML());
        for (File file : getRuleFiles()) {
            kfs.write("src/main/resources/" + file.getName(),ResourceFactory.newClassPathResource(RULES_PATH + File.separator + file.getName(), "UTF-8"));
        }
        KieBuilder kb = ks.newKieBuilder(kfs);
        kb.buildAll();
        if (kb.getResults().hasMessages(Message.Level.ERROR)) {
            throw new RuntimeException("Build Errors:\n"+ kb.getResults().toString());
        }
        KieContainer kContainer = ks.newKieContainer(rid);
        kContainer.updateToVersion( rid );
        KieSession kieSession = kContainer.newKieSession("ksession");
        kieSession.fireAllRules();
    }
}
```

通过指定规则文件方式实现动态规则时，其中 ResourceFactory 资料文件工厂类是做了很大贡献的，它实现了读取规则文件的方式，具体的可以看其都有哪些方式可以提供使用，如图 13-4 所示，通过加载 Class、绝对路径、URL 等，还能做一些其他方面的设置。

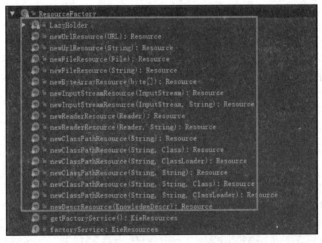

图13-4 ResourceFactory的组织结构

ResourceFactory 类的主要目的是用来获取规则文件的，真正的规则引擎的执行还是 KieHelper

工具类，不管是校验或是构建都离不开 KieHelper 类。

3. 通过代码生成kjar

生成 kjar 的动态规则与拼接字符串的方式有相似之处，核心思想与 Workbench 构建 kjar 包相似，kjar 的核心是 kmodule.xml 配置文件，而 Kie 为开发人员提供了这样的 API。动态创建 kmodule.xml 配置文件，其内容如下。

```
KieServices ks = KieServices.Factory.get();
KieFileSystem kfs = ks.newKieFileSystem();//kie文件系统
KieModuleModel kModuleModel = ks.newKieModuleModel();
        kModuleModel.newKieBaseModel("kiemodulemodel")
                    .newKieSessionModel("ksession");
        kfs.writeKModuleXML(kModuleModel.toXML());
        System.out.println(kModuleModel.toXML());
```

这段代码的核心是 newKieFileSystem 与 newKieModuleModel，其实在 KieHelper 中也是这样操作的，只是 KieHelper 使用的是默认值，而通过 KieFileSystem 细节更方便一些。

上述代码简单地描述了如何通过 API 创建 Kmodule.xml 虚拟文件，根据此思路，对 kjar 动态规则进行详细说明，创建 kjar 类似于 Workbench 构建生成的 kjar 包，所需要创建的文件有 Pom.xml、Kmodule.xml。

在项目 drools 中创建一个包目录，如在 comTwo.sf 中创建 kjar 包目录，本节的 Java 文件均在 kjar 目录下。创建 ResourceWrapper.java 文件，其内容如下。

```
package comTwo.sf.kjar;

import org.kie.api.io.Resource;

public class ResourceWrapper {
    private Resource resource;
    private String   targetResourceName;
    public ResourceWrapper(Resource resource, String targetResourceName) {
        this.resource = resource;
        this.targetResourceName = targetResourceName;
    }

    public Resource getResource() {
        return resource;
    }

    public void setResource(Resource resource) {
        this.resource = resource;
    }

    public String getTargetResourceName() {
        return targetResourceName;
```

```
    }

    public void setTargetResourceName(String targetResourceName) {
        this.targetResourceName = targetResourceName;
    }
}
```

创建 DroolsUtils.java 文件，该类为创建 Kie 所需要的工具类，也是核心，其内容如下。

```
package comTwo.sf.kjar;

import org.drools.compiler.kie.builder.impl.InternalKieModule;
import org.kie.api.KieServices;
import org.kie.api.builder.KieBuilder;
import org.kie.api.builder.KieFileSystem;
import org.kie.api.builder.ReleaseId;
import org.kie.api.builder.model.KieBaseModel;
import org.kie.api.builder.model.KieModuleModel;
import org.kie.api.builder.model.KieSessionModel;
import org.kie.api.conf.EqualityBehaviorOption;
import org.kie.api.conf.EventProcessingOption;

public class DroolsUtils {

    public static InternalKieModule createKieJar(KieServices ks,
ReleaseId releaseId, ResourceWrapper resourceWrapper) {
        KieFileSystem kfs = createKieFileSystemWithKProject(ks, true);
        kfs.writePomXML(getPom(releaseId));
        kfs.write("src/main/resources/" + resourceWrapper.
getTargetResourceName(), resourceWrapper.getResource());
        KieBuilder kieBuilder = ks.newKieBuilder(kfs);
        if (!kieBuilder.getResults().getMessages().isEmpty()) {
            System.out.println(kieBuilder.getResults().getMessages());
            throw new IllegalStateException("Error creating
KieBuilder.");
        }
        return (InternalKieModule) kieBuilder.getKieModule();
    }
    /**
     * 创建默认的kbase和stateful的kiesession
     *
     * @param ks
     * @param isdefault
     * @return
     */
    public static KieFileSystem createKieFileSystemWithKProject(KieServ
ices ks, boolean isdefault) {
        KieModuleModel kproj = ks.newKieModuleModel();
        KieBaseModel KieBaseModel1 = kproj.newKieBaseModel("KBase").
setDefault(isdefault)
```

```java
                .setEqualsBehavior(EqualityBehaviorOption.EQUALITY).set
EventProcessingMode(EventProcessingOption.STREAM);
        // Configure the KieSession.
        KieBaseModel1.newKieSessionModel("KSession").
setDefault(isdefault)
                .setType(KieSessionModel.KieSessionType.STATEFUL);
        KieFileSystem kfs = ks.newKieFileSystem();
        kfs.writeKModuleXML(kproj.toXML());
        System.out.println(kproj.toXML());
        return kfs;
    }
    /**
     * 创建kjar的pom
     *
     * @param releaseId
     * @param dependencies
     * @return
     */
    public static String getPom(ReleaseId releaseId, ReleaseId...
dependencies) {
        String pom = "<?xml version=\"1.0\" encoding=\"UTF-8\"?>\n"
                + "<project xmlns=\"http://maven.apache.org/POM/4.0.0\"
xmlns:xsi=\"http://www.w3.org/2001/XMLSchema-instance\"\n"
                + "         xsi:schemaLocation=\"http://maven.apache.
org/POM/4.0.0 http://maven.apache.org/maven-v4_0_0.xsd\">\n"
                + "    <modelVersion>4.0.0</modelVersion>\n" + "\n" + "    
<groupId>" + releaseId.getGroupId()
                + "</groupId>\n" + "    <artifactId>" + releaseId.
getArtifactId() + "</artifactId>\n" + "    <version>"
                + releaseId.getVersion() + "</version>\n" + "\n";
        if (dependencies != null && dependencies.length > 0) {
            pom += "<dependencies>\n";
            for (ReleaseId dep : dependencies) {
                pom += "<dependency>\n";
                pom += "  <groupId>" + dep.getGroupId() + "</groupId>\
n";
                pom += "  <artifactId>" + dep.getArtifactId() + "</
artifactId>\n";
                pom += "  <version>" + dep.getVersion() + "</version>\
n";
                pom += "</dependency>\n";
            }
            pom += "</dependencies>\n";
        }
        pom += "</project>";
        return pom;
    }
}
```

创建 DynamicDrlTest.java 文件，该类为测试类。其内容如下。

```java
package comTwo.sf.kjar;

import org.drools.compiler.kie.builder.impl.InternalKieModule;
import org.kie.api.KieServices;
import org.kie.api.builder.KieRepository;
import org.kie.api.builder.ReleaseId;
import org.kie.api.runtime.KieContainer;
import org.kie.api.runtime.KieSession;
import org.kie.internal.io.ResourceFactory;

public class DynamicDrlTest {
    private static final String RULESFILE_NAME = "rules.drl";
    /**
     * 规则文件内容（可以从数据库中加载）
     */
    private static final String ruleCoutext = "package rules rule \"Hello World 2\" when eval(true) then System.out.println(\"Test, Drools!\"); end";

    public static void main(String[] args) throws Exception {
        KieServices kieServices = KieServices.Factory.get();
        //指定kjar包
        final ReleaseId releaseId = kieServices.newReleaseId("rule", "test", "1.0.0");
        // 创建初始化的kjar
        InternalKieModule kJar = DroolsUtils.createKieJar(kieServices, releaseId,
                new ResourceWrapper(ResourceFactory.newByteArrayResource(ruleCoutext.getBytes()), RULESFILE_NAME));
        KieRepository repository = kieServices.getRepository();
        repository.addKieModule(kJar);
        KieContainer kieContainer = kieServices.newKieContainer(releaseId);
        KieSession session = kieContainer.newKieSession();
        try {
            session.fireAllRules();
        } catch (Exception e) {
        } finally {
            session.dispose();
        }
        System.out.println("------------");
    }
}
```

执行 DynamicDrlTest 的主函数，结果如图 13-5 所示。

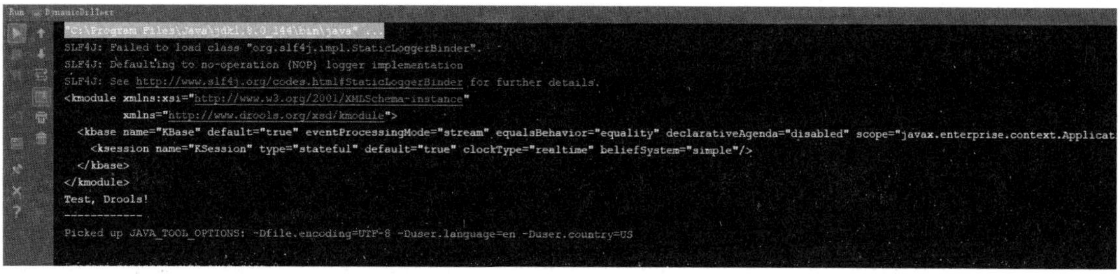

图13-5 使用kjar包的动态规则效果

4. 模仿官方文档方式

官方文档也提供了一种方式，写法比通过代码生成 kjar 要更简单，其中原理都是相似的。创建 InitRuleParam.java 文件，目录为 comTwo/sf，其内容如下。

```java
package comTwo.sf;

import com.pojo.Person;
import org.kie.api.KieServices;
import org.kie.api.builder.KieBuilder;
import org.kie.api.builder.KieFileSystem;
import org.kie.api.builder.ReleaseId;
import org.kie.api.builder.model.KieModuleModel;
import org.kie.api.builder.model.KieSessionModel.KieSessionType;
import org.kie.api.runtime.KieContainer;
import org.kie.api.runtime.StatelessKieSession;

import java.io.BufferedReader;
import java.io.File;
import java.io.FileReader;
import java.util.ArrayList;
import java.util.HashMap;
import java.util.List;
import java.util.Map;

public class InitRuleParam {
    //可自己配置日志功能
    private final static String PATH = "src/main/resources/rules";
    private final static String PATHt = "rules";
    private final static String GROUPID = "com.plus";
    private final static String ARTIFACTID = "plusRules";
    private final static String VERSION = "1.0";

    public static KieContainer kContainer;
    public static KieServices kieServices;
    public static KieFileSystem kfs;
    public static KieModuleModel kmm;

    public static Map<String, StatelessKieSession> mapKieSession = new
```

```java
HashMap<String, StatelessKieSession>();
    static Map<String, String> fileMap = new HashMap<>();
    static KieContainer kc;


    public InitRuleParam() {
        System.out.println("开始初始化ksession");
        long star1 =System.currentTimeMillis();
        try {
            // 动态加载 drl (把 drl 内容读取出来拼接动态生成 kiesession, 相当于从数据库读取字符串 )
            Map<String, String> map = getFile("D:\\project\\drools\\src\\main\\resources\\rules\\isKieSession");
            loadGeneral(map);
        } catch (Exception e) {
            e.printStackTrace();
        }
        long end =System.currentTimeMillis();
        System.err.println("初始总用时间:"+(end-star1));
        System.out.println("初始化成功，总用时: "+(end-star1));

        init();
    }

    /**
     * 测试
     */
    public void init() {
        System.out.println("(3) 测试...");
        for (final Map.Entry<String, StatelessKieSession> entry : mapKieSession.entrySet()) {
            new Thread(new Runnable() {
                @Override
                public void run() {
                    StatelessKieSession kieSession = null;
                    try {
                        System.out.println("(3) 取出ksession");
                        kieSession = entry.getValue();
                        List<String> myGlobalList = new ArrayList<>();  // 违反规则, global
                        kieSession.setGlobal("myGlobalList", myGlobalList);

                        System.err.println(kieSession);
                        Person person = new Person("张三",30);
                        // 无状态
                        System.out.println("(3) 省略了有状态的insert fireAllRules");
                        /**
```

```
                            * 无状态会话不支持迭代调用，调用execute（…）的动作是
一个单独的方法，将内部实例化KieSession，
                            * 添加所有用户数据并执行用户命令，调用fireAllRules，然
后调用dispose()
                            */
                            kieSession.execute(person);
                            System.out.println("(3) 获取违反规则长度为
"+myGlobalList.size());
                            System.err.println(myGlobalList.size());
                    } catch (Exception e2) {
                        e2.printStackTrace();
                    }
                }
            }).start();
        }
    }

    /**
     * 把字符串动态生成 kiesession
     * @param map
     */
    private  void loadGeneral(Map<String, String> map) {
        System.out.println("(2) 动态生成kiesession...");
        kieServices = KieServices.Factory.get();
        kfs = kieServices.newKieFileSystem();
        // 创建 pom.xml
        ReleaseId rId = kieServices.newReleaseId(GROUPID, ARTIFACTID, VERSION);
        kfs.generateAndWritePomXML(rId);
        System.out.println("(2) 创建pom.xml文件成功");
        // 创建 kmoudle.xml
        kmm = kieServices.newKieModuleModel();
        System.out.println("(2) 创建kmoudle.xml文件成功");

        // 规则写入本地
        if (null != map && map.size() != 0) {
            for (Map.Entry<String, String> entry : map.entrySet()) {
                String path = PATH + "/" + entry.getKey() + "/" + entry.getKey() + ".drl";
                kfs.write(path, entry.getValue());
                System.err.println(path);
                kmm.newKieBaseModel(entry.getKey() + "z").addPackage(PATHt + "." + entry.getKey())
                        .newKieSessionModel(entry.getKey()).setType(KieSessionType.STATELESS);
            }
        }
        System.out.println("(2) 组建kmoudle.xml文件成功");
```

```java
        kfs.writeKModuleXML(kmm.toXML());
        KieBuilder kbd = kieServices.newKieBuilder(kfs);
        kbd.buildAll();
        System.out.println("(2) 编写kmoudle.xml文件成功,内容为"+kmm.
toXML());

        kc = kieServices.newKieContainer(rId);

        kc.updateToVersion(rId);

        if (null != map && map.size() != 0) {
            for (Map.Entry<String, String> entry : map.entrySet()) {
                StatelessKieSession kSession = null;
                kSession = kc.newStatelessKieSession(entry.getKey());
                mapKieSession.put(entry.getKey(), kSession);
            }
        }
        System.out.println("(2) 动态生成Ksession,存入mapKieSession成功,长
度为"+mapKieSession.size()+"…");
    }

    /**
     * 读取 drl 内容,文件名作为 key ,内容作为 value
     *
     * @param frold
     * @return
     * @throws Exception
     */
    private Map<String, String> getFile(String frold) throws Exception
{
        System.out.println("(1) 读取drl文件,存入fileMap…");
        File f = new File(frold);
        File fa[] = f.listFiles();
        for (int i = 0; i < fa.length; i++) {
            File fs = fa[i];
            if (!fs.isDirectory()) {
                BufferedReader br = new BufferedReader(new
FileReader(new File(fs.toString()))); 
                StringBuilder result = new StringBuilder();
                String s = null;
                while ((s = br.readLine()) != null) {// 使用readLine方法,
一次读一行
                    result.append(System.lineSeparator() + s);
                }
                fileMap.put(fs.getName().replace(".drl", ""), result.
toString());
                System.out.println("(1) 存入fileMap,key: "+fs.getName().
replace(".drl", "")+"    value:"+result.toString());
            } else {
```

```
            System.err.println(fs.getName());
            System.out.println("(1) 修改路径不正确："+fs.getName());
        }
    }

    System.out.println("(1) 读取drl文件，存入fileMap完成，map长度为："+fileMap.size()+"...");
    return fileMap;
    }

    public  static void main(String[] args) {
        new InitRuleParam();
    }
}
```

执行 InitRuleParam 的主函数，结果如图 13-6 所示。

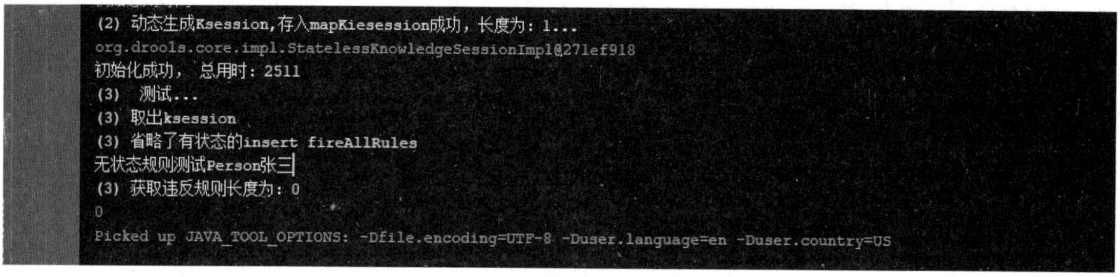

图13-6　模仿官方文档的动态规则

模仿官方文档的动态规则总结如下。

KieServices 对象得到一个 KieContainer，然后根据 Session Name 来新建一个 KieSession，最后通过 KieSession 来运行规则，该接口提供了很多方法，可以通过这些方法访问 Kie 关于构建和运行的相关对象，如可以获取 KieContainer，利用 KieContainer 来访问 KieBase 和 KieSession 等信息；可以获取 KieRepository 对象，利用 KieRepository 来管理 KieModule 等。它是一个中心，通过它获取的各种对象来完成规则构建、管理和执行等操作。

KieContainer：KieContainer 就是一个 KieBase 的容器。

KieBase：KieBase 就是一个知识仓库，包含了若干的规则、流程、方法等，在 Drools 中主要就是规则和方法，KieBase 本身并不包含运行时的数据，如果需要执行 KieBase 中的规则，就需要根据 KieBase 创建 KieSession。

KieSession：KieSession 就是一个与 Drools 引擎打交道的会话，其基于 KieBase 创建，它包含运行时的数据，包含"事实 Fact"，并对运行时的数据事实进行规则运算。通过 KieContainer 创建 KieSession 是一种较为方便的做法，其本质是从 KieBase 中创建出来的。KieSession 就是应用程序与规则引擎进行交互的会话通道。创建 KieBase 是一个成本非常高的事情，它会建立知识（规则、流程）仓库，而创建 KieSession 则是一个成本非常低的事情，所以 KieBase 会建立缓存，而

KieSession 则并不是必须要缓存的。

维护 KieSession StatelessSession 没有持久化上下文，也不提供多少高层的生命周期语义。特别是无状态 Session 不实现第一级 Cache，也不与第二级缓存或者查询缓存交互。它不实现事务化写入，也不实现脏读数据检查。使用 Stateless Session 进行的操作甚至不级联到关联实例。

Stateless Session 忽略集合类 (Collections)。通过 Stateless Session 进行的操作不触发 Hibernate 的事件模型和拦截器。

实现动态规则可以分为两部分，第一部分是通过 KieFileSystem 创建出虚拟文件，该文件包含了"规则文件"、虚拟 kmodule.xml 文件及 KieBase 的 packages，通过构建生成知识库 (规则库)。第二部分就是与实际操作规则相关的 API，包括知识库（规则库）、KieSession 等。

了解了动态规则的实现方式，也明白了什么是动态规则，解释为什么要使用动态规则就容易多了，动态规则又可以称为动态业务，规则引擎的核心目的之一就是用来处理项目中的业务。例如，风险控制系统，这类的项目往往包含了太多的业务规则，以传统的方式开发，都会将其逻辑业务写到代码中，也称为硬编码方式。这样做无形中会为系统增加累赘，变更需求，代码需要重新讨论，开发、测试、上线可能会错过最好的时间段，而且还要承担重启服务器所带来的后果。

第14章

多线程中的Drools

Drools 规则引擎技术指南

实际开发过程中，规则引用一般都会作为一个平台出现，如常见的电商类促销规则，金融类风险控制、反欺诈，医疗类的合理用药等，这些平台都有一个共同的特性，那就是处于一个高并发的状态下。下面就通过多线程方式来模拟高并发状态下的规则引擎并进行总结。

讲到多线程就不得不多提一个概念，即 Multithreaded rule engine，在官方文档中，有介绍多线程规则引擎的内容，也明确说明了该功能还处于实验阶段，如图 14-1 所示。

图14-1　Multithreaded注意事项

该功能默认处于关闭状态，在 KieBaseConfiguration 中设置，该功能由 KieServices 提供，查阅官方给出的代码[①]，如图 14-2 所示。

```
KieServices ks = KieServices.Factory.get();
KieBaseConfiguration kieBaseConf = ks.newKieBaseConfiguration();
kieBaseConf.setOption(MultithreadEvaluationOption.YES);
KieBase kieBase = kieContainer.newKieBase(kieBaseConf);
```

图14-2　开启Multithreaded 说明

通过 KieServices 创建 KieBaseConfiguration，通过命名的方式打开，创建 KieBean 时导入，或者通过系统属性 drools.multithreadEvaluation = true，如图 14-3 所示是 MultithreadEvaluation 的部分源码。

```
/**
 * An Enum for multithread evaluation option.
 *
 * drools.multithreadEvaluation = &lt;true|false&gt;
 *
 * DEFAULT = false
 */
public enum MultithreadEvaluationOption implements SingleValueKieBaseOption {

    YES(true),
    NO(false);

    /**
     * The property name for the multithread evaluation option
     */
    public static final String PROPERTY_NAME = "drools.multithreadEvaluation";

    private boolean value;

    MultithreadEvaluationOption( final boolean value ) { this.value = value; }

    /**
     * {@inheritDoc}
     */
    public String getPropertyName() {
        return PROPERTY_NAME;
    }

    public boolean isMultithreadEvaluation() {
        return this.value;
    }
```

图14-3　MultithreadEvaluation 的部分源码

---
① 图 14-2 是在使用 Kmodule.xml 或在代码中使用 KieServices 时的赋值方式，使用 KieHelper 则不需要。

目前这样的多线程规则不支持 queries、salience、agenda-group，如果有这些属性存在规则中，则会自动切换回单线程模式，根据官方所表达的意思，在编译过程将会发出警告信息。还有一些特殊的情况，就是不使用自定义的排序，规则也有一些默认的排序，可能会执行 myRuleFile1，也有可能会执行 myRuleFile2，还有可能执行一部分 myRuleFile1 与 myRuleFile2，总而言之，规则整个执行下来的过程之和，符合调用规则的次数。

测试多线程在规则中使用是比较麻烦的，因为存在相同的规则库、相同的 KieSession，不同的规则库、不同的 KieSession，有状态的 KieSession 或是无状态的 StatelessKieSession 等。针对这一问题，分析了 8 种不同的情况，其中有些结论是相同的，有些结论非常相似，下面先将相同与相似的总结如下。

（1）同一个规则库中，同逻辑路径下的规则名是不能重复的，这符合 kmodule.xml 规范。

（2）不同 KieHelper 不同 KieSession（有状态），规则名是能重复的，这符合 kmodule.xml 规范要求。

（3）"helper1.build(MultithreadEvaluationOption.YES).newKieSession();"并无实际效果。规则没有并发执行。

（4）有状态创建的时间比无状态创建的时间要长。

（5）同 KieHelper 不同 KieSession（有状态），不同 KieHelper 不同 KieSession（有状态），同 KieHelper 同 StatelessKieSession（无状态），同 KieHelper 不同 StatelessKieSession（无状态），这 4 种在多线程执行下，规则可能会被激活。

不同的结论会针对性地在测试用例下方做出总结。在 Drools 项目下创建 comThree 包，本节中所有的测试文件均在该目录下。

创建 RuleValueFile.java 文件，其内容如下。

```
package comThree;

public class RuleValueFile {
    //规则文件
    public static final String myRuleFile1 = "package rules " +
"import com.pojo.Person;" +    "import com.pojo.School;" +
        "rule \"myRule1\" " +
        " salience 10" +
        "   when " +
        "       $p:Person(name==\"张三\")" +
        "       $s:School(className==\"北大\")" +
        "   then " +
        "       $p.setName(\"李四\");" +
        "       $s.setClassName(\"清华\");" +
        "       update($p);" +
        "       update($s);" +
        "end              " +
        "rule \"myRule2\" " +
```

```
"          when " +
"              $p:Person(name==\"李四\")" +
"              $s:School(className==\"清华\")" +
"          then " +
"              System.out.println(\"规则myRuleFile1被调用\");" +
"end";
public static final String myRuleFile2 = "package rules " + "import com.pojo.Person;" +       "import   com.pojo.School;" +
"rule \"myRule3\" " +
"          when " +
"              $p:Person(name==\"张三\")" +
"              $s:School(className==\"北大\")" +
"          then " +
"              $p.setName(\"王王\");" +
"              $s.setClassName(\"张大\");" +
"              update($p);" +
"              update($s);" +
"end                            " +
"rule \"myRule4\" " +
"          when " +
"              $p:Person(name==\"王王\")" +
"              $s:School(className==\"张大\")" +
"          then " +
"     System.out.println(\"规则myRuleFile2调用\");" +
"end";
}
```

14.1 同KieHelper 同KieSession（有状态）

同 KieHelper 与同 KieSession（有状态）是指在代码中使用同一个 KieHelper 来构建规则库，并创建同一个有状态的 KieSession。

创建 DrlKieSessionOne.java 文件，其内容如下。

```
package comThree;

import com.pojo.Person;
import com.pojo.School;
import org.kie.api.io.ResourceType;
import org.kie.api.runtime.KieSession;
import org.kie.internal.conf.MultithreadEvaluationOption;
import org.kie.internal.utils.KieHelper;

import java.util.concurrent.ExecutorService;
import java.util.concurrent.Executors;
```

```java
import java.util.concurrent.Semaphore;

import static comThree.RuleValueFile.myRuleFile1;
import static comThree.RuleValueFile.myRuleFile2;

public class DrlKieSessionOne {
    private static int thread_num = 1000;//线程数,设置同时并发线程数
    private static int client_num = 100;//访问次数

    public static void main(String[] args) throws InterruptedException {

        long start = System.currentTimeMillis();
        KieHelper helper1 = new KieHelper();
        long end = System.currentTimeMillis();
        System.out.println("输出创建KieHelper用的毫秒是="+(end - start));

        long startAddrule = System.currentTimeMillis();
        //分别将规则myRuleFile2 myRuleFile1 加载到虚拟文件中
        helper1.addContent(myRuleFile1, ResourceType.DRL);
        helper1.addContent(myRuleFile2, ResourceType.DRL);
        long endAddrule = System.currentTimeMillis();
        System.out.println("导入规则所用到的毫秒是="+(endAddrule - startAddrule));
        long startNewKieSession = System.currentTimeMillis();
        KieSession kieSession = helper1.build(MultithreadEvaluationOption.YES).newKieSession();
        long endNewKieSession = System.currentTimeMillis();
        System.out.println("创建有状态KieSession所用到的毫秒是="+(endNewKieSession - startNewKieSession));

        ExecutorService exec = Executors.newCachedThreadPool();
        final Semaphore semp = new Semaphore(thread_num);
        for (int index = 0; index < client_num; index++) {
            Runnable run = () -> {
                try {
                    semp.acquire();
                    long startrule = System.currentTimeMillis();
                    Person person = new Person();
                    person.setName("张三");
                    person.setAge((int) (Math.random() * 100));
                    School school = new School();
                    school.setClassName("北大");
                    school.setClassCount("40");
                    kieSession.insert(person);
                    kieSession.insert(school);
                    int s1 = kieSession.fireAllRules();
                    System.out.println("kieSession执行规则"+s1);
                    System.out.println("修改后的"+person.getName());
```

```
                    //kieSession.dispose();//注释的原因  同一个KieSession
（有状态）是线程以外创建的KieSession，所以必须要注释
                    long endrule = System.currentTimeMillis();
                    System.out.println("规则执行所用到的毫秒是="+(endrule
- startrule));
                    System.out.println("==============================分
割线======================================");
                    semp.release();
                } catch (Exception e) {
                    e.printStackTrace();
                }
            };
            exec.execute(run);
        }
        exec.shutdown();
    }
}
```

结论：测试并使用同 KieHelper 同 KieSession（有状态）线程数特别多时，规则可能有一部分是不会执行的，如图 14-4 所示。

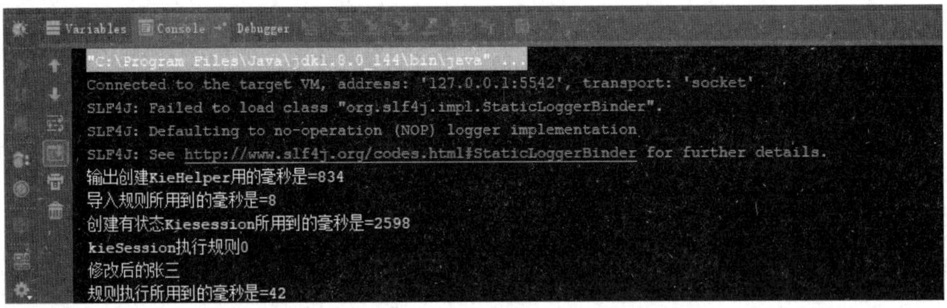

图14-4　同KieHelper同KieSession（有状态）

创建规则调用后，输出的结果则只有一个李四，并没有出现迭代的效果，如图 14-5 所示。

图14-5　同KieHelper同KieSession（有状态）

通过上述测试用例的执行过程得出区别于其他情况的总结如下。

规则与规则之间的 Fact 事实对象会相互作用，也就是说规则输出修改后的值可以是李四，也可以是王王，但只有一个规则文件会执行，并受规则优先级属性的影响，如图 14-6 所示。

```
=============================分割线=============================
kieSession执行规则2
修改后的李四
规则执行所用到的毫秒是=404
=============================分割线=============================
kieSession执行规则2
修改后的王王
规则执行所用到的毫秒是=406
=============================分割线=============================
```

图14-6　同KieHelper不同KieSession（有状态）

14.2　同KieHelper 不同KieSession（有状态）

同 KieHelper 与不同 KieSession（有状态）是指在代码中使用同一个 KieHelper 来构建规则库，并创建两个有状态的 KieSession。

创建 DrlKieSessionTwo.java 文件，其内容如下。

```java
package comThree;

import com.pojo.Person;
import com.pojo.School;
import org.kie.api.io.ResourceType;
import org.kie.api.runtime.KieSession;
import org.kie.internal.utils.KieHelper;

import java.util.concurrent.ExecutorService;
import java.util.concurrent.Executors;
import java.util.concurrent.Semaphore;

import static comThree.RuleValueFile.myRuleFile1;
import static comThree.RuleValueFile.myRuleFile2;

public class DrlKieSessionTwo {
    private static int thread_num = 2000;//线程数,设置同时并发线程数
    private static int client_num = 100;//访问次数

    public static void main(String[] args) {

        long start = System.currentTimeMillis();
        KieHelper helper1 = new KieHelper();
        long end = System.currentTimeMillis();
        System.out.println("输出创建KieHelper用的毫秒是=" + (end - start));

        long startAddrule = System.currentTimeMillis();
        //分别将规则myRuleFile2 myRuleFile1 加载到虚拟文件中
```

```java
            helper1.addContent(myRuleFile1, ResourceType.DRL);
            helper1.addContent(myRuleFile2, ResourceType.DRL);
            long endAddrule = System.currentTimeMillis();
            System.out.println("导入规则所用到的毫秒是=" + (endAddrule - startAddrule));

            ExecutorService exec = Executors.newCachedThreadPool();
            final Semaphore semp = new Semaphore(thread_num);
            for (int index = 0; index < client_num; index++) {
                Runnable run = () -> {
                    try {
                        semp.acquire();
                        long startNewKieSession = System.currentTimeMillis();
                        KieSession kieSession = helper1.build().newKieSession();
                        long endNewKieSession = System.currentTimeMillis();
                        System.out.println("创建有状态KieSession所用到的毫秒是=" + (endNewKieSession - startNewKieSession));
                        long startrule = System.currentTimeMillis();
                        Person person = new Person();
                        person.setName("张三");
                        person.setAge((int) (Math.random() * 100));
                        School school = new School();
                        school.setClassName("北大");
                        school.setClassCount("40");
                        kieSession.insert(person);
                        kieSession.insert(school);
                        int s1 = kieSession.fireAllRules();
                        System.out.println("kieSession执行规则" + s1);
                        System.out.println("修改后的"+person.getName());
                        kieSession.dispose();//这里可注释可不注释，因为每次KieSession都是不同的
                        long endrule = System.currentTimeMillis();
                        System.out.println("规则执行所用到的毫秒是=" + (endrule-startrule));
                        System.out.println("=============================分割线======================================");
                        semp.release();
                    } catch (Exception e) {
                        e.printStackTrace();
                    }
                };
                exec.execute(run);
            }
            exec.shutdown();
        }
    }
```

14.3 不同KieHelper 不同KieSession（有状态），KieSession只创建一次

不同 KieHelper 肯定是不同的 KieSession，区别在于 KieSession 只创建一次，测试过程中不做修改。

创建 DrlKieSessionThree.java 文件，其内容如下。

```
package comThree;

import com.pojo.Person;
import com.pojo.School;
import org.kie.api.io.ResourceType;
import org.kie.api.runtime.KieSession;
import org.kie.internal.conf.MultithreadEvaluationOption;
import org.kie.internal.utils.KieHelper;

import java.util.concurrent.ExecutorService;
import java.util.concurrent.Executors;
import java.util.concurrent.Semaphore;

import static comThree.RuleValueFile.myRuleFile1;
import static comThree.RuleValueFile.myRuleFile2;
public class DrlKieSessionThree {
    private static int thread_num = 20;//线程数,设置同时并发线程数
    private static int client_num = 100;//访问次数

    public static void main(String[] args) {

        long start = System.currentTimeMillis();
        KieHelper helper1 = new KieHelper();
        KieHelper helper2 = new KieHelper();
        long end = System.currentTimeMillis();
        System.out.println("输出创建KieHelper用的毫秒是="+(end - start));

        long startAddrule = System.currentTimeMillis();
        //分别将规则myRuleFile2 myRuleFile1 加载到虚拟文件中
        helper1.addContent(myRuleFile1, ResourceType.DRL);
        helper2.addContent(myRuleFile2, ResourceType.DRL);//*******
        long endAddrule = System.currentTimeMillis();
        System.out.println("导入规则所用到的毫秒是="+(endAddrule - startAddrule));

        long startNewKieSession = System.currentTimeMillis();
        KieSession kieSession1 = helper1.build(MultithreadEvaluationOption.YES  ).newKieSession();
```

```java
        KieSession kieSession2 = helper2.build(MultithreadEvaluationOption.YES ).newKieSession();
        long endNewKieSession = System.currentTimeMillis();
        System.out.println("创建有状态KieSession所用到的毫秒是="+(endNewKieSession - startNewKieSession));

        ExecutorService exec = Executors.newCachedThreadPool();
        final Semaphore semp = new Semaphore(thread_num);
        for (int index = 0; index < client_num; index++) {
            Runnable run = () -> {
                try {
                    semp.acquire();
                    long startrule = System.currentTimeMillis();
                    Person person = new Person();
                    person.setName("张三");
                    person.setAge((int) (Math.random() * 100));
                    School school = new School();
                    school.setClassName("北大");
                    school.setClassCount("40");
                    kieSession1.insert(person);
                    kieSession2.insert(person);
                    kieSession1.insert(school);
                    kieSession2.insert(school);
                    int s1 = kieSession1.fireAllRules();
                    int s2 = kieSession2.fireAllRules();
                    System.out.println("kieSession1执行规则"+s1);
                    System.out.println("kieSession2执行规则"+s2);
                    System.out.println("修改后的"+person.getName());
                    //kieSession1.dispose();//注释的原因   同一个KieSession有状态 是线程以外创建的KieSession 所以必须要注释
                    //kieSession2.dispose();//注释的原因   同一个KieSession有状态 是线程以外创建的KieSession 所以必须要注释
                    long endrule = System.currentTimeMillis();
                    System.out.println("规则执行所用到的毫秒是="+(endrule-startrule));
                    System.out.println("==============================分割线======================================");
                    semp.release();
                } catch (Exception e) {
                    e.printStackTrace();
                }
            };
            exec.execute(run);
        }
        exec.shutdown();
    }
}
```

通过上述测试用例的执行过程得出区别于其他情况的总结如下。

（1）线程数特别多时，规则可能有一部分是不会执行的，这与同 KieHelper 同 KieSession 是一样的。

（2）规则与规则之间的 Fact 事实对象会相互作用，虽然在代码中分别将 Person 对象和 School 对象嵌入两个不同的 KieSession 中，但由于 myRulefile1 先执行，导致 Fact 对象发现变化，myRuleFile2 规则无法满足条件。

（3）虽然 KieSession 是不同的，KieHelper 也是不同的，myRuleFile2 与 myRuleFile1 都会被执行，但结果可能存在迭代。

14.4 不同KieHelper 不同KieSession（有状态），KieSession在线程代码中创建

不同 KieHelper 肯定是不同的 KieSession，KieSession 的创建放在线程中。

创建 DrlKieSessionTwo.java 文件，其内容如下。

```
package comThree;

import com.pojo.Person;
import com.pojo.School;
import org.kie.api.io.ResourceType;
import org.kie.api.runtime.KieSession;
import org.kie.internal.conf.MultithreadEvaluationOption;
import org.kie.internal.utils.KieHelper;

import java.util.concurrent.ExecutorService;
import java.util.concurrent.Executors;
import java.util.concurrent.Semaphore;

import static comThree.RuleValueFile.myRuleFile1;
import static comThree.RuleValueFile.myRuleFile2;

public class DrlKieSessionTwo {
    private static int thread_num = 2000;//线程数,设置同时并发线程数
    private static int client_num = 100;//访问次数

    public static void main(String[] args) {

        long start = System.currentTimeMillis();
        KieHelper helper1 = new KieHelper();
        KieHelper helper2 = new KieHelper();
```

```java
            long end = System.currentTimeMillis();
            System.out.println("输出创建KieHelper用的毫秒是=" + (end - start));

            long startAddrule = System.currentTimeMillis();
            //分别将规则myRuleFile2 myRuleFile1 加载到虚拟文件中
            helper1.addContent(myRuleFile1, ResourceType.DRL);
            helper1.addContent(myRuleFile2, ResourceType.DRL);
            helper2.addContent(myRuleFile1, ResourceType.DRL);
            helper2.addContent(myRuleFile2, ResourceType.DRL);
            long endAddrule = System.currentTimeMillis();
            System.out.println("导入规则所用到的毫秒是=" + (endAddrule - startAddrule));

            ExecutorService exec = Executors.newCachedThreadPool();
            final Semaphore semp = new Semaphore(thread_num);
            for (int index = 0; index < client_num; index++) {
                Runnable run = () -> {
                    try {
                        semp.acquire();
                        long startNewKieSession = System.currentTimeMillis();
                        KieSession kieSession1 = helper1.build(  ).newKieSession();
                        KieSession kieSession2 = helper2.build(  ).newKieSession();
                        long endNewKieSession = System.currentTimeMillis();
                        System.out.println("创建有状态KieSession所用到的毫秒是="+ (endNewKieSession - startNewKieSession));
                        long startrule = System.currentTimeMillis();
                        Person person = new Person();
                        person.setName("张三");
                        person.setAge((int) (Math.random() * 100));
                        School school = new School();
                        school.setClassName("北大");
                        school.setClassCount("40");
                        kieSession1.insert(person);
                        kieSession1.insert(school);
                        kieSession2.insert(person);
                        kieSession2.insert(school);
                        int s1 = kieSession1.fireAllRules();
                        int s2 = kieSession2.fireAllRules();
                        System.out.println("kieSession1执行规则" + s1);
                        System.out.println("kieSession2执行规则" + s2);
                        System.out.println("修改后的"+person.getName());
                        kieSession1.dispose();//这里可注释 可不注释，因为每次KieSession1都是不同的
                        kieSession2.dispose();//这里可注释 可不注释，因为每次KieSession2都是不同的
                        long endrule = System.currentTimeMillis();
```

```
                System.out.println("规则执行所用到的毫秒是=" + (endrule-
startrule));
                System.out.println("==============================分
割线===================================");
                    semp.release();
                } catch (Exception e) {
                    e.printStackTrace();
                }
            };
            exec.execute(run);
        }
        exec.shutdown();
    }
}
```

通过上述测试用例的执行过程得出区别于其他情况的总结如下。

（1）虽然 KieSession 是不同的，KieHelper 也是不同的，但同一个 Fact 会受到影响，与 insert 是否提前无关。

（2）KieHelper 创建的 KieBase 都是默认的，所以有可能是 KieBase 的 packages 所至，就当前例子而言 myRuleFile2 规则不会被执行。

（3）"helper1.build(MultithreadEvaluationOption.YES).newKieSession();" 会触发多线程效果，但同一个 Fact 对象的结果会以最后一次调用规则的 KieSession 返回值为准。

14.5 同KieHelper 同StatelessKieSession（无状态）

同 KieHelper 与同 StatelessKieSession（无状态）是指在代码中使用同一个 KieHelper 来构建规则库，并创建同一个无状态的 StatelessKieSession。

创建 DrlKieSessionFives.java 文件，其内容如下。

```
package comThree;

import com.pojo.Person;
import com.pojo.School;
import org.kie.api.io.ResourceType;
import org.kie.api.runtime.StatelessKieSession;
import org.kie.internal.command.CommandFactory;
import org.kie.internal.utils.KieHelper;

import java.util.ArrayList;
import java.util.List;
import java.util.concurrent.ExecutorService;
import java.util.concurrent.Executors;
```

```java
import java.util.concurrent.Semaphore;

import static comThree.RuleValueFile.myRuleFile1;
import static comThree.RuleValueFile.myRuleFile2;

public class DrlKieSessionFives {
    private static int thread_num = 200;//线程数,设置同时并发线程数
    private static int client_num = 10;//访问次数

    public static void main(String[] args) {

        long start = System.currentTimeMillis();
        KieHelper helper1 = new KieHelper();
        long end = System.currentTimeMillis();
        System.out.println("输出创建KieHelper用的毫秒是=" + (end - start));

        long startAddrule = System.currentTimeMillis();
        //分别将规则myRuleFile2 myRuleFile1 加载到虚拟文件中
        helper1.addContent(myRuleFile1, ResourceType.DRL);
        helper1.addContent(myRuleFile2, ResourceType.DRL);

        long endAddrule = System.currentTimeMillis();
        System.out.println("导入规则所用到的毫秒是=" + (endAddrule - startAddrule));
        long startNewKieSession = System.currentTimeMillis();
        StatelessKieSession kieSession = helper1.build().newStatelessKieSession();
        long endNewKieSession = System.currentTimeMillis();
        System.out.println("创建无状态StatelessKieSession所用到的毫秒是=" + (endNewKieSession - startNewKieSession));

        ExecutorService exec = Executors.newCachedThreadPool();
        final Semaphore semp = new Semaphore(thread_num);
        for (int index = 0; index < client_num; index++) {
            Runnable run = () -> {
                try {
                    semp.acquire();
                    long startrule = System.currentTimeMillis();
                    Person person = new Person();
                    person.setName("张三");
                    person.setAge((int) (Math.random() * 100));
                    School school = new School();
                    school.setClassName("北大");
                    school.setClassCount("40");
                    List cmds = new ArrayList();
                    cmds.add(CommandFactory.newInsert(person, "p1"));
                    cmds.add(CommandFactory.newInsert(school, "s1"));

kieSession.execute(CommandFactory.newBatchExecution(cmds));
```

```
                    System.out.println(person.getName());
                    long endrule = System.currentTimeMillis();
                    System.out.println("规则执行所用到的毫秒是=" + (endrule
- startrule));
                    System.out.println("=============================分
割线=======================================");
                    semp.release();
                } catch (Exception e) {
                    e.printStackTrace();
                }
            };
            exec.execute(run);
        }
        exec.shutdown();
    }
}
```

14.6 同KieHelper 不同StatelessKieSession（无状态）

同 KieHelper 与不同 StatelessKieSession（无状态）是指在代码中使用同一个 KieHelper 来构建规则库，并创建两个无状态的 StatelessKieSession。

创建 DrlKieSessionSix.java 文件，其内容如下。

```
package comThree;

import com.pojo.Person;
import com.pojo.School;
import org.kie.api.io.ResourceType;
import org.kie.api.runtime.StatelessKieSession;
import org.kie.internal.command.CommandFactory;
import org.kie.internal.conf.MultithreadEvaluationOption;
import org.kie.internal.utils.KieHelper;

import java.util.ArrayList;
import java.util.List;
import java.util.concurrent.ExecutorService;
import java.util.concurrent.Executors;
import java.util.concurrent.Semaphore;

import static comThree.RuleValueFile.myRuleFile1;
import static comThree.RuleValueFile.myRuleFile2;

public class DrlKieSessionSix {
    private static int thread_num = 200;//线程数,设置同时并发线程数
```

```java
    private static int client_num = 10;//访问次数

    public static void main(String[] args) {

        long start = System.currentTimeMillis();
        KieHelper helper1 = new KieHelper();
        long end = System.currentTimeMillis();
        System.out.println("输出创建KieHelper用的毫秒是=" + (end - start));

        long startAddrule = System.currentTimeMillis();
        //分别将规则myRuleFile2 myRuleFile1 加载到虚拟文件中
        helper1.addContent(myRuleFile1, ResourceType.DRL);
        helper1.addContent(myRuleFile2, ResourceType.DRL);
        long endAddrule = System.currentTimeMillis();
        System.out.println("导入规则所用到的毫秒是=" + (endAddrule - startAddrule));

        ExecutorService exec = Executors.newCachedThreadPool();
        final Semaphore semp = new Semaphore(thread_num);
        for (int index = 0; index < client_num; index++) {
            Runnable run = () -> {
                try {
                    semp.acquire();
                    long startNewKieSession = System.currentTimeMillis();
                    StatelessKieSession kieSession = helper1.build(MultithreadEvaluationOption.YES).newStatelessKieSession();
                    long endNewKieSession = System.currentTimeMillis();
                    System.out.println("创建无状态StatelessKieSession所用到的毫秒是=" + (endNewKieSession - startNewKieSession));
                    long startrule = System.currentTimeMillis();
                    Person person = new Person();
                    person.setName("张三");
                    person.setAge((int) (Math.random() * 100));
                    School school = new School();
                    school.setClassName("北大");
                    school.setClassCount("40");
                    List cmds = new ArrayList();
                    cmds.add(CommandFactory.newInsert(person, "p1"));
                    cmds.add(CommandFactory.newInsert(school, "s1"));
                    kieSession.execute(CommandFactory.newBatchExecution(cmds));
                    System.out.println(person.getName());
                    long endrule = System.currentTimeMillis();
                    System.out.println("规则执行所用到的毫秒是=" + (endrule-startrule));
                    System.out.println("=============================分割线=====================================");
                    semp.release();
                } catch (Exception e) {
```

```
                e.printStackTrace();
            }
        };
        exec.execute(run);
    }
    exec.shutdown();
}
```

通过上述测试用例的执行过程得出区别于其他情况的总结如下。

规则与规则之间的 Fact 事实对象会相互作用，这一点与同 KieHelper 同 KieSession（有状态）的结论是一样的。

14.7 不同KieHelper 不同StatelessKieSession（无状态），StatelessKieSession只创建一次

不同 KieHelper 肯定是不同的 StatelessKieSession，但在线程中则是唯一的，只创建一次不做修改的 StatelessKieSession。

创建 DrlKieSessionSeven.java 文件，其内容如下。

```java
package comThree;

import com.pojo.Person;
import com.pojo.School;
import org.kie.api.io.ResourceType;
import org.kie.api.runtime.StatelessKieSession;
import org.kie.internal.command.CommandFactory;
import org.kie.internal.utils.KieHelper;

import java.util.ArrayList;
import java.util.List;
import java.util.concurrent.ExecutorService;
import java.util.concurrent.Executors;
import java.util.concurrent.Semaphore;

import static comThree.RuleValueFile.myRuleFile1;
import static comThree.RuleValueFile.myRuleFile2;

public class DrlKieSessionSeven {
    private static int thread_num = 200;//线程数,设置同时并发线程数
    private static int client_num = 10;//访问次数
```

```java
    public static void main(String[] args) {

        long start = System.currentTimeMillis();
        KieHelper helper1 = new KieHelper();
        KieHelper helper2 = new KieHelper();
        long end = System.currentTimeMillis();
        System.out.println("输出创建KieHelper用的毫秒是="+(end - start));

        long startAddrule = System.currentTimeMillis();
        //分别将规则myRuleFile2 myRuleFile1 加载到虚拟文件中
        helper1.addContent(myRuleFile1, ResourceType.DRL);
        helper2.addContent(myRuleFile2, ResourceType.DRL);//*******
        long endAddrule = System.currentTimeMillis();
        System.out.println("导入规则所用到的毫秒是="+(endAddrule - startAddrule));

        long startNewKieSession = System.currentTimeMillis();
        StatelessKieSession kieSession2 = helper1.build(    ).newStatelessKieSession();
        StatelessKieSession kieSession1 = helper1.build(    ).newStatelessKieSession();
        long endNewKieSession = System.currentTimeMillis();
        System.out.println("创建有状态KieSession所用到的毫秒是="+(endNewKieSession - startNewKieSession));

        ExecutorService exec = Executors.newCachedThreadPool();
        final Semaphore semp = new Semaphore(thread_num);
        for (int index = 0; index < client_num; index++) {
            Runnable run = () -> {
                try {
                    semp.acquire();
                    long startrule = System.currentTimeMillis();
                    Person person = new Person();
                    person.setName("张三");
                    person.setAge((int) (Math.random() * 100));
                    School school = new School();
                    school.setClassName("北大");
                    school.setClassCount("40");
                    //设计两个List 同时将实体放在集合中
                    List cmds1 = new ArrayList();
                    List cmds2 = new ArrayList();
                    cmds1.add( CommandFactory.newInsert(person,"p1") );

                    cmds2.add( CommandFactory.newInsert(person,"p2") );

                    cmds1.add( CommandFactory.newInsert(school,"s1") );

                    cmds2.add( CommandFactory.newInsert(school,"s2")
```

```
);
kieSession1.execute(CommandFactory.newBatchExecution(cmds1));
kieSession2.execute(CommandFactory.newBatchExecution(cmds2));

                    System.out.println(person.getName());
                    long endrule = System.currentTimeMillis();
                    System.out.println("规则执行所用到的毫秒是="+(endrule-startrule));
                    System.out.println("=============================分割线=====================================");
                    semp.release();
                } catch (Exception e) {
                    e.printStackTrace();
                }
            };
            exec.execute(run);
        }
        exec.shutdown();
    }
}
```

14.8 不同KieHelper不同StatelessKieSession（无状态），StatelessKieSession在线程代码中创建

不同 KieHelper 肯定是不同的 StatelessKieSession（无状态），StatelessKieSession 的创建放在线程中。

创建 DrlKieSessionEight.java 文件，其内容如下。

```
package comThree;

import com.pojo.Person;
import com.pojo.School;
import org.kie.api.io.ResourceType;
import org.kie.api.runtime.StatelessKieSession;
import org.kie.internal.command.CommandFactory;
import org.kie.internal.utils.KieHelper;

import java.util.ArrayList;
import java.util.List;
import java.util.concurrent.ExecutorService;
import java.util.concurrent.Executors;
import java.util.concurrent.Semaphore;
```

```java
import static comThree.RuleValueFile.myRuleFile1;
import static comThree.RuleValueFile.myRuleFile2;

public class DrlKieSessionEight {
    private static int thread_num = 200;//线程数,设置同时并发线程数
    private static int client_num = 100;//访问次数

    public static void main(String[] args) {

        long start = System.currentTimeMillis();
        KieHelper helper1 = new KieHelper();
        KieHelper helper2 = new KieHelper();
        long end = System.currentTimeMillis();
        System.out.println("输出创建KieHelper用的毫秒是=" + (end - start));

        long startAddrule = System.currentTimeMillis();
        //分别将规则myRuleFile2 myRuleFile1 加载到虚拟文件中
        helper2.addContent(myRuleFile2, ResourceType.DRL);//*******
        helper1.addContent(myRuleFile1, ResourceType.DRL);
        long endAddrule = System.currentTimeMillis();
        System.out.println("导入规则所用到的毫秒是=" + (endAddrule - startAddrule));

        ExecutorService exec = Executors.newCachedThreadPool();
        final Semaphore semp = new Semaphore(thread_num);
        for (int index = 0; index < client_num; index++) {
            Runnable run = () -> {
                try {
                    semp.acquire();
                    long startNewKieSession = System.currentTimeMillis();
                    StatelessKieSession kieSession1 = helper1.build().newStatelessKieSession();
                    StatelessKieSession kieSession2 = helper2.build().newStatelessKieSession();
                    long endNewKieSession = System.currentTimeMillis();
                    System.out.println("创建有状态KieSession所用到的毫秒是=" + (endNewKieSession - startNewKieSession));
                    long startrule = System.currentTimeMillis();
                    Person person = new Person();
                    person.setName("张三");
                    person.setAge((int) (Math.random() * 100));
                    School school = new School();
                    school.setClassName("北大");
                    school.setClassCount("40");
                    List cmds1 = new ArrayList();
                    List cmds2 = new ArrayList();
                    cmds1.add(CommandFactory.newInsert(school, "s1"));
                    cmds1.add(CommandFactory.newInsert(person, "p1"));
```

```
                    cmds2.add(CommandFactory.newInsert(person, "p2"));
                    cmds2.add(CommandFactory.newInsert(school, "s2"));
kieSession2.execute(CommandFactory.newBatchExecution(cmds2));
kieSession1.execute(CommandFactory.newBatchExecution(cmds1));
                    long endrule = System.currentTimeMillis();
                    System.out.println(person.getName());
                    System.out.println("规则执行所用到的毫秒是=" + (endrule-startrule));
                    System.out.println("==============================分割线======================================");
                    semp.release();
                } catch (Exception e) {
                    e.printStackTrace();
                }
            }
        };
        exec.execute(run);
    }
    exec.shutdown();
}
```

通过上述测试用例的执行过程得出区别于其他情况的总结如下。

虽然 StatelessKieSession 是不同的，KieHelper 也是不同的，执行的顺序是根据 execute 来判定的。

通过上述的测试用例中，其实与 MultithreadEvaluationOption.YES 关系并不大，KieSession 不论是有状态的还是无状态的，都会受规则中 Fact 事实对象是否发生变化而影响，所以说规则本身是不会受这种多线程的影响的。这就好比在数据库表中设置锁，但 Drools 是基于内存数据库的，所以它的效率很高。不同的规则库，所操作的 KieSession 也是不同的，既然 Fact 是通过 Java 创建并 insert 到规则中的，Fact 对象会相互作用，这也就导致了有时只能执行一个规则。

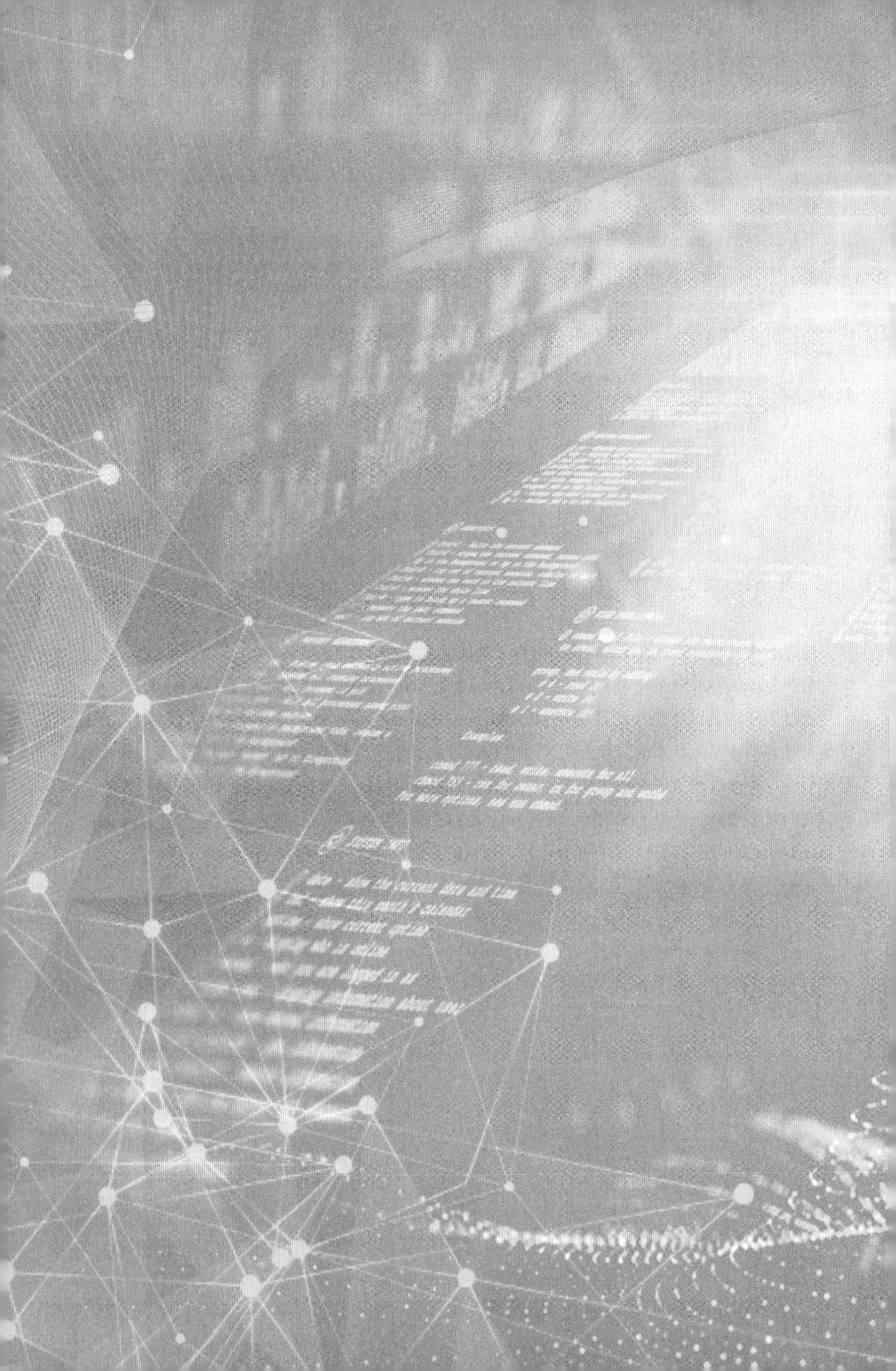

Drools 规则引擎技术指南

第五篇

源码篇

第15章
Drools源码分析

第 15 章 Drools 源码分析

研究 Drools 源码是深入了解 Drools 规则引擎的重要途径之一，从源码能了解 Drools 规则引擎是如何在项目中创建、构建和使用的。分析源码是一个很枯燥的事情，但阅读源码又是一个可以跟大师学习的机会，了解大师为什么会这样写代码。

分析 Drools 源码，可以从两个方面着手，一方面是本书中最常用的 kmodule.xml 配置文件方式，另一方面是讲述动态规则中常使用的 API，其实它们的原理与思想都是一样的。阅读源码还是需要从基础开始，拿一个非常简单的例子来进行说明。找到 Drools 项目中的 RulesHello.java 文件，其内容如下（注意加粗部分）。

```java
package com.rulesHello;

import com.pojo.Person;
import org.kie.api.KieServices;
import org.kie.api.runtime.KieContainer;
import org.kie.api.runtime.KieSession;
public class RulesHello {
    public static void main(String[] args) {
        KieServices kss = KieServices.Factory.get();
        KieContainer kc = kss.getKieClasspathContainer();
        KieSession ks = kc.newKieSession("testhelloworld");
        Person person = new Person();           person.setName("张三");
        person.setAge(30);
        ks.insert(person);
        int count = ks.fireAllRules();
        System.out.println("总执行了" + count + "条规则");
        System.out.println("输出修改后的Person age" + person.getAge()+person);
        ks.dispose();
    }
}
```

15.1 KieServices 分析

KieServices 组织结构如图 15-1 所示。

它是一个线程安全的单例，可以通过 KieServices kieServices = KieServices.Factory.get(); 对 KieServices 提供的工厂进行引用，其源码如下。

```java
/**
 * Returns a reference to the KieServices singleton
 */
static KieServices get() {
    return Factory.get();
```

```
}

/**
 * A Factory for this KieServices
 */
class Factory {

    private static class LazyHolder {
        private static KieServices INSTANCE = ServiceRegistry.
getInstance().get(KieServices.class);
    }

    /**
     * Returns a reference to the KieServices singleton
     */
    public static KieServices get() {
        return LazyHolder.INSTANCE;
    }
}
```

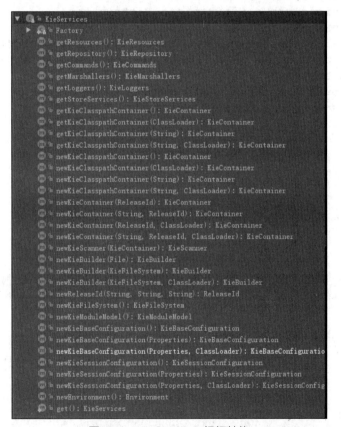

图15-1 KieServices组织结构

这是一个单例模式,是KieServices接口中的内部类。在它的LazyHolder类型私有类中

"ServiceRegistry.getInstance().get(KieServices.class);"使用了该引用。它的继承关系如图 15-2 所示，其源码如下：

```java
/**
 * Internal Interface
 *
 */
public interface ServiceRegistry extends Service {
    static ServiceRegistry getInstance() {
        return ServiceRegistryImpl.LazyHolder.INSTANCE;
    }

    <T> T get(Class<T> cls);
}
```

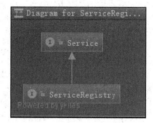

图15-2　ServiceRegistry继承关系

这是一个内部接口，它有一个内部实现类 ServiceRegistryImpl，其源码如下。

```java
package org.kie.api.internal.utils;

import java.util.Map;

/**
 * This is an internal class, not for public consumption.
 */
public class ServiceRegistryImpl
        implements
        ServiceRegistry {
    private Map<String, Object> registry;
    调用该类的LazyHolder时会创建ServiceRegistryImpl的实例
    static class LazyHolder {
        static final ServiceRegistryImpl INSTANCE = new ServiceRegistryImpl();
    }
    构建函数，它使用了ServiceDiscoveryImpl，也是一个单例
    public ServiceRegistryImpl() {
        registry = ServiceDiscoveryImpl.getInstance().getServices();
    }
    重启
    public synchronized void reset() {
        ServiceDiscoveryImpl.getInstance().reset();
```

```
    }
    public synchronized void reload() {
        registry = ServiceDiscoveryImpl.getInstance().getServices();
    }
```
判断是否为某个类的实例。如果没有则返回NULL
```
    public synchronized <T> T get(Class<T> cls) {
        Object service = this.registry.get( cls.getName() );
        return cls.isInstance( service ) ? (T) service : null;
    }
}
```
ServiceRegistryImpl 的组织结构如图 15-3 所示。

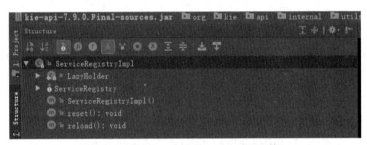

图15-3　ServiceRegistryImpl组织结构

进入 ServiceDiscoveryImpl 类中，它是一个实现类。进行了这么多层的封装，终于找到了 KieServices 做事的部分。它负责将 Kie 相关的类动态加载进来，主要是通过 getServices() 方法，其源码如下：

```
public synchronized Map<String, Object> getServices() {
    if (!sealed) {
        if (kiecConfDiscoveryAllowed) {
            Enumeration<URL> confResources = null;
            try {
                confResources = getClassLoader().getResources(path);
            } catch (Exception e) {
                throw new IllegalStateException("Discovery started, but no kie.conf's found");
            }
            if (confResources != null) {
                while (confResources.hasMoreElements()) {
                    registerConfs( getClassLoader(), confResources.nextElement() );
                }
            }
            buildMap();
        }

        cachedServices = Collections.unmodifiableMap( cachedServices );
```

```
            sealed = true;
        }
        return cachedServices;
    }

    private ClassLoader getClassLoader() {
        if (classloader == null) {
            classloader = Thread.currentThread().getContextClassLoader();
            if (classloader == null) {
                classloader = ClassLoader.getSystemClassLoader();
            }
        }
        return classloader;
    }
```

加粗的部分是比较核心的，内容比较简单。加载成功后并返回，用的也是单例的方式，所以这种加载方式也只会加载一次。

完成初始化操作后，再来看看 KieServices 还提供什么样的服务，其服务函数列表如下。

```
加载资料文件，并返回一个KieResources
KieResources getResources();
共享的存储库，是一个单例，返回KieRepository
KieRepository getRepository();
命名工厂，返回一个KieCommands
KieCommands getCommands();
返回一个KieMarshallers
KieMarshallers getMarshallers();
一个KieRuntimeLogger工厂，返回KieLoggers 处理日志
KieLoggers getLoggers();
加载KieSession 返回一个 KieStoreServices
KieStoreServices getStoreServices();
获取操作KieBase容器，返回一个KieContainer，它还可以指不同的参数来进行控制
KieContainer getKieClasspathContainer();
KieContainer getKieClasspathContainer(ClassLoader classLoader);
KieContainer getKieClasspathContainer(String containerId);
KieContainer getKieClasspathContainer(String containerId, ClassLoader classLoader);
创建一个新的KieContainer，就算该值存在
KieContainer newKieClasspathContainer();它还可以指定不同的参数来进行创建
KieContainer newKieClasspathContainer(ClassLoader classLoader);
KieContainer newKieClasspathContainer(String containerId);
KieContainer newKieClasspathContainer(String containerId, ClassLoader classLoader);
功能与newKieClasspathContainer相似，可通过指定ReleaseId来创建，它也可以指定不同的参数进行创建
KieContainer newKieContainer(ReleaseId releaseId);
KieContainer newKieContainer(String containerId, ReleaseId releaseId);
KieContainer newKieContainer(ReleaseId releaseId, ClassLoader classLoader);
```

```
KieContainer newKieContainer(String containerId, ReleaseId releaseId,
ClassLoader classLoader);
创建扫描器,自动检测更新新的KieModule
KieScanner newKieScanner(KieContainer kieContainer);
创建一个新的KieBuilder,构建指定文件夹中包含的KieModule
KieBuilder newKieBuilder(File rootFolder);
支持虚拟文件的创建
KieBuilder newKieBuilder(KieFileSystem kieFileSystem);
KieBuilder newKieBuilder(KieFileSystem kieFileSystem, ClassLoader
classLoader);
使用指定的groupId、artifactId和version创建新的ReleaseId
ReleaseId newReleaseId(String groupId, String artifactId, String
version);
创建新的KieFileSystem
KieFileSystem newKieFileSystem( );
创建新的KieModuleModel
KieModuleModel newKieModuleModel();
设置KieModuleModel中KieBase属性,在它新建的基础上。
KieBaseConfiguration newKieBaseConfiguration();
KieBaseConfiguration newKieBaseConfiguration(Properties properties);
KieBaseConfiguration newKieBaseConfiguration(Properties properties,
ClassLoader classLoader);
设置KieModuleModel中KieSession属性,在它新建的基础上。
KieSessionConfiguration newKieSessionConfiguration();
KieSessionConfiguration newKieSessionConfiguration(Properties properties);
KieSessionConfiguration newKieSessionConfiguration(Properties properties,
ClassLoader classLoader);
实例化,返回环境
Environment newEnvironment();
```

实例中,采用 getKieClasspathContainer 的获取方式,找到其对应源码,看看它都做了什么事,实例源码如下。

```
KieContainer kc = kss.getKieClasspathContainer();
```

它的实现类是 KieServicesImpl,其源码如下。

```
/**
 * Returns KieContainer for the classpath
 */
public KieContainer getKieClasspathContainer() {
    return getKieClasspathContainer( null, findParentClassLoader() );
}
```

getKieClasspathContainer() 方法是核心力。它有两个值,而在本实例中,因为它是获取 kmodule.xml 的所有配置下 KieBase 的 packages,所以它有一个默认参数。

```
public static ClassLoader findParentClassLoader() {
    ClassLoader parent = Thread.currentThread().getContextClassLoader();
    if (parent == null) {
        parent = ClassLoader.getSystemClassLoader();
```

```
    }
    if (parent == null) {
        parent = ProjectClassLoader.class.getClassLoader();
    }
    return parent;
}
```

看看 getKieClasspathContainer() 方法都做了什么事，其源码如下。

```
public KieContainer getKieClasspathContainer(String containerId,
ClassLoader classLoader) {
    if ( classpathKContainer == null ) {
        // these are heavy to create, don't want to end up with two
        synchronized ( lock ) {
            if ( classpathKContainer == null ) {
                classpathClassLoader = classLoader;
                if (containerId == null) {
                    classpathKContainerId = UUID.randomUUID().toString();
                } else {
                    classpathKContainerId = containerId;
                }
                classpathKContainer = newKieClasspathContainer(classp-
athKContainerId, classLoader);
            } else if (classLoader != classpathClassLoader) {
                throw new IllegalStateException("There's already
another KieContainer created from a different ClassLoader");
            }
        }
    } else if (classLoader != classpathClassLoader) {
        throw new IllegalStateException("There's already another
KieContainer created from a different ClassLoader");
    }

    if (containerId != null && !classpathKContainerId.
equals(containerId)) {
        throw new IllegalStateException("The default global singleton
KieClasspathContainer was already created with id
"+classpathKContainerId);
    }

    return classpathKContainer;
}
```

首先它会判断容器是否为空。如果为空，则进行创建；如果不为空，则根据传入的加载类与已加载的类进行比较，如果不相等，则抛 IllegalStateException 的异常。然后再进行判断，指定的 containerId 与以创建的 classpathKContainerId 是否相等。如果不相等也会抛一个 IllegalStateException 的异常。

核心处理在同步块中调用 newKieClasspathContainer 的代码。但它做了一些前提操作，首先它进行上锁操作，然后进行 classpathKContainer 是否为空的判断，这符合线程安全的单例模式设计。

```
synchronized ( lock ) {
    if ( classpathKContainer == null ) {
        classpathClassLoader = classLoader;
        if (containerId == null) {
            classpathKContainerId = UUID.randomUUID().toString();
        } else {
            classpathKContainerId = containerId;
        }
        classpathKContainer = newKieClasspathContainer(classpathKContainerId, classLoader);
    } else if (classLoader != classpathClassLoader) {
        throw new IllegalStateException("There's already another KieContainer created from a different ClassLoader");
    }
}
```

将 classLoader 赋值给 classpathClassLoader，如果判断没有指定 containerId，则生成 UUID，反之将 containerId 赋值给 classpathKContainerId。newKieClasspathContainer 是这个单例中的核心部分，其源码如下。

```
public KieContainer newKieClasspathContainer(String containerId, ClassLoader classLoader) {
    if (containerId == null) {
        KieContainerImpl newContainer = new KieContainerImpl(UUID.randomUUID().toString(), new ClasspathKieProject(classLoader, listener), null);
        return newContainer;
    }
    if ( kContainers.get(containerId) == null ) {
        KieContainerImpl newContainer = new KieContainerImpl(containerId, new ClasspathKieProject(classLoader, listener), null);
        KieContainer check = kContainers.putIfAbsent(containerId, newContainer);
        if (check == null) {
return newContainer;
        } else {
            newContainer.dispose();
            throw new IllegalStateException("There's already another KieContainer created with the id "+containerId);
        }
    } else {
        throw new IllegalStateException("There's already another KieContainer created with the id "+containerId);
    }
}
```

进入该方法，又进行了一次判断。它其实是属于一个公共方法的。如果 containerId 参数为空，则再次进行 UUIT 指定，创建一个新的 KieContainer 并返回。判断 kContainers 中是否包含 containerId 的值，如果为空，则进行创建；如果不为空，则该 containerId 已经存在，会抛一个 IllegalStateException 异常。kContainers 是一个线程安全的全局常量 ConcurrentMap 类型的值。代码 KieContainerImpl 是核心，在本方法中，通过 KieContainerImpl 构造了函数创建，而代码中 new ClasspathKieProject(classLoader, listener) 也做了很多事，其源码如下。

```
ClasspathKieProject(ClassLoader parentCL,
WeakReference<KieServicesEventListerner> listener) {
    this.kieRepository = KieServices.Factory.get().getRepository();
    this.listener = listener;
    this.parentCL = parentCL;
}
```

它是 ClasspathKieProject.java 类的构造函数，而其功能是加载 kmodule.xml 配置路径上的所有类，每个生成 KieModule 的类都会添加到 KieRepository 中。它调用了 KieServices.Factory.get().getRepository()。返回来继续看 KieContainerImpl() 方法，其源码如下。

```
public KieContainerImpl(String containerId, KieProject kProject,
KieRepository kr) {
    this.kr = kr;
    this.kProject = kProject;
    this.containerId = containerId;
    kProject.init();
    initMBeans(containerId);
}
```

创建成功后，将新创建对象放在 kContainers 中，但要注意，这里的 Map 用的是 putIfAbsent 方式，容器就算初始化成功了。

15.2 KieContainer 分析

完成了容器的初始化操作后，就需要看看 KieContainer 做了什么事情，通过分析 KieServices 得知，它创建了 KieContainer，但 KieContainer 又能完成什么事情呢。先看一下它的组织结构，如图 15-4 所示。

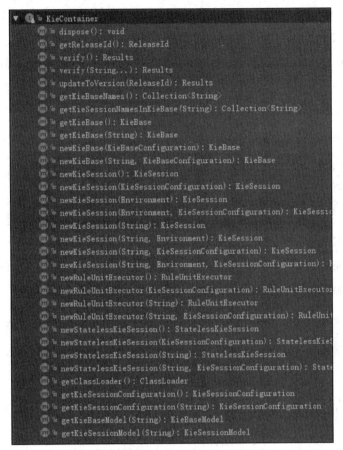

图15-4 KieContainer组织结构

可以看出 KieContainer 基本上都是操作规则库 KieBase 与会话 KieSession 的。其中，有比较常用的 newKieSession(String)，其他函数的说明如下。

```
清除当前所在容器的KieSession
void dispose();
返回ReleaseId
ReleaseId getReleaseId();
创建KieModule并构建所有KieBase，它还有可以指定的方式，
Results verify();
Results verify(String... kBaseNames);
切换修改版本号
Results updateToVersion(ReleaseId version);
返回所有可以用的KieBase的名称
Collection<String> getKieBaseNames();
返回指定KieBase，在当前KieContainer中定义所有KieSession的名称
Collection<String> getKieSessionNamesInKieBase(String kBaseName);
返回一个默认值KieBase，
KieBase getKieBase();
返回一个指定名称的KieBase
KieBase getKieBase(String kBaseName);
```

根据配置的默认值，创建一个KieBase
KieBase newKieBase(KieBaseConfiguration conf);
根据指定的KBaseName进行创建。
KieBase newKieBase(String kBaseName, KieBaseConfiguration conf);
创建一个默认的KieSession
KieSession newKieSession();
根据配置的默认值，创建一个KieSession
KieSession newKieSession(KieSessionConfiguration conf);
根据指定的环境，为当前容器创建默认的KieSession
KieSession newKieSession(Environment environment);
根据指定的环境，为当前容器创建默认的KieSession
KieSession newKieSession(Environment environment, KieSessionConfiguration conf);
根据指定的KieSessionName创建KieSession
KieSession newKieSession(String kSessionName);
根据指定环境，为当前容器创建指定名称的KieSession
KieSession newKieSession(String kSessionName, Environment environment);
根据配置，创建指定名称的KieSession
KieSession newKieSession(String kSessionName, KieSessionConfiguration conf);
根据指定环境和指定配置，创建指定名称的KieSession
KieSession newKieSession(String kSessionName, Environment environment, KieSessionConfiguration conf);
创建默认的RuleUnitExecutor
RuleUnitExecutor newRuleUnitExecutor();
根据配置，创建默认的RuleUnitExecutor
RuleUnitExecutor newRuleUnitExecutor(KieSessionConfiguration conf);
根据指定的Kisession，创建RuleUnitExecutor
RuleUnitExecutor newRuleUnitExecutor(String kSessionName);
根据配置指定KieSession，创建RuleUnitExecutor
RuleUnitExecutor newRuleUnitExecutor(String kSessionName, KieSessionConfiguration conf);
创建默认的无状态KieSession
StatelessKieSession newStatelessKieSession();
根据配置创建默认无状态的KieSession
StatelessKieSession newStatelessKieSession(KieSessionConfiguration conf);
创建指定名称的无状态KieSession
StatelessKieSession newStatelessKieSession(String kSessionName);
根据配置，创建指定名称的无状态KieSession
StatelessKieSession newStatelessKieSession(String kSessionName, KieSessionConfiguration conf);
返回当前容器中的ClassLoader
ClassLoader getClassLoader();
返回默认的KieSession的KieSessionConfiguration
KieSessionConfiguration getKieSessionConfiguration();
返回指定KieSession名称的KieSessionConfiguration
KieSessionConfiguration getKieSessionConfiguration(String kSessionName);
返回指定KBaseName的KieBaseModel

```
KieBaseModel getKieBaseModel( String kBaseName );
返回指定kSessionName的KieBaseModel
KieSessionModel getKieSessionModel( String kSessionName );
```

看过 KieContainer 的函数列表后,再来看实例中的使用,本章开始的例子是通过使用指定 KieSession 名称的方式来创建有状态的 KieSession 的,其实例源码如下。

```
KieSession ks = kc.newKieSession("testhelloworld");
```

它的实现类是 KieContainerImpl.java,具体的 newKieSession 内容如下。

```
public KieSession newKieSession(String kSessionName) {
    return newKieSession(kSessionName, null, null);
}
```

通过源代码可以得出,通过 KieSession 名称创建 KieSession 进行了封装,通过重载的方式进行操作,实现操作还是用它的 newKisession 方法,其源码如下。

```
public KieSession newKieSession(String kSessionName, Environment
environment, KieSessionConfiguration conf) {
    KieSessionModelImpl kSessionModel = kSessionName != null ?
                                (KieSessionModelImpl)
getKieSessionModel(kSessionName) :
                                (KieSessionModelImpl)
findKieSessionModel(false);

    if ( kSessionModel == null ) {
        log.error("Unknown KieSession name: " + kSessionName);
        return null;
    }
    if (kSessionModel.getType() == KieSessionModel.KieSessionType.STATE-
LESS) {
        throw new RuntimeException("Trying to create a stateful
KieSession from a stateless KieSessionModel: " + kSessionModel.
getName());
    }
    KieBase kBase = getKieBase( kSessionModel.getKieBaseModel().
getName() );
    if ( kBase == null ) {
        log.error("Unknown KieBase name: " + kSessionModel.
getKieBaseModel().getName());
        return null;
    }

    KieSession kSession = kBase.newKieSession( conf != null ? conf :
getKieSessionConfiguration( kSessionModel ), environment );
    if (isJndiAvailable()) {
        wireSessionComponents( kSessionModel, kSession );
    }
    registerLoggers(kSessionModel, kSession);
```

```
        ((StatefulKnowledgeSessionImpl) kSession).initMBeans(containerId,
((InternalKnowledgeBase) kBase).getId(), kSessionModel.getName());

    kSessions.put(kSessionModel.getName(), kSession);
    return kSession;
}
```

分析上述的代码，首先它会判断 KieSessionname 是否存在，不为空时，则通过 getKieSessionModel 到配置文件中进行查询，其源码如下。

```
public KieSessionModel getKieSessionModel(String kSessionName) {
    return kProject.getKieSessionModel(kSessionName);
}
```

如果为空时，则去配置文件中查找默认的值，对有状态的、无状态的进行查询，其源码如下。

```
private KieSessionModel findKieSessionModel(boolean stateless) {
    KieSessionModel defaultKieSessionModel = stateless ? kProject.getDefaultStatelessKieSession() : kProject.getDefaultKieSession();
    if (defaultKieSessionModel == null) {
        throw new RuntimeException(stateless ? "Cannot find a default StatelessKieSession" : "Cannot find a default KieSession");
    }
    return defaultKieSessionModel;
}
```

比较完成后会生成 KieSessionModelImpl 实例，然后进行判断，如果 KieSessionModel 为空，则在日志中记录信息，返回一个空，所以要注意，这是初始化者常犯的一个错误，前提是已经读到源码篇，这里讲述的是为什么会产生空指针异常。

其次就是判断创建 KieSession 是什么类型，如果创建出来的 KieSession 是无状态的，则会抛一个 RuntimeException 异常，说明当前指定的或默认的 KieSession 类型与返回值类型不匹配。

最后就是创建 KieBase 的过程，因为 KieSession 会话是通过 KieBase 创建的。虽然很多实例是直接通过 newKieSession 的，但阅读源码后，可以知道它是在 newKieSession 中获取的，其源码如下。

```
KieBase kBase = getKieBase( kSessionModel.getKieBaseModel().getName() );
if ( kBase == null ) {
    log.error("Unknown KieBase name: " + kSessionModel.getKieBaseModel().getName());
    return null;
}

KieSession kSession = kBase.newKieSession( conf != null ? conf : getKieSessionConfiguration( kSessionModel ), environment );
if (isJndiAvailable()) {
    wireSessionComponents( kSessionModel, kSession );
}
registerLoggers(kSessionModel, kSession);
```

```
((StatefulKnowledgeSessionImpl) kSession).initMBeans(containerId,
((InternalKnowledgeBase) kBase).getId(), kSessionModel.getName());
kSessions.put(kSessionModel.getName(), kSession);
```

通过 KieSessionModle 的 getKieBaseModel 的名称获取 KieBase，KieSessions 进行 newKieSession 创建、验证等。KieSession 是一个线程安全的 Map，所以当前容器的 KieSession 变化时，或者是在多线程、高并发下线程也是安全的。但有状态的一般都会被清空，防止数据迭代。

15.3 KieSession分析

KieSession 是使用最多的功能之一，它分为有状态、无状态，本章将详细地对 KieSessionAPI 进行说明。KieSession 是一个接口，它继承了 StatefulRuleSession、StatefulProcessSession、CommandExecutor、KieRuntime 接口。

KieSession 接口本身并没有提供很多方法，使用时基本都是其实现类进行执行调用的。它的继承关系如图 15-5 所示。

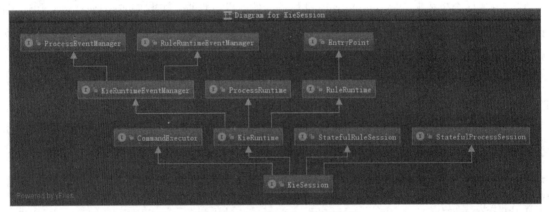

图15-5　KieSession继承关系

找到 EntryPoint（入口点）接口，它的组织结构如图 15-6 所示。

EntryPoint 有几个常用的函数，如 insert、delete、update 等。但它只是一个接口，它的实现类是在 StatefulKnowledgeSessionImpl 类中，找到 StatefulKnowledgeSessionImpl 类，查阅其继承关系图，确实可以实现很多接口，而且它还有本章所讲到的 KieSession 接口，从这个类中可以看到，该类包括了所有 Kie 的操作。

图15-6　EntryPoint的组织结构

insert 所调用的规则是这个方法，它会先去检查当前对象是否存活，然后再进行 insert 操作，其源码如下。

```java
public FactHandle insert(final Object object,
                         final boolean dynamic,
                         final RuleImpl rule,
                         final TerminalNode terminalNode) {
    checkAlive();
    return this.defaultEntryPoint.insert(object,
                                         dynamic,
                                         rule,
                                         terminalNode);
}
```

fireAllRules() 方法是 StatefulRuleSession 接口文件中的接口方法，实际还是在 StatefulKnowledgeSessionImpl 实现类中操作 fireAllRules() 方法，其源码如下。

```java
public int fireAllRules(final AgendaFilter agendaFilter,
                        int fireLimit) {
    checkAlive();
    try {
        startOperation();
        return internalFireAllRules(agendaFilter, fireLimit);
    } finally {
        endOperation();
    }
}
```

Dispose 方法是清空当前会话，其源码如下。

```java
public void dispose() {
    if (!agenda.dispose(this)) {
        return;
    }
    if (logger != null) {
```

```
        try {
            logger.close();
        } catch (Exception e) { /* the logger was already closed,
swallow */ }
    }

    for (WorkingMemoryEntryPoint ep : this.entryPoints.values()) {
        ep.dispose();
    }
    this.ruleRuntimeEventSupport.clear();
    this.ruleEventListenerSupport.clear();
    this.agendaEventSupport.clear();
    for (KieBaseEventListener listener : KieBaseEventListeners) {
        this.kBase.removeEventListener(listener);
    }
    if (processRuntime != null) {
        this.processRuntime.dispose();
    }
    if (timerService != null) {
        this.timerService.shutdown();
    }
    if (this.workItemManager != null) {
((org.drools.core.process.instance.WorkItemManager)this.
workItemManager).dispose();
    }
    this.kBase.disposeStatefulSession( this );
    if (this.mbeanRegistered.get()) {
DroolsManagementAgent.getInstance().unregisterKnowledgeSessionUnderName
(mbeanRegisteredCBSKey, this);
    }
}
```

通过源码发现，它把所有相关 KieSession 的内存数据都进行了 clear 操作，这样做是为了使用有状态时防止数据迭代，使用有状态时，如果不是要使用到原数据，最好在代码末尾添加 "kiesession.dispose();" 代码。

15.4 KieBase分析

KieBase 是规则库，这里也称为知识库，它本身不包含执行时的数据，如 Fact 事实对象。但它是所有应用程序知识库的存储库，它可以进行定义、销毁知识库的操作，包含规则、流程、函数、数据类型的操作，以及自己的创建等。它的组织结构如图 15-7 所示。

可以看出它包含了 newKieSession 的操作，还有 remove***()、get***() 的操作，分别用来操作 package、rule 与流程等相关的函数。

它通过 KnowledgeBaseImpl 实现类来进行规则库的实际操作，它的继承关系如图 15-8 所示。

图15-7　KieBase组织结构

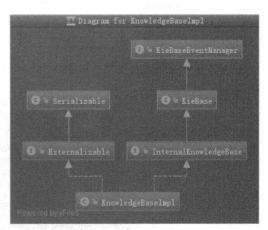

图15-8　KnowledgeBaseImpl继承关系

在 KnowledgeBaseImpl 实现类中，又有很多的方法，如 getRule、getQuery、getGlobals 等，都是日常开发过程中用来监听或指定一些特殊操作时所用的函数。其实在规则库中，用到的基本上都是在规则本质的操作，而并不像 KieSession 进行数据的传送，这说明 Drools 研究人员对其做的分离很充分，各司其职。它只能对所负责的地方进行维护。

15.5　KieFileSystem分析

KieFileSystem 是一个内存文件系统，可以用来虚拟构建一个类似 kmodule.xml 的文件，在使用 KIE 时，要创建规则库、知识会话都必须通过 kmodule.xml 配置文件来指定规则相关的文件。但这并不利于操作动态规则，所以在动态规则中，并没有使用这种方式，而是通过 KieFileSystem 自定义一个虚拟的 kmodule.xml 文件。KieFileSystem 组织结构如图 15-9 所示。

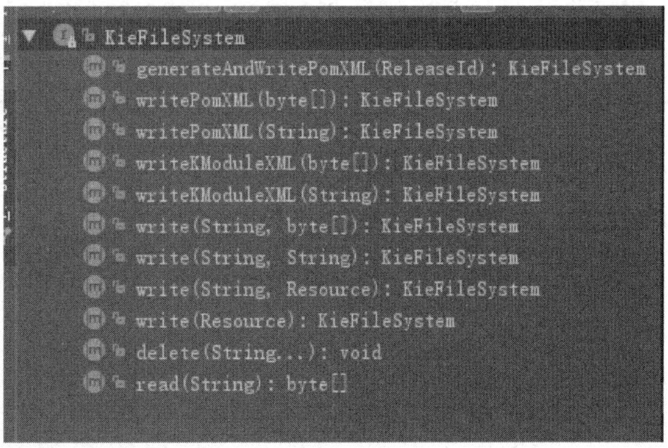

图15-9　KieFileSystem组织结构

KieFileSystem 的功能不仅可以创建 kmodule.xml 虚拟文件，还可以构建 kjar 的 pom.xml 文件。这两个方法的使用，在动态规则中已进行了说明。

15.6 KieHelper分析

KieHelper 是一个工具类，这个类是官方文档所没有提到的，它的功能与官方所提供的不通过 kmodule.xml 访问规则的方式非常相似，它在 org.kie.internal.utils 路径下，其组织结构如图 15-10 所示。

图15-10　KieHelper组织结构

通过 KieHelper 这个工具类的源码分析，可以很容易地理解动态规则的使用。

```
//构建函数
public KieHelper()
public KieHelper( KnowledgeBuilderOption... options )
//创建KieBase
public KieBase build( KieBaseConfiguration KieBaseConf )
public KieBase build(KieBaseOption... options)
//获取容器
public KieContainer getKieContainer()
//规则校验
public Results verify()
//设置类加载器
public KieHelper setClassLoader(ClassLoader classLoader)
//设置Kie的模块模板
public KieHelper setKieModuleModel(KieModuleModel kieModel)
//添加规则内容，并指定类型
```

```java
public KieHelper addContent(String content, ResourceType type)
//添加规则内容,并指定文件名,该文件名是在资源目录下
public KieHelper addContent(String content, String name)
//通过类的路径进行添加,并指定名称或编码
public KieHelper addFromClassPath(String name)
public KieHelper addFromClassPath(String name, String encoding)
//加载资源内容,可通过指定类型
public KieHelper addResource(Resource resource)
public KieHelper addResource(Resource resource, ResourceType type)
//生成资源名称,可通过指定类型
private String generateResourceName(ResourceType type)
```

该类在构建时会对 KIE 进行初始化,加载 KieServices、KieFileSystem。它使用的方式与在动态规则中讲到的内容很相似,它与自定义封装的类区别在于 KieBase 与 KieSession 是默认的。同样,在第一次构建规则库时需要发费的时间会很长。

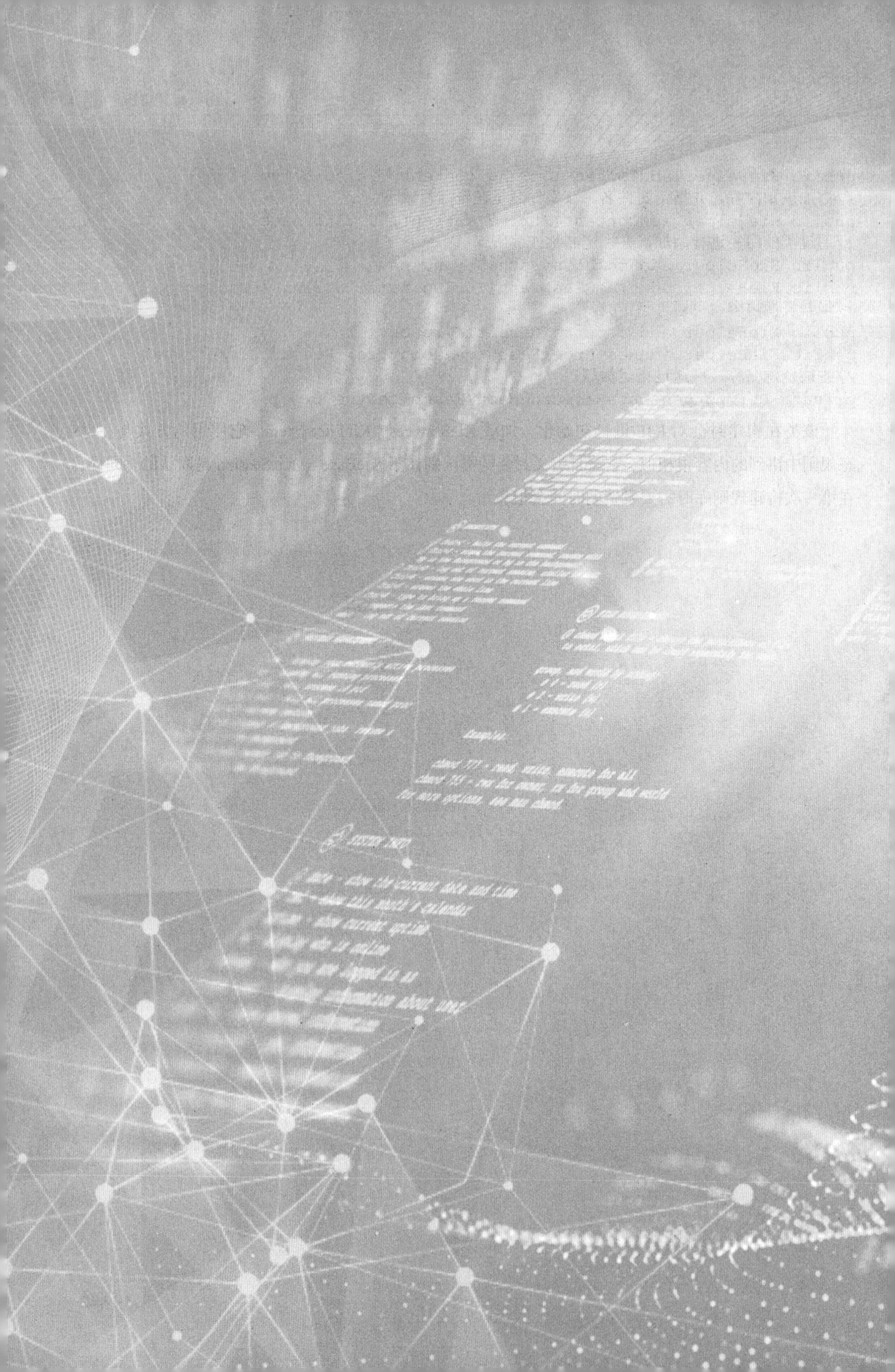

Drools 规则引擎技术指南

第六篇

扩展篇

第16章 Drools扩展说明

16.1　规则引擎优化方案

Drools 规则引擎的优化是在学习了前面的内容就会想到的问题。Drools 在 5.x 版本中采用了 RETE 的网络算法，该算法有一个好处就是节点是共享的，规则在执行过程中不关心规则的个数，所以规则的优化就显得非常重要了。需要借鉴决策表特殊用法中的思想来考虑这个问题。

1. 语法优化

规则体中的 When 部分，是用来判断当前规则是否满足条件，如果遇到这种情况，就需要考虑进行优化了。例如，代码如下：

```
...
rule "可优化的规则"
    when
        $s:School() OR
        $p:Person();
    then
        ***
end
...
```

很明显，上述规则体中的 When 部分，是一个逻辑或的关系，但强调过 Drools 的算法，规则执行时并不关心规则的数量，使用逻辑或反而会影响规则的执行效率，其优化方案如下。

```
...
rule "已优化的规则1"
    when
        $s:School()
    then
        ***
End
rule "已优化的规则2"
    when
        $p: Person ()
    then
        ***
end
...
```

代码看似很简单，但它隐含了很多的问题，设计规则中，一旦出现存在 OR 关键字的情况，就需要根据具体的规则业务将规则设计成多个。原因有如下几点。

（1）使用逻辑或（OR）会存在短路机制。

（2）影响规则执行的效果。

（3）规则算法的不合理应用。

如果是手写规则代码，这样优化是没有问题的，但根据实战应用来看，这样的实现是一个非常复杂的过程。在 Drools 整合 Spring Boot 的实战项目中，使用了模板引擎，这样的优化对模板来说是一个比较大的挑战。灵活性往往代表着复杂度，虽然从上述代码中看不出优化后的效果，但是应对风控系统、金融行业中的业务规则，量级不只是书中所提到的这些简单规则。

语法的优化只是从 When 整体逻辑判断做了一个优化，基础篇中提到过常用的约束条件，就如同写 SQL 语句，尽可能先比较精确值，然后是范围值，最后是模糊值。例如，代码如下：

```
...
rule "可优化的规则"
    when
        $p:Person(name matches "张.*",age>=30,className=="一班");
    then
        ***
end
...
```

规则代码这样写是没有问题的，运行时结果也是正常的，但一般不会这样去写，好比 SQL 查询语句的优化，首先想到的是比较符的优化，这里也是同样的道理。其优化方案如下：

```
...
rule "以优化的规则"
    when
        $p:Person(className=="一班", age>=30, name matches "张.*");
    then
        ***
end
...
```

Drools 规则引擎的核心离不开规则库，规则库是存放在内存中的，所以它有一个核心思想是用空间换时间，牺牲了本地服务器内存，来换取更快的执行速度，语法的优化同样是至关重要的。

规则语法的整合优化：规则在执行时，会有一个类似预处理的过程，规则只执行满足条件的部分（如果使用 accumulate 可以测试），而并不是因为规则多，其实规则多少对整合性能影响并不是很大，只是在生成规则库时需要大量的时间。所以优化规则语法的原则如下。

（1）规则拆分原则：将规则进行拆分，避免出现 OR 的情况。

（2）规则比较原则：将区间或模糊查询的方式排在比较值的后面。

（3）规则简单原则：尽量避免出现过于复杂的比较值。

（4）规则结果原则：then 中避免出来 if eles。

2. 实战中的优化

规则的执行离不开 KieSession，而 KieSession 的创建离不开规则库。规则库构建成功后是存放在内存中的，但构建规则库是一个相对比较慢的过程，例如，当前进程中第一次构建规则库的时间约为 1300ms，虽然说第一次构建规则库的时间比较长，但也可以在项目启动时就去创建规则库，可以避免一些风险。项目初始化时加载规则，是实战中的优化方案之一。

当前进程中的规则库被初始化后，第二次构建的时间就会大大缩短，但消耗的时间为 200ms 左右，且对内存也是一个不小的开销。时间上、空间上都并不是一个可行的方案。因为实战项目中，业务人员不可能实时或频繁地变更业务，也就是说在代码中的规则库并不会频繁的构建。所以这时就有了一个优化方案，通过创建本地缓存。那么本地缓存应该如何创建，可以采用 Spring 的一些思想，在操作 Drools 规则引擎的地方创建一个私有全局静态变量的 ConcurrentHashMap，其线程是安全的。它的 Key 存放着业务场景的标识，其 Value 存放着一个 JavaBean，JavaBean 的内容如下。

```java
import org.kie.api.KieBase;
import org.kie.api.runtime.KieSession;
import org.kie.api.runtime.StatelessKieSession;
/**
 *
 * 功能描述：规则封装类，用于规则的执行，
 *
 * @auther: lai.zhihui
 * @date: 2018/10/18 16:51
 */
public class RuleExecute{
    //无状态的KieSession
    private StatelessKieSession statelessKieSession;
    //有状态的KieSession
    private KieSession kieSession;
    //存入KieBase
    private KieBase KieBase;
…此处省略get set方法 读者不要忘记加…
}
…
public class DroolsService {
    //日志配置
    private Logger LOGGER = LoggerFactory.getLogger(DroolsService.class);
    //全局静态常量信息
    private static Map<String, RuleExecute> ruleExecuteMap = new ConcurrentHashMap<>();
…
```

通过 JavaBean 可以看出，在构建规则库时，将当前规则 KieBase、StatelessKieSession、KieSession 放在对象且全局变量 Map 中。使用时只需要将 Map 对应 Key 的 Value 值执行规则就可以了。这样做的好处就是只有当业务人员操作变更系统业务时才会进行重构规则库，减少了构建规则库所带来的内存压力和响应效率。当然如果是需要实时变更业务规则的，那么可以忽略这个优化方案。但是规则库的频繁构建侧面证明了当前的设计是存在问题的。通过本地缓存方案，是规则执行的优化方案之一。

为什么不放在 Redis 缓存中或其他第三方提供的缓存中呢？下面就来说明其原因，使用 Redis 缓存还可以实现集群间的数据共享，解决了数据同步的问题。

第一点：KieBase 是规则库，它只存活在当前进程中。也就是说它是一个内存级的数据，内存数据是无法共享的。

第二点：KieBase 是不可被序列化/反序列化的，无法放在缓存中。

通过以上两点，可以得出：要实现集群但数据无法共享时，只能通过一些手段来让集群中规则库保持最终一致性。在设计过程中，也有人这样问过，既然规则库无法通过第三方缓存中实现数据统一，那是不是将规则内容放在缓存中就好了？让集群中的每一台服务器都去读取这个缓存并获取内容，构建其规则库。这种做法也是一种思路，但并不赞同这样做。首先，编辑的规则内容和业务数据一定会持久化，通常的作法是通过数据库；其次，有一个前提条件就是业务人员不可能实时去变更业务规则，也就是说业务服务器不频繁去调用数据库，当只有业务规则变更时才会去调用数据库，这对数据库来说并没有压力，而且规则管理服务器只关系数据库，而不需要再去更新缓存。如果有频繁或实时变更业务的需求，就需要考虑两点：规则业务设计的是否合理，增加服务器的硬件配置。

16.2 规则实战架构

学习过 Drools 规则引擎后，大部分程序员都在关注这样一个问题，规则引擎在实战中的应用，如何结合现有的技术框架来整合 Drools，这也成了很多程序员的难题。规则引擎技术本身是个相对冷门的技术，而且现有的规则引擎大部分都已经被商业化。Drools 作为一个开源的规则引擎技术，资料是少之又少。作者从 2016 年 6 月开始接触并研究这门技术，经过几个实战项目的经验，总结出 Drools 规则引擎在实战中的架构流程。

实战中一般会采用两套服务系统，一套是规则的管理系统，另一套是规则业务系统。业务系统是一个集群，管理系统是对业务人员统一操作的服务系统。这样做的好处是实现业务定制化，具体业务由业务操作员来选定，做到可动态发布、及时修改，使业务操作员不在依赖 IT 部门，快速完成业务的变更。图 16-1 所示为平台架构流程。

第 16 章 Drools 扩展说明

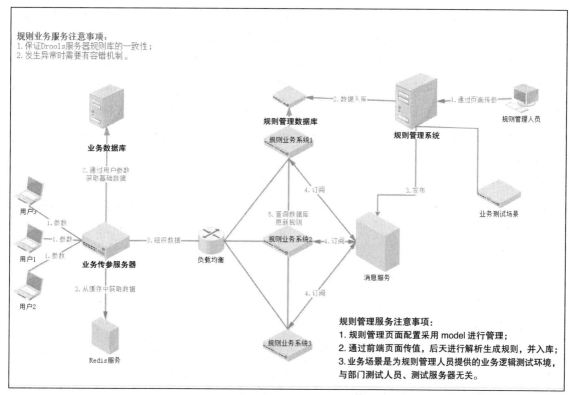

图16-1　Drools实战项目中的平台架构流程

可以看出，规则业务系统是一个集群，是用户调用规则的系统。通过 Nginx 做负载均衡与反向代理，减轻业务应用的压力。规则管理系统是用户生成规则的管理项目，管理系统将业务员生成的业务数据，业务规则存放在规则管理数据库中，这里的规则管理数据库是结构化数据库，可以是 MySql、Oracle 等。规则管理服务器还提供了操作业务的功能，如发布业务、删除业务、变更业务等。这里的做法是通过消息服务实现异步操作业务的功能。消息服务通过 RabbitMq 做中间介，采用发布/订阅的方式进行信息同步。发布端为规则管理系统，订阅端为规则业务系统。Drools 提供的 API 只能对其语法进行校验，但业务员并不了解当前所配置的业务规则是否是正确的，换而言之，业务员并不知道业务逻辑是否满足预定的规则，这里建议实现一个业务测试场景系统，该系统只为进行测试业务逻辑的合理性，是交互业务操作员使用的，注意，该模块与测试环境是不一样的。

负载均衡前的业务参数服务器是用来组合规则可识别的数据所要完成的事，理论上，规则业务系统是对外提供接口的，是一个具有通用性的平台化项目。所以在接收参数方面，是有一定的约定的，也是俗称的约定大约配置。

通过图 16-1 所表现出来的思想，进行一个扩展性的说明，将每一个规则业务系统看成是一类型的业务。并且每一类型的业务即是一个集群，通过规则管理系统实现一套由业务员操作的庞大业务系统，且非常符合平台化系统的结构流程。

业务系统引用规则引擎后，给业务人员带来了极大方便，与此同时在架构设计、代码开发层面就提高了一定的复杂度。灵活性与复杂性是成正比的，一套非常灵活的系统是产品化的前提，这也就是国内使用了 Drools 做底层的规则引擎平台成为商业化产品的原因之一。

涉及集群就一定会涉及分布式锁、分布式事务的问题，在实战的优化中提到通过本地缓存来控制规则的执行，从而减少频繁更新规则库所带来的性能问题，规则库是存放在当前服务器所在的进程中的，无法共享。集群规则库的一致性、本地缓存"事件管理"管理都成为了问题，不管是采用强一致性或最终一致性，都需要设计人员去不断地完善。与其他类分布式项目一样，容错机制是必须要存在的。

在实际开发过程中，项目经理一般情况都不会给程序员太多的时间去对平台进行完善，并不是说项目经理不懂技术，不懂学习成本，而是项目经理需要把控进度。Drools 为什么在国内发展的比较慢，根据分析有以下几点。

（1）学习成本高。

（2）可学习的开源项目少。

（3）中文资料少。

（4）多数成了商业化产品。

面对这样的压力，程序员就不得不先做出一个功能来，哪怕当前的功能还不完善，但可以证明这门技术是可行的。在有限的时间中完成一项有难度的项目，是需要制订一些计划的。首先，程序员必须要了解 Drools 可以做什么；其次，通过 Workbench 提供的页面实现一套可动态配置的简易项目，来证明 Drools 的可行性；最后，学习 Drools 的基础语法，通过了解规则实战架构完成定制化平台开发。

1. 实战技术选型

说到架构图就会讲到技术选型，通常用表 16-1 所示的技术来搭建 Drools 项目的软件和硬件设施。

表16-1 搭建Drools项目的软件和硬件设施

功 能	选用技术
硬件设备	ecs（阿里云服务器）8核16GB的配置
系统版本	CentOS 7
反向代理/负载均衡	Nginx
数据库	Mysql、Postgre
缓存	Redis
消息队列	rabbitMq

续表

功　能	选用技术
后端技术	JDK 1.8
	Spring Boot 2.0.1.RELEASE
	Drools 7.10
	log4j2 日志
	阿里JSON包
	Mybatis
交互	前后端分离/统一提供接口
规则生成	模板引擎ST4 4.0.8/自定制模板（参考Workbench）
开发工具	IntelliJ IDEA（个人习惯）
版本控制	GitHub

2．数据库表设计

（1）项目系统类的表需以 Sys_*** 开始，如 Sys_OperateLog。

（2）项目业务类的表需根据当前业务的缩写开始，如 Rule_Content。

（3）项目业务类的表关联根据当前主表进行扩展，如 Rule_Content_Type。

（4）命名规则：文件名采用驼峰式（类名首字母必须大写，类首行注释写名该类的相关业务、作者）；方法/变量采用驼峰式（方法/方法参数需要写明用途，方法上要写明作者）。

（5）工具类统一管理。

（6）日志风格，输出地址统一管理。

（7）配置文件需放在 resources 中，按业务进行文件夹创建。

16.3　规则引擎项目的定位

规则引擎在项目中的位置是设计人员必须清楚的，规则引擎并不是万能的，被使用的原因是它可以让业务更透明、简单，解放生产力，而一个完整的项目体系可远远不止这些。图 16-2 所示是一个相对全面的组织结构，而且符合现在流行的微服务架构。

决策层中包含了模型管理、规则配置、知识库、规则库、规则调用、业务管理。通过此图要说明的问题是，规则引擎是整合项目中的一个部分，它是需要结合其他服务来协作的，如权限管理，业务操作员受权限的影响，其操作的范围也就有所不同。所以定位决定了设计的范围。

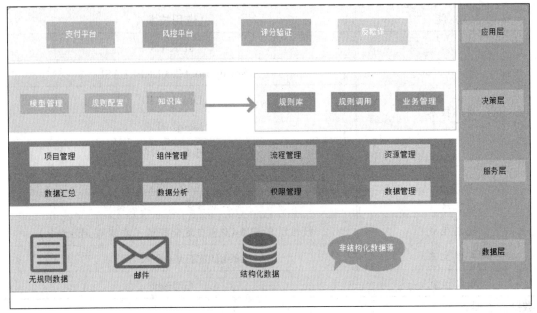

图16-2 项目组织结构

16.4 规则引擎实战应用思想

规则引擎主要处理的是业务规则，常用于风控系统、医疗系统、促销系统等，所涉及的领域也是比较广泛的。读者在使用规则引擎时常常没有思路，不知道规则引擎可以做什么、怎么做、最终的目的是什么。介于开发设计人员的种种烦恼，这里先提出以下几点来开阔你的思维方式。

1. 规则流

规则文件是规则文件，流程文件是流程文件。两者是相互独立的文件，并没有冲突。

2. 动态规则"文件？"

规则内容是规则内容，文件是文件，内容可以是 String，可以是规则所支持的类型，规则内容不关心它的存储介质，只能被规则库加载构建。

规则文件是需要编译的，但并非是 class 方式的编译，修改规则文件不会造成规则本身的变化，动态规则方案中可以提供更有效的思路。

3. 设计产品的定位

在规则引擎的实战应用中，一般面临的都是规则的可变性，也就是多次讲到的动态规则（动态代码），规则引擎的开发定位很重要，这里的定位是指项目面对的使用者，程序员可以采用 Workbench，它可以快速开发，但对开发人员有较大的依赖；业务操作员，可以采用定制化页面，但开发周期要长一些。

规则引擎项目的开发过程是比较麻烦的，语法稍有不慎，就会变成死循环，或者影响业务的正常运行，所以掌握语法至关重要。

4. 动态规则的两种主流

第一种：Workbench+Kie-Server 集群 + 负载均衡。

Workbench+Kie-Server 方式，Workbench 是定位，WEB 是 IDE，通过这类 IDE 都可以完成动态规则的实现，这种方式是程序员相对喜欢的方式。

第二种：String+DB 的应用。

String+DB 规则内容入库方式，操作数据库，实现定制化，作为 Web 项目，操作数据库是正常不过的，这里通过页面进行配置，生成规则，业务人员这样的技术小白自定制规则，更适合互联网这样的项目应用。例如，金融系统，它的业务变化快，效果要求高。但业务开发一套"类似"Workbench 的页面操作规则系统，注意点比较多，页面的设置、数据表的设计等。重要的是，语法的测试及业务场景的测试都是要有的，因此真正实现起来也是非常困难的。

16.5 规则引擎日志输出

使用 Drools 规则引擎时，如何查看 Drools 的日志是一个问题，在第 10 章的事件监听中有提到过关于 Drools 日志的一些使用。本节的重要内容是如何生成日志文件。在第 15.1 节中，有接口"KieLoggers getLoggers();"，它的功能是输出 Kie 运行时的日志信息，进入这个接口查看源代码，其中内容如下。

```
KieRuntimeLogger newFileLogger(KieRuntimeEventManager session,
                    String fileName);
KieRuntimeLogger newFileLogger(KieRuntimeEventManager session,
                    String fileName,
                    int maxEventsInMemory);
KieRuntimeLogger newThreadedFileLogger(KieRuntimeEventManager session,
                    String fileName,
                    int interval);
KieRuntimeLogger newConsoleLogger(KieRuntimeEventManager session);
```

Drools 规则引擎的日志操作共分为 3 种方式，即指定输出日志文件、控制台输出及线程操作的日志，这 3 种日志中有两种是可以创建日志的，分别是 newThreadedFileLogger 与 newFileLogger，而 newConsoleLogger 是输出在控制台。其中 newFileLogger、newConsoleLogger 有两个方法，区别在于指定日志事件的最大数量，默认为 1000；newThreadedFileLogger 是指线程的日志文件，而 newConsoleLogger 是指在控制台输出不指定文件。使用 Kie 日志时必须要加载日志引用，无论是通过 Jar 包引用还是通过 pom.xml，否则日志或监听都是无法生效的。

1. newFileLogger

任意编译一个执行规则 Java，在方法中添加操作日志代码，其内容如下。

```
...
KieServices kss = KieServices.Factory.get();
kss.getLoggers().newFileLogger(ks,"./droolsLog");
...
```

执行规则调用代码，会在指定目录创建一个名为 droolsLog.log 的日志文件，内容是类似 XML 文件的以标签为分隔的日志执行过程，其截取部分内容为：

```
...
<org.drools.core.audit.event.ActivationLogEvent>
    <type>4</type>
    <activationId>test002 [1]</activationId>
    <rule>test002</rule>
    <declarations>$p=Person{name=&apos;张三&apos;, age=30, className=&apos;null&apos;}</declarations>
    <factHandleIds>1</factHandleIds>
</org.drools.core.audit.event.ActivationLogEvent>
...
```

上述日志内容的规则是"test002"，所操作的对象是"Person"，内容是"name=张三，age=30…"。

返回 Java 代码中，newFileLogger 默认值为 1000，如果第三个参数的值设置的特别小，也就是说操作最大内存的事件值比较小时，规则日志会默认生成多个 droolsLog.log 文件，不同的是日志文件会进行编辑，如：droolsLog1.log、droolsLog2.log。

使用 newFileLogger 时要注意其与 newConsoleLogger 控制台输出日志是有冲突的，如在 newFileLogger 之后使用 newConsoleLogger 则会导致日志文件无法被正常创建。

2. newConsoleLogger

任意编译一个执行规则 Java，在方法中添加操作日志代码，其内容为：

```
...
KieServices kss = KieServices.Factory.get();
kss.getLoggers().newConsoleLogger (ks);
...
```

执行规则调用代码，会在控制台输出每一个被匹配成功后的规则操作日志，如图 16-3 所示。

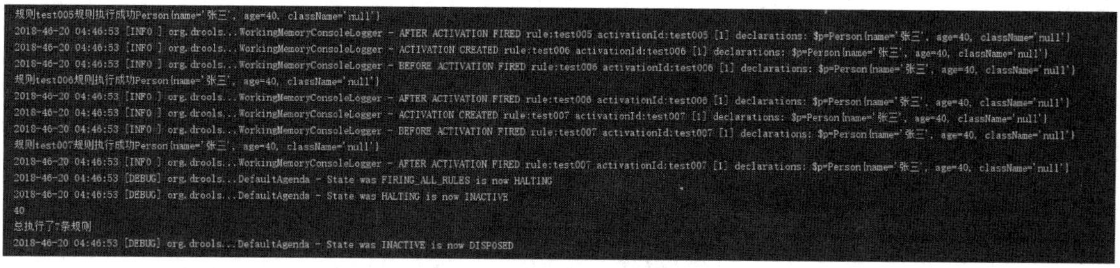

图16-3 规则控制台日志输出

还有一种方式也可以实现类似控制台输出的效果，就是前面讲过的事件监听的另一种用法，代码中添加事件监听器，其内容如下。

```
KieServices kss = KieServices.Factory.get();
...
ks.addEventListener(new DebugAgendaEventListener());
ks.addEventListener(new DebugRuleRuntimeEventListener());
...
```

执行规则调用代码，输出结果与图16-3所示大致相同。

3. newThreadedFileLogger

任意编译一个执行规则Java，在方法中添加操作日志代码，其内容为：

```
...
KieServices kss = KieServices.Factory.get();
kss.getLoggers().newThreadedFileLogger(ks,"./Threaded",1000);
...
```

执行规则调用代码，功能与newFileLogger相似，也是在指定目录下创建日志文件，生成的日志文件内容也大致相同，但newThreadedFileLogger是记录线程中的一些信息，如WorkingMemoryLog的一些操作信息。

参考文献

［1］袁鸣凯.jboss 规则引擎 KIE Droolshttp://blog.csdn.net/lifetragedy/article/details/60755213，2017-10-12.

［2］Maciej Swiderski.Installing KIE Server and Workbench on same server.https://mswiderski.blogspot.com/2015/10/installing-Kie-Server-and-Workbench-on.html，2016-10-15.

［3］李 震.Installing KIE Server and Workbench on same server. http://dyingbleed.com/drools-13/，2016-08-12.

［4］朱胜智.Drools7.0.0.Final 规则引擎教程. https://blog.csdn.net/wo541075754/article/details/76695415，2017-12-11.

［5］Drools 官方文档. Drools Documentation.https://docs.jboss.org/drools/release/7.10.0.Final/drools-docs/html_single/index.html，2018-12-14.

［6］蔡从洋.Drools6.4 动态加载规则之（三）kie-wb 与 Kie-Server 的集群应用.https://blog.csdn.net/caicongyang/article/details/53056890，2016-11-13.